Genetic Engineering of Plants for Stress Tolerance

Genetic Engineering of Plants for Stress Tolerance

Guest Editor

Moxian Chen

MDPI

Basel • Beijing • Wuhan • Barcelona • Belgrade • Novi Sad • Cluj • Manchester

Guest Editor
Moxian Chen
Research and Development
Center for Fine Chemicals
Guizhou University
Guiyang
China

Editorial Office
MDPI AG
Grosspeteranlage 5
4052 Basel, Switzerland

This is a reprint of the Special Issue, published open access by the journal *International Journal of Molecular Sciences* (ISSN 1422-0067), freely accessible at: https://www.mdpi.com/journal/ijms/special_issues/2L0PDKAW8Y.

For citation purposes, cite each article independently as indicated on the article page online and as indicated below:

Lastname, A.A.; Lastname, B.B. Article Title. *Journal Name* **Year**, *Volume Number*, Page Range.

ISBN 978-3-7258-6514-7 (Hbk)
ISBN 978-3-7258-6515-4 (PDF)
https://doi.org/10.3390/books9978-3-7258-6515-4

Contents

International Journal of
Molecular Sciences

MDPI

Editorial

Special Issue "Genetic Engineering of Plants for Stress Tolerance"

Moxian Chen

State Key Laboratory of Green Pesticide, Key Laboratory of Green Pesticide and Agricultural Bioengineering, Ministry of Education, Research and Development Center for Fine Chemicals, Guizhou University, Guiyang 550025, China; cmx2009920734@gmail.com

1. Advances in Molecular Understanding of Plant Stress Tolerance

Plants are continuously exposed to diverse environmental challenges such as drought, salinity, and temperature extremes, which restrict their growth and yield potential. With the accelerating pace of climate change, improving stress tolerance in plants has become an urgent global priority. Genetic engineering, combined with high-throughput omics and synthetic biology, has transformed our ability to dissect the complex networks governing plant adaptation. Integrative omics approaches, spanning metabolomics, transcriptomics, and phosphoproteomics, have enabled a system-level understanding of stress responses. For example, analyses of maize have revealed transcriptional regulation and metabolic specialization in alkaloid biosynthesis and ureide metabolism that underlie their ecological adaptation [1]. Similarly, combined metabolomic and transcriptomic profiling of wheat during the reproductive stage under drought stress has identified amino acid metabolism, heat shock proteins, and transporter systems as key contributors to thermal tolerance [2]. Complementary phosphoproteomic studies in maize have demonstrated that phosphorylation events mediated by MAPK signaling play crucial roles in early heat stress responses [3]. These studies collectively emphasize that stress tolerance arises from intricate molecular coordination across transcriptional, metabolic, and post-translational levels. Moreover, recent discoveries highlight the critical role of alternative splicing in modulating plant developmental transitions under environmental stress. A striking example is the identification of VRF1 alternative splicing as a molecular switch that regulates stress-induced early flowering in Arabidopsis [4]. The study demonstrated that isoform ratios of VRF1 determine whether plants prioritize stress tolerance or reproductive escape, offering a new mechanistic insight into plant adaptation strategies.

2. Genetic Engineering, Synthetic Modules, and Regulatory Mechanisms

Alongside omics-driven insights, genetic engineering and synthetic biology have opened new avenues for developing crops with enhanced stress resilience. The creation of synthetic multi-gene modules exemplifies this progress. High expression of the class II *TPS* gene osTPS8 in rice significantly improved salinity tolerance by enhancing osmotic adjustment, activating antioxidant defense systems, and upregulating stress-related genes [5]. This demonstrates how synthetic modules integrating protective functions from extremophiles can be used to fortify plants against multiple stress factors simultaneously.

Genome-wide identification of transcription factor families has also advanced our understanding of regulatory networks involved in stress responses. The characterization of the TCP gene family in *Cenchrus fungigraminus* revealed genes responsive to drought and cold stress, with several members associated with growth and developmental regulation [6].

Beyond plant-specific mechanisms, cross-disciplinary studies have revealed parallels between plant stress adaptation and resistance mechanisms across biological systems. For instance, recent reviews on drug target discovery and membrane protein-mediated resistance provide valuable perspectives for plant biotechnology [7,8]. These works illustrate how convergent evolution and membrane-associated signaling contribute to adaptive resistance in organisms ranging from microbes to plants, highlighting shared molecular logic that can inspire the design of novel resistance-breaking strategies in crops. Moreover, alternative splicing has emerged as a vital post-transcriptional mechanism that fine-tunes gene expression during plant–fungal symbioses, balancing immune responses and beneficial interactions [9]. Together, these findings highlight that plant resilience depends not only on the activation of stress-related genes but also on sophisticated transcriptional and post-transcriptional regulation.

3. Evolutionary Insights and Future Perspectives

Exploring stress adaptation in naturally resilient species continues to provide valuable lessons for crop improvement. Investigations of high-altitude plants from the Qinghai–Tibet Plateau revealed that long-term adaptation to low temperatures, intense UV radiation, and nutrient-poor soils involves integrated physiological and morphological adjustments, including enhanced antioxidant activity and photosynthetic efficiency [10]. When viewed together with molecular insights such as VRF1-mediated flowering control and membrane protein-driven resistance evolution, it becomes clear that plants employ both evolutionary and engineered strategies to optimize survival under stress. These insights exemplify the evolutionary templates that modern biotechnology seeks to emulate through targeted engineering.

Collectively, the studies in this Special Issue provide a multidimensional perspective on plant stress tolerance, demonstrating that sustainable agricultural productivity depends on the integration of molecular, synthetic, and ecological approaches. Omics-based discoveries guide the identification of key pathways, while synthetic biology offers the tools to reprogram plants for enhanced performance under stress. Future research should focus on validating engineered traits under field conditions, where multiple stresses often occur concurrently, and on exploring developmental stage-specific and combined stress responses.

Integrating multi-omics data with ecological modeling will also be crucial for deploying stress-tolerant varieties in suitable environments. The convergence of these efforts will accelerate the creation of crops capable of maintaining productivity and stability under climate variability.

The guest editors express their sincere gratitude to all authors, reviewers, and the editorial team of the *International Journal of Molecular Sciences* for their invaluable contributions. Collectively, these papers mark an important step forward in applying molecular science and biotechnology to ensure agricultural sustainability in a changing world.

Funding: This research received no external funding.

Conflicts of Interest: The author declares no conflicts of interest.

List of Contributions

1. Liu, R.; Xu, K.; Li, Y.; Zhao, W.; Ji, H.; Lei, X.; Ma, T.; Ye, J.; Zhang, J.; Du, H.; et al. Investigation on the Potential Functions of ZmEPF/EPFL Family Members in Response to Abiotic Stress in Maize. *Int. J. Mol. Sci.* **2024**, *25*, 7196. https://doi.org/10.3390/ijms25137196.
2. Li, Y.; Jiang, S.; Hong, Y.; Yao, Z.; Chen, Y.; Zhu, M.; Ding, J.; Li, C.; Zhu, X.; Xu, W.; et al. Transcriptomic and Hormonal Changes in Wheat Roots Enhance Growth under Moderate Soil Drying. *Int. J. Mol. Sci.* **2024**, *25*, 9157. https://doi.org/10.3390/ijms25179157.

3. Farooq, M.A.; Zeeshan Ul Haq, M.; Zhang, L.; Wu, S.; Mushtaq, N.; Tahir, H.; Wang, Z. Transcriptomic Insights into Salt Stress Response in Two Pepper Species: The Role of MAPK and Plant Hormone Signaling Pathways. *Int. J. Mol. Sci.* **2024**, *25*, 9355. https://doi.org/10.3390/ijms25179355.
4. Zhu, N.; Qian, Y.; Song, L.; Yu, Q.; Sheng, H.; Li, Y.; Zhu, X. Regulating Leaf Photosynthesis and Soil Microorganisms through Controlled-Release Nitrogen Fertilizer Can Effectively Alleviate the Stress of Elevated Ambient Ozone on Winter Wheat. *Int. J. Mol. Sci.* **2024**, *25*, 9381. https://doi.org/10.3390/ijms25179381.
5. Yan, X.; Wu, X.; Sun, F.; Nie, H.; Du, X.; Li, X.; Fang, Y.; Zhai, Y.; Zhao, Y.; Fan, B.; et al. Cloning and Functional Study of *AmGDSL1* in *Agropyron mongolicum*. *Int. J. Mol. Sci.* **2024**, *25*, 9467. https://doi.org/10.3390/ijms25179467.
6. Kang, L.; Liu, J.; Zhu, H.; Liao, L.; Ye, M.; Wei, Y.; Liu, N.; Ke, Q.; Kim, H.S.; Kwak, S.-S.; et al. StEPF2 and StEPFL9 Play Opposing Roles in Regulating Stomatal Development and Drought Tolerance in Potato (*Solanum tuberosum* L.). *Int. J. Mol. Sci.* **2024**, *25*, 10738. https://doi.org/10.3390/ijms251910738.
7. Liu, R.; Ma, T.; Li, Y.; Lei, X.; Ji, H.; Du, H.; Zhang, J.; Cao, S.-K. Genomic Identification and Expression Analysis of Regulator of Chromosome Condensation 1-Domain Protein Family in Maize. *Int. J. Mol. Sci.* **2024**, *25*, 11437. https://doi.org/10.3390/ijms252111437.
8. Tian, Y.; Wang, X.; Huang, H.; Deng, X.; Zhang, B.; Meng, Y.; Wu, L.; Chen, H.; Zhong, Y.; Chen, W. Genome-Wide Identification of the DnaJ Gene Family in Citrus and Functional Characterization of *ClDJC24* in Response to Citrus Huanglongbing. *Int. J. Mol. Sci.* **2024**, *25*, 11967. https://doi.org/10.3390/ijms252211967. Correction in *Int. J. Mol. Sci.* **2025**, *26*, 411. https://doi.org/10.3390/ijms26010411.
9. Zhang, Y.; Qin, J.; Hou, J.; Liu, C.; Geng, S.; Qin, M.; Li, W.; Dai, Z.; Wu, Z.; Lei, Z.; et al. Identification of the Brassinazole-Resistant (*BZR*) Gene Family in Wheat (*Triticum aestivum* L.) and the Molecular Cloning and Functional Characterization of *TaBZR2.1*. *Int. J. Mol. Sci.* **2024**, *25*, 12545. https://doi.org/10.3390/ijms252312545.
10. Zhang, K.-L.; Leng, Y.-N.; Hao, R.-R.; Zhang, W.-Y.; Li, H.-F.; Chen, M.-X.; Zhu, F.-Y. Adaptation of High-Altitude Plants to Harsh Environments: Application of Phenotypic-Variation-Related Methods and Multi-Omics Techniques. *Int. J. Mol. Sci.* **2024**, *25*, 12666. https://doi.org/10.3390/ijms252312666.
11. Xiong, G.; Cui, D.; Tian, Y.; Schwarzacher, T.; Heslop-Harrison, J.S.; Liu, Q. Genome-Wide Identification of the Lectin Receptor-like Kinase Gene Family in *Avena sativa* and Its Role in Salt Stress Tolerance. *Int. J. Mol. Sci.* **2024**, *25*, 12754. https://doi.org/10.3390/ijms252312754.
12. Ali, A.; Khan, N.M.; Jiang, Y.; Zhou, G.; Wan, Y. Comprehensive Genome-Wide Identification and Expression Profiling of *bHLH* Transcription Factors in *Areca catechu* Under Abiotic Stress. *Int. J. Mol. Sci.* **2024**, *25*, 12936. https://doi.org/10.3390/ijms252312936.
13. Xing, B.; Li, P.; Li, Y.; Cui, B.; Sun, Z.; Chen, Y.; Zhang, S.; Liu, Q.; Zhang, A.; Hao, L.; et al. Integrated Transcriptomic and Metabolomic Analysis of *G. hirsutum* and *G. barbadense* Responses to *Verticillium* Wilt Infection. *Int. J. Mol. Sci.* **2025**, *26*, 28. https://doi.org/10.3390/ijms26010028.

References

1. Arif, M.; Helal, M.M.U.; Bilal, M.S.; Shi, W.; Wang, Z.; Xu, R.; Ghorbani, A.; Chen, M.; Li, L. Transcriptomic and metabolomic profiling reveal the close implication of alkaloid biosynthesis and ureide metabolism to drought stress tolerance in maize. *J. Agric. Food Chem.* **2025**, *73*, 15416–15428. [CrossRef] [PubMed]
2. Mega, R.; Kim, J.; Tanaka, H.; Ishii, T.; Abe, F.; Okamoto, M. Metabolic and transcriptomic profiling during wheat seed development under progressive drought conditions. *Sci. Rep.* **2023**, *13*, 15001. [CrossRef] [PubMed]
3. Dunayevich, P.; Baltanás, R.; Clemente, J.A.; Couto, A.; Sapochnik, D.; Vasen, G. Heat-stress triggers MAPK crosstalk to turn on the hyperosmotic response pathway. *Sci. Rep.* **2018**, *8*, 15168. [CrossRef] [PubMed]
4. Chen, M.; Tian, Y.; Zhu, F.; Fan, T.; Yan, H.; Sun, P.; Li, M.; Hou, X.; Lin, P.; Song, Y.; et al. Alternative splicing of VRF1 acts as a molecular switch to regulate stress-induced early flowering. *Cell Rep.* **2024**, *43*, 114918. [CrossRef] [PubMed]
5. Vishal, B.; Krishnamurthy, P.; Ramamoorthy, R.; Kumar, P.P. OsTPS8 controls yield-related traits and confers salt stress tolerance in rice by enhancing suberin deposition. *New Phytol.* **2019**, *221*, 1369–1386. [CrossRef] [PubMed]

6. Ke, S.; He, X.; Zhang, M.; Luo, L.; Chen, Y.; Cui, Y.; Huang, Y. Genome-wide identification and stress-responsive expression profiling of the TCP gene family in *Cenchrus fungigraminus* under drought and cold stress. *BMC Genom.* **2025**, *26*, 626. [CrossRef] [PubMed]
7. Jia, Z.C.; Yang, X.; Wu, Y.K.; Li, M.; Das, D.; Chen, M.X.; Wu, J. The Art of Finding the Right Drug Target: Emerging Methods and Strategies. *Pharmacol. Rev.* **2024**, *76*, 896–914. [CrossRef] [PubMed]
8. Yang, X.; Li, M.; Jia, Z.; Liu, Y.; Wu, S.; Chen, M.; Hao, G.; Yang, Q. Unraveling the secrets: Evolution of resistance mediated by membrane proteins. *Drug Resist. Updates* **2024**, *77*, 101140. [CrossRef] [PubMed]
9. Zhao, X.; Muhammad Aslam, M.; Chen, M.; Das, D. Plant–Fungi Mutualism, Alternative Splicing, and Defense Responses: Balancing Symbiosis and Immunity. *Int. J. Mol. Sci.* **2024**, *26*, 5197. [CrossRef] [PubMed]
10. Sun, P.; Hao, R.; Fan, F.; Wang, Y.; Zhu, F. Adaptation of High-Altitude Plants to Plateau Abiotic Stresses: A Case Study of the Qinghai-Tibet Plateau. *Int. J. Mol. Sci.* **2024**, *26*, 2292. [CrossRef] [PubMed]

International Journal of
Molecular Sciences

MDPI

Article

Investigation on the Potential Functions of ZmEPF/EPFL Family Members in Response to Abiotic Stress in Maize

Rui Liu [1,2], Keli Xu [1], Yu Li [1], Wanqing Zhao [1], Hongjing Ji [1], Xiongbiao Lei [1], Tian Ma [1], Juan Ye [1], Jianhua Zhang [2,3], Hewei Du [1,*] and Shi-Kai Cao [1,3,*]

1 School of Life Science, Yangtze University, Jingzhou 434025, China; liuxiaoshuang_6@163.com (R.L.); xukeli2024@163.com (K.X.); tinashecool@163.com (Y.L.); m15062656201@163.com (W.Z.); jjing0105@163.com (H.J.); leixiongbiao627@163.com (X.L.); 19164047603@163.com (T.M.); 18882839330@163.com (J.Y.)

2 Department of Biology, Hong Kong Baptist University, Hong Kong, China; jzhang@hkbu.edu.hk

3 State Key Laboratory of Agrobiotechnology, The Chinese University of Hong Kong, Shatin, Hong Kong

* Correspondence: 521052@yangtzeu.edu.cn (H.D.); shikaicao@yangtzeu.edu.cn (S.-K.C.)

Abstract: Maize is an important crop used for food, feed, and fuel. Abiotic stress is an important factor affecting maize yield. The *EPF/EPFL* gene family encodes class-specific secretory proteins that play an important role in the response to abiotic stress in plants. In order to explore and utilize the EPF/EPFL family in maize, the family members were systematically identified, and their chromosomal localization, physicochemical properties, *cis*-acting element prediction in promoters, phylogenetic tree construction, and expression pattern analysis were carried out using bioinformatics techniques. A total of 18 ZmEPF/EPFL proteins were identified in maize, which are mostly alkaline and a small portion acidic. Subcellular localization results showed that ZmEPF6, ZmEPF12, and ZmEPFL2 are localized in the nucleus and cytoplasm. Analysis of *cis*-acting elements revealed that members of the ZmEPF/EPFL family contain regulatory elements such as light response, anoxic, low temperature, and hormone response regulatory elements. RT-qPCR results showed that these family members are indeed responding to cold stress and hormone treatments. These results of this study provide a theoretical basis for improving the abiotic stress resistance of maize in future research.

Keywords: abiotic stress; EPF/EPFL; gene family; hormone; bioinformatic analysis; maize

1. Introduction

Agricultural ecosystems have been adversely affected by abiotic stress, such as high temperature, low temperature, drought, salt, and so on [1]. Stomata is one of the key factors for plants to resist adversity stresses, whose size and density change with the environment [2]. By altering stomatal density, pore size, and switching time, carbon dioxide uptake in plant photosynthesis can be optimized, and external abiotic stresses can be resisted [3]. A large number of ROS accumulated in guard cells promote stomatal closure through activation of calcium channels in the plasma membrane in response to a variety of abiotic stresses [4]. Plants could increase drought tolerance by regulating stomatal closure and reducing water transpiration [5,6].

The fossil record indicates that terrestrial plants had stomatal-like structures on their surface over 400 million years ago [7]. This feature emerged independently in newly evolved terrestrial plant strains [8]. Stomata are composed of guard cells and the pores formed around them, which exist on the surface of most terrestrial plants and act as microscopic pressure-driven valves [7]. Each stoma is surrounded by two highly specialized cells called Guard Cells, which require stereotypical asymmetric and symmetric cell division [9]. Its cell development process strictly follows the cell division program, ensuring that there is at least one interval of cell separation between each mature stoma [10].

In order to facilitate the correct development of stomata and the absence of intercellular separation between stomata, some extracellular and plasma membrane-bound proteins are essential for the coordinated development of stomata and signaling between Epidermal Pavement Cells [11]. These signals include EPF (EPIDERMAL PATTERNING FACTOR) and EPFL (EPIDERMAL PATTERNING FACTOR-Like) signal peptides, leucine-rich repetitive ER (ERECTA) series membrane receptor kinases ERECTA/ERECTA-LIKE1 (ERL1), and Leu-rich repetitive membrane protein TMM (Too Many Mouths) [10]. During stomatal development, ER family proteins regulate the direction of differentiation of MMC (Meristemoid Mother Cells) and are regulated by EPF [12].

So far, members of the *EPF/EPFL* family have been identified in several species, such as moss (*Physcomitrium patens*), barley (*Hordeum vulgare*), rice (*Oryza sativa*), wheat (*Triticum aestivum* L.), and *Arabidopsis thaliana* [13–17]. For example, moss has conserved homologs of EPF, which function together to facilitate the correct development of stomata [13]. Plants develop stomata in a manner that employs multiple transcription factors and EPF ligands to more precisely control subsequent differentiation [18]. In barley, overexpression of HvEPF1 restricts the access of HvEPF1 to the stomatal developmental pathway and significantly enhances water use efficiency, drought tolerance, and soil water retention performance [14]. In rice, overexpression of OsEPF1 and OsEPF2 could reduce the stomatal density and stomatal conductance in leaves, which led to improved water use efficiency [15, 19]. Moreover, several studies have demonstrated that EPF/EPFL family members are involved in the stomatal development process in *Arabidopsis thaliana*. *EPF1* encodes a small secretory protein that is expressed in stomatal precursor cells and controls stomatal development by regulating asymmetric cell division in the MMC [10]. *EPF2* regulates the stomatal density of mature leaves and controls the number of cells entering the stomata during leaf development [20]. Elevated expression of EPFL9 increases stomatal density and aggregation [21]. However, the control of stomatal aggregation by EPFL9 and EPF1, as well as the regulation of stomatal density by EPFL9 and EPF2, are independent of each other [22].

Besides involvement in stomatal regulation, EPFL members are also involved in plant growth, fertility, and yield regulation. For instance, OsEPFL1 has the function of elongating awns in rice [23]. OsEPFL2 is involved in the development of awns and large grains, regulating grain length and grain weight [24]. Furthermore, OsEPFL6, OsEPFL7, OsEPFL8, and OsEPFL9 negatively regulate the number of spikelets per spike. The *osepfl6 osepfl7 osepfl9* triple mutants significantly increased grain yield without affecting spikelet fertility, and specific inhibition of the OsEPFL6-OsER1, OsEPFL7-OsER1, and OsEPFL9-OsER1 ligand receptors could optimize rice panicle architecture [25].

Although the number of EPF/EPFL homologs has been identified in several species, there is still a lack of systematic identification and related research reports on the members of the EPF/EPFL family in maize. In this paper, we conducted a series of bioinformatics analyses to systematically identify and analyze the members of the *EPF/EPFL* gene family in maize. These results contribute to a better understanding of the evolutionary history and functional mechanisms of the EPF/EPFL family in maize.

2. Results

2.1. Gene Structure Analysis and Protein-Conserved Motif and Domain Prediction

To identify the EPF/EPFL family members in maize, we downloaded the maize genome DNA, cDNA, protein sequence, and maize annotation files from the Ensembl Plants website and performed the protein blast using the Arabidopsis EPF/EPFL family members. A total of 18 EPF/EPFL members were identified in maize, named ZmEPF1-15 and ZmEPFL1-L3, respectively (Figure 1). To visualize the gene structure of maize EPF/EPFL family members, we used TBtools-II (V2.042) software to draw a schematic diagram of gene structure. All 17 maize *EPF/EPFL* genes contain one or two introns, except for *ZmEPF7*, which is an intronless gene (Figure 1A).

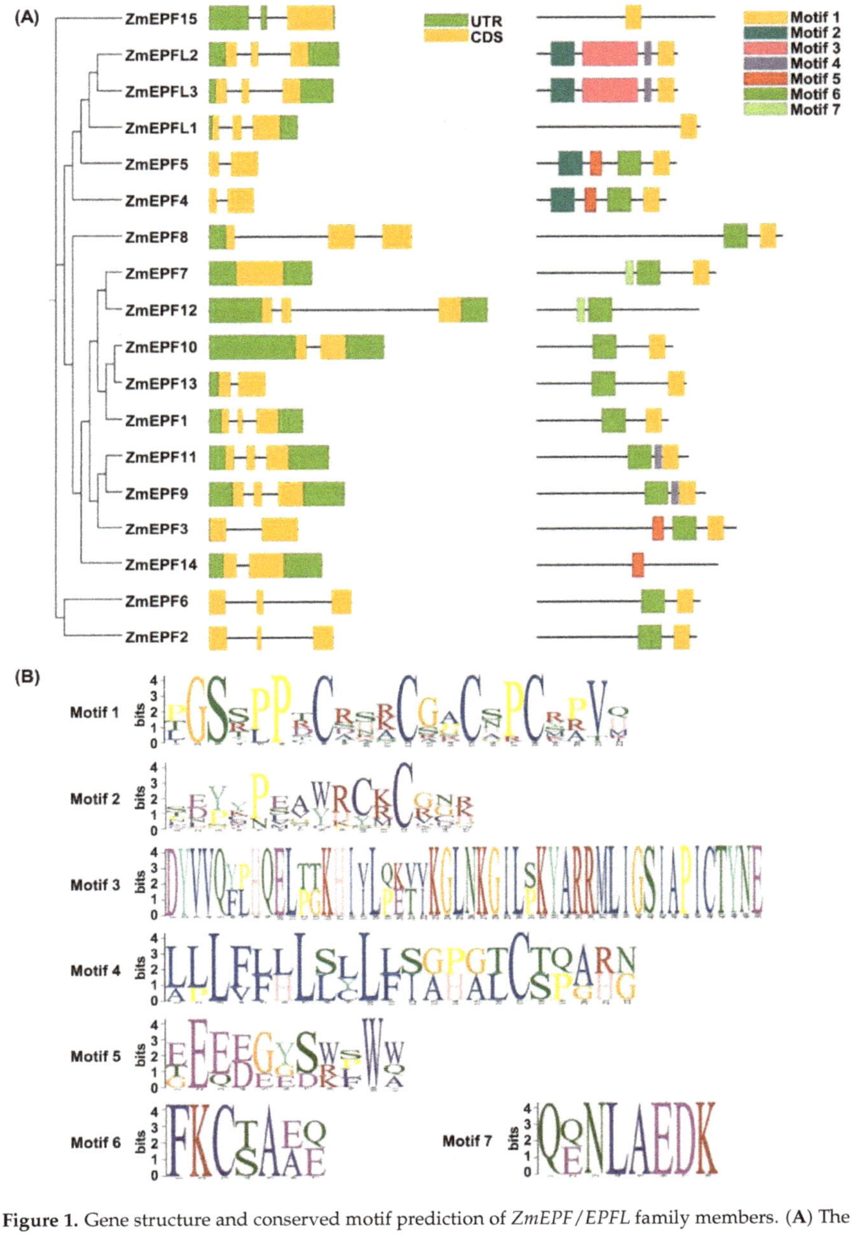

Figure 1. Gene structure and conserved motif prediction of *ZmEPF/EPFL* family members. (**A**) The phylogenetic relationship, gene structure, and conserved motifs of cDNA sequences are listed on the picture's left, middle, and right sides, respectively. Gray lines represent introns, green squares represent UTR (untranslated region), and yellow squares represent exons (middle). Gray lines represent proteins expressed by family members; green, red, gray, and yellow boxes represent different motifs (right). (**B**) The conserved sequences and main protein domains.

Next, a prediction of the protein-conserved domain and motif of all maize EPF/EPFL family members was performed. Results of conserved motif prediction showed that there are seven motifs in the maize EPF/EPFL family members, from motif 1 to 7 (Figure 1B). Three family members have one motif, seven members contain two motifs, three members have three motifs, and four members have four motifs (Figure 1A). Prediction of domain showed that there are three conserved domains in the maize EPF/EPFL family members, including the EPF domain, the stomagen-like superfamily domain, and the PLN03207 domain (Figure S1). Among the maize EPF/EPFL family members, 12 members contain EPF domains. ZmEPFL1 contains a stomagen-like superfamily domain, and ZmEPFL2 and L3 contain PLN03207 domains (Figure S1).

2.2. Phylogenetic Relationship Analysis

To analyze the phylogenetic relationships of the ZmEPF/EPFL family, phylogenetic tree construction of EPF/EPFL proteins was completed in the three species of *Arabidopsis thaliana*, *Oryza sativa*, and *Zea mays* (including 11 Arabidopsis proteins, 12 rice proteins, and 18 maize proteins) using the MEGA 7 (V7.0.26) software (Figure 2). Six subfamilies were classified, from Class I to Class VI subfamilies. There are five members in the Class I subfamily, five in Class II, two in Class III, three in Class IV, three in Class V, and zero in Class VI, respectively (Figure 2).

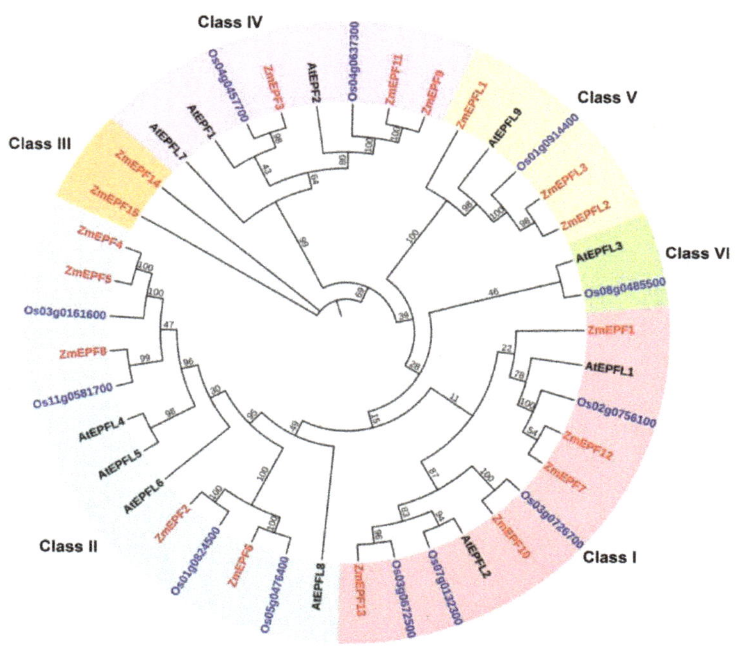

Figure 2. Phylogenetic tree of EPFs and EPFLs in *Zea mays*, *Oryza sativa*, and *Arabidopsis thaliana*. Red font indicates EPF/EPFL family members in *Zea mays*, black font indicates EPF/EPFL family members in *Arabidopsis thaliana*, and blue font indicates EPF/EPFL family members in *Oryza sativa*, and different color backgrounds highlight different classes: pink background highlights Class I, blue background highlights Class II, yellow background highlights Class III, purple background highlights Class IV, light yellow background highlights Class V, and green background highlights Class VI.

In addition, we performed the collinearity analysis of EPFs and EPFLs in different species (*Arabidopsis thaliana*, *Zea mays*, and *Oryza sativa*) (Figures S2 and S3). Results showed that there are three homologous genes between *Arabidopsis thaliana* and *Zea mays*, while there are 12 homologous genes between *Zea mays* and *Oryza sativa* (Figure S2). In *Zea mays*, there are five homologous pairs of *EPF/EPFL* genes, including *ZmEPF2* and *6*, *ZmEPF4* and *5*, *ZmEPF7* and *12*, *ZmEPF9* and *11*, *ZmEPFL2* and *L3* (Figure S3). These results suggested that maize EPF/EPFL family members are more closely related to rice compared with Arabidopsis, consisting of the phylogenetic tree.

2.3. Physicochemical Property Prediction

To get information on the physicochemical properties of ZmEPF/EPFL family members, a comprehensive analysis of the ZmEPF/EPFL family was performed, including subcellular localization, protein length, molecular mass, theoretical isoelectric point (pI), instability index, and grand average of hydropathicity (GRAVY) (Table 1). As shown in Table 1, the relative protein length of the ZmEPF/EPFL members ranges from 113 to 216 amino acids, the relative molecular weight is from 12,115.97 to 23,070.52 Da, and the pI is between 6.81 and 10.7. Only one family member has an isoelectric point less than 7, and the rest have isoelectric points greater than 7. The instability indexes (II) were computed to range from 46.20 to 76.84, indicating that all family members may be unstable. The GRAVY ranges from −0.032 to −0.680, indicating that all the family members may be hydrophilic proteins.

Table 1. Prediction of physicochemical properties of the maize EPF/EPFL family members.

Gene ID	Protein Name	Subcellular Localization	Protein Length (aa)	Molecular Mass (Da)	Theoretical pI	Instability Index	Grand Average of Hydropathicity
Zm00001eb121050	ZmEPF1	Nucleus	115	12,115.97	9.30	54.00	−0.203
Zm00001eb149400	ZmEPF2	Plasma membrane, Nucleus	140	15,319.77	9.44	71.03	−0.309
Zm00001eb423950	ZmEPF3	Nucleus	175	18,452.86	7.57	61.90	−0.069
Zm00001eb403370	ZmEPF4	Nucleus	113	12,325.37	9.91	69.28	−0.174
Zm00001eb004650	ZmEPF5	Plasma membrane, Nucleus	122	13,264.23	9.39	75.99	−0.313
Zm00001eb290530	ZmEPF6	Chloroplast, Nucleus, Cytoplasm	143	14,827.99	9.49	54.09	−0.031
Zm00001eb194040	ZmEPF7	Chloroplast, Nucleus	157	16,999.45	9.13	55.22	−0.265
Zm00001eb200540	ZmEPF8	Chloroplast, Nucleus	216	23,070.52	10.7	76.84	−0.278
Zm00001eb431780	ZmEPF9	Plasma membrane	148	15,291.78	8.07	46.20	0.194
Zm00001eb055900	ZmEPF10	Nucleus	119	13,069.02	9.78	63.50	−0.680
Zm00001eb069150	ZmEPF11	Plasma membrane	133	13,454.54	8.43	52.04	0.180
Zm00001eb255380	ZmEPF12	Nucleus	143	15,175.36	9.04	56.26	−0.267
Zm00001eb032920	ZmEPF13	Chloroplast	131	13,665.28	9.98	59.87	−0.582
Zm00001eb217540	ZmEPF14	Nucleus	159	17,542.06	8.36	55.92	−0.233
Zm00001eb352020	ZmEPF15	Nucleus, Cytoplasm	156	16,344.11	10.00	52.98	−0.724
Zm00001eb174960	ZmEPFL1	Plasma membrane	143	15,131.95	6.81	61.21	−0.376
Zm00001eb143570	ZmEPFL2	Chloroplast, Nucleus, Cytoplasm	123	13,623.8	8.83	47.80	−0.062
Zm00001eb364920	ZmEPFL3	Plasma membrane	123	13,615.88	8.92	52.74	−0.032

Subcellular localization prediction revealed that ZmEPF/EPFL family members may localize in the nucleus, plasma membrane, chloroplast, or cytoplasm (Table 1). To further study the characteristics of these proteins, we fused them with pGWB6-ZmEPF6, ZmEPF12, and ZmEPFL2, transformed tobacco leaf epidermis through Agrobacterium-mediated methods, and observed fluorescence signals using fluorescence confocal microscopy, respectively. Results showed that the green fluorescence signals of pGWB6-ZmEPF6, 12, and L2 were observed in the nucleus, cytoplasm, cell wall, or cell membrane (Figure 3), partially consistent with the results of subcellular localization prediction (ZmEPF6: chloroplast, nucleus, and cytoplasm; ZmEPF12: nucleus; ZmEPFL2: chloroplast, nucleus, and cytoplasm). Regarding whether ZmEPF6 and L2 have chloroplast localization signals, this needs to be further investigated by constructing vectors in which GFP is fused to the C-terminus of the target proteins.

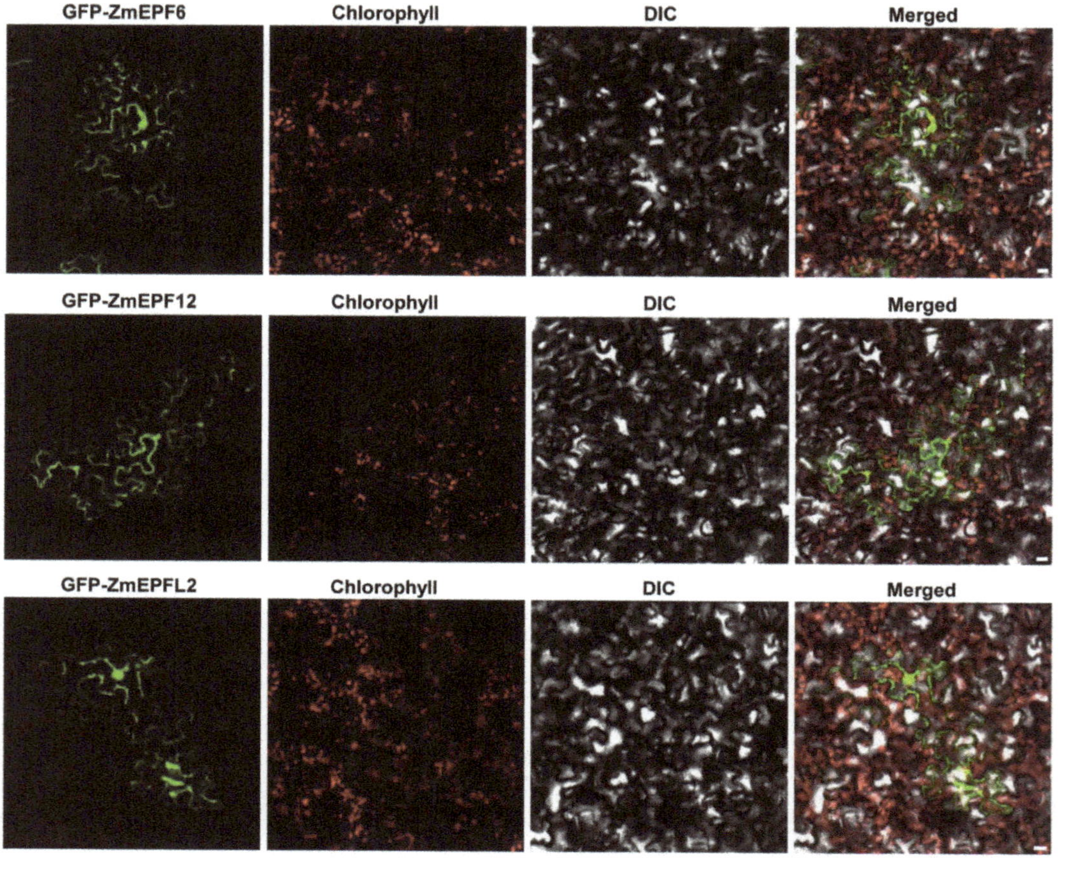

Figure 3. The subcellular localization of ZmEPF6, ZmEPF12, and ZmEPFL2 in *N. benthamiana*. ZmEPF6, ZmEPF12, and ZmEPFL2 fused with N-terminal GFP were infiltrated in *N. benthamiana*, respectively. Scale bar, 20 μm.

2.4. Identification and Chromosome Distribution

Based on genome annotation information, we drew the chromosome distribution map of 18 maize EPF/EPFL family members using the TBtools-II software. Results showed that all 18 *ZmEPF/EPFL* members were distributed on nine chromosomes randomly from chromosome 1 to 10, except for 7 (Figure 4).

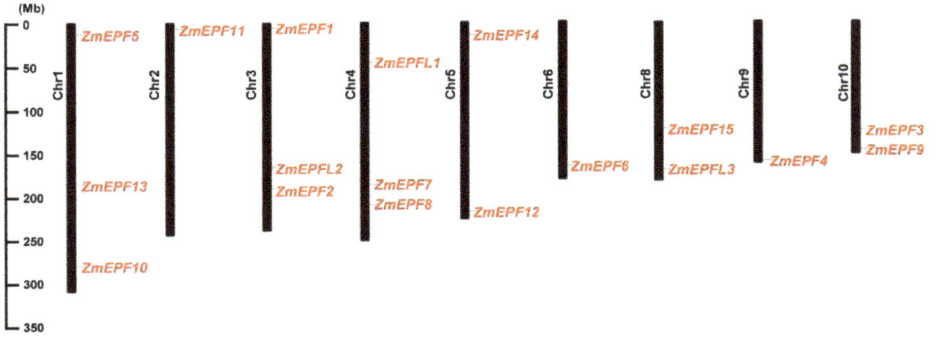

Figure 4. Gene distribution of ZmEPF/EPFL family members in chromosomes. The black bars represent the chromosomes, the gray lines indicate the location of the genes on the chromosomes, and the length of the chromosomes is shown on the left scale.

2.5. Analysis of cis-Acting Elements in Promoters

In order to understand which factors may be related to ZmEPF/EPFL family members, we performed a *cis*-acting element analysis of the promoter region of ZmEPF/EPFL family members. Results showed that the promoters of the family members contain nineteen *cis*-element (Figure 5), including abscisic acid responsive element (ten members), anoxic specific inducible element (ten members), defense and stress responsive element (six members), endosperm expression regulatory element (five members), gibberellin responsive element (nine members), light responsive regulatory element (eighteen members), low temperature responsive element (five members), MeJA responsive regulatory element (eighteen members), meristem expression regulatory element (eleven members), auxin responsive regulatory element (three members), drought inducible element (seven members), Light responsive element (thirteen members), salicylic acid responsive element (eight members), anaerobic inducible regulatory element (nine members), auxin responsive element (eight members), zein metabolism regulatory element (eight members), AT-rich DNA binding protein binding element (one member), seed specific regulatory element (four members), and cell cycle regulatory element (one member).

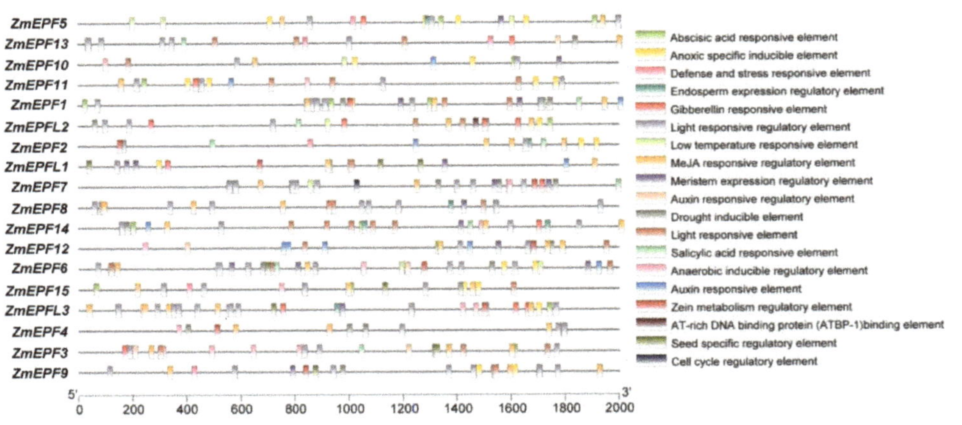

Figure 5. *Cis*-element analysis in promoters of ZmEPF/EPFL family members. Gray lines represent promoters corresponding to family members, various colored boxes represent different *cis*-elements, the *cis*-element corresponding to each colored box is shown on the right side of the diagram, and promoter lengths are shown on the scale below.

2.6. Expression Pattern Analysis

To investigate the expression pattern of ZmEPF/EPFL family members, a spatiotemporal expression heatmap was established based on the transcriptome data (Figure 6). As shown in Figure 6, *ZmEPFL1* exhibits high expression levels in embryos (24 DAP), *ZmEPF3, 9, 11,* and *L2* exhibit high expression level in leaf zone 2 (stomatal) comparing with other members. *ZmEPF2, 7, 12,* and *L3* exhibit relatively moderate expression levels in internode 7–8, leaf zone 2 (stomatal), and vegetative meristem (16–19 days). ZmEPF15 exhibits a medium expression level in embryos (twenty-four DAP) and whole seeds (four DAP).

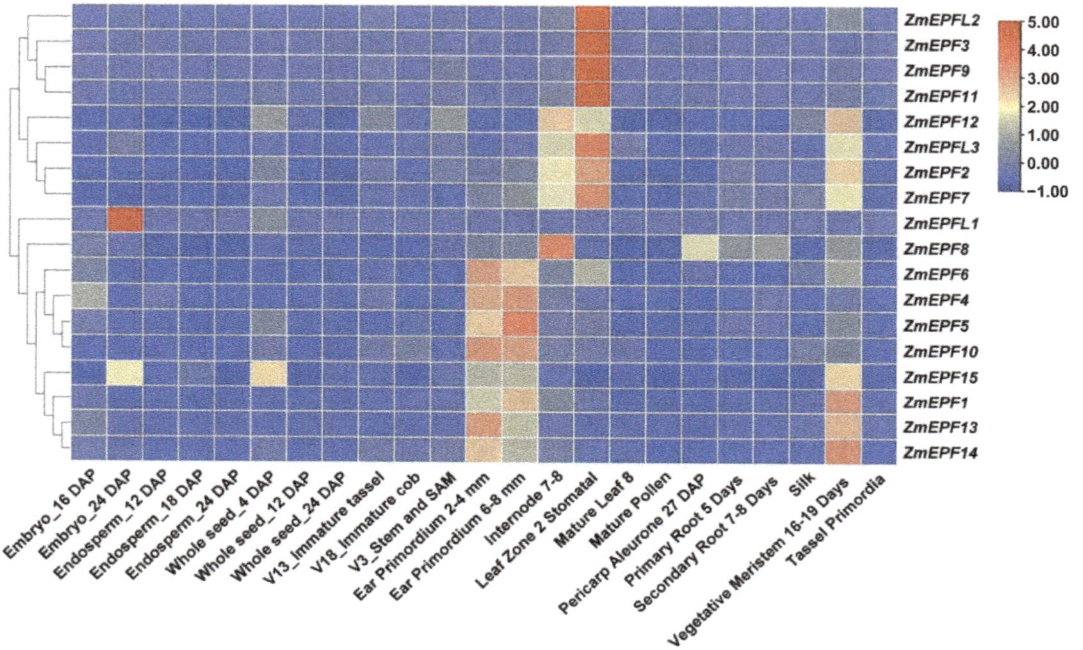

Figure 6. Expression pattern of *ZmEPF/EPFL* family members in different tissues. The horizontal coordinates represent the tissues from which the material was taken, and the different colors indicate the transcript levels of the genes of each member of the family.

2.7. Expression Levels of ZmEPF/EPFL Genes in Response to Abiotic Stresses

To investigate the potential response to abiotic stresses of ZmEPF/EPFL family members, expression heatmaps under different abiotic stresses were established, including cold stress, heat stress, salt stress, UV stress, and water deficit stress (Figure 7). Results of heatmaps showed that expression of six members increased under cold stress, including *ZmEPF3, 5, 8, 13, 15,* and *L1*. The expression of two members (*ZmEPF13* and *L1*) increased under heat stress. Under salt stress, only two members (*ZmEPF11* and *12*) showed increased expression. Under UV stress, seven members exhibited increased expression levels, including *ZmEPF1, 2, 10, 11, 12, 14,* and *15* (Figure 7A).

Furthemore, the transcription level of ZmEPF/EPFL family members was also analyzed under different degrees of water deficit (Figure 7B). There were no significant differences in the expression of the four family members in the presence or absence of a water deficit, including *ZmEPF3, 13, 15,* and *L1*. After 6 h under mild water deficit stress, the transcription levels of five members increased, including *ZmEPF7, 8, 10, 12,* and *L3*. After 24 h under mild water deficit stress, the transcription levels of seven members increased, including *ZmEPF6, 7, 9, 11, 12, L2,* and *L3*. Compared with a mild water deficit, more family members responded to a severe water deficit. After 6 h under severe water deficit stress, the transcription levels of seven members increased, including *ZmEPF2*, *7*, *8*, *10*, *12*, *L2*, and *L3*. After 24 h under severe water deficit stress, the transcription levels of ten members increased, including *ZmEPF2*, *6*, *7*, *8*, *9*, *10*, *11*, *12*, *L2*, and *L3*.

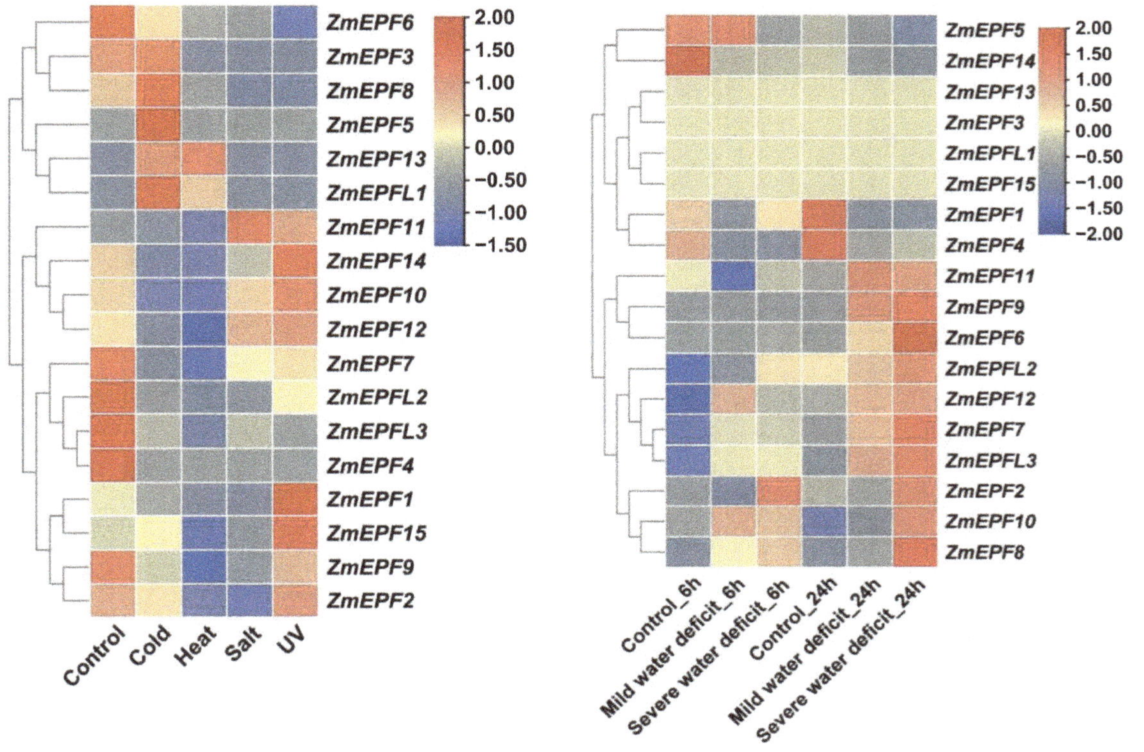

(**A**) (**B**)

Figure 7. Expression pattern of *ZmEPF/EPFL* family members under different stresses. (**A**) Expression pattern of *ZmEPFs/EPFLs* under cold, heat, salt, or UV stress. (**B**) Expression pattern of *ZmEPFs/EPFLs* under water deficit stress.

We detected the expression levels of 11 *ZmEPF/EPFL* family members under drought, heat, and cold treatments by RT-qPCR, including *ZmEPF2*, *6*, *7*, *8*, *9*, *10*, *12*, *14*, *15*, *L2*, and *L3* (Figure 8). Results showed that the expression of all these genes was reduced under drought stress. The expression of *ZmEPF7*, *8*, and *15* was induced under heat stress, and *ZmEPF7*, *10*, *12*, *14*, and *L2* were significantly reduced under cold stress (Figure 8). These results suggested that *ZmEPF/EPFL* members differentially respond to drought, heat, and cold treatments.

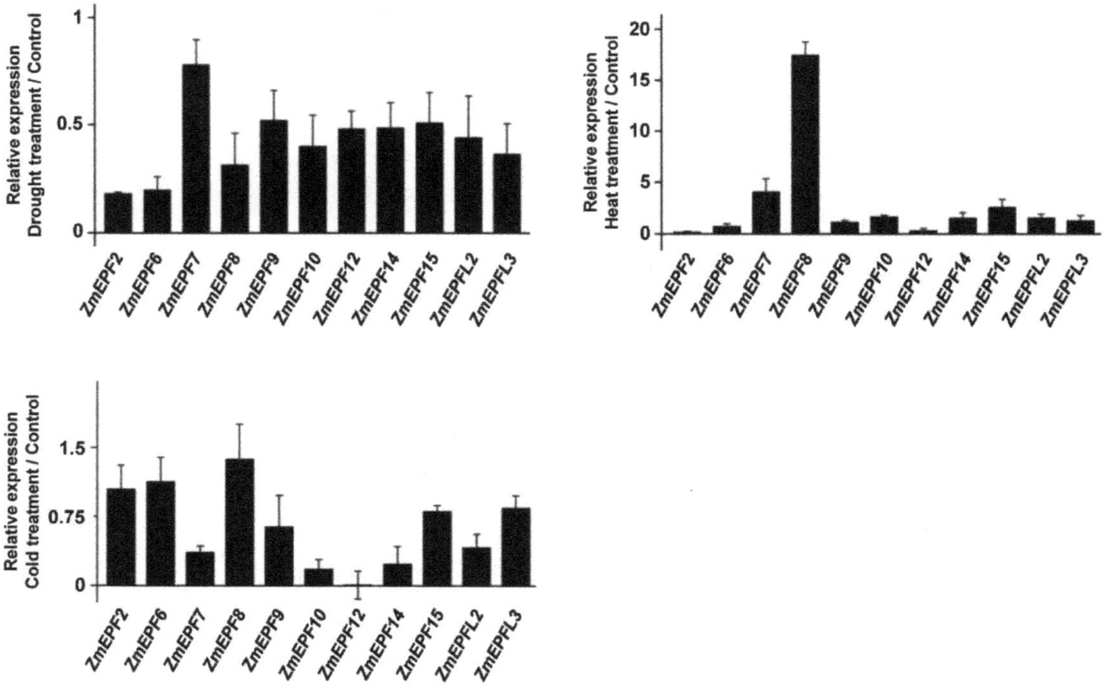

Figure 8. Expression level of *ZmEPF/EPFL* members under drought, heat, and cold treatments. Expression levels of *ZmEPF2, 6, 7, 8, 9, 10, 12, 14, 15, L2,* and *L3* were detected by RT-qPCR under drought, heat, and cold treatments. Data are means (±SE) of three biological replicates.

In addition, the expression levels of 11 *ZmEPF/EPFL* family members under different hormone treatments were detected by RT-qPCR (Figure 9), including *ZmEPF2, 6, 7, 8, 9, 10, 12, 14, 15, L2,* and *L3*. Results showed that the expression of *ZmEPF6, 9, 10, 14, and L3* was significantly reduced, while *ZmEPF8* was induced under SA treatment. Under GA treatment, expression of *ZmEPF10, 12, 14, L2,* and *L3* was reduced, while *ZmEPF8* was induced. Under ABA treatment, except for *ZmEPF7*, the expression of *ZmEPF8* was obviously induced by ABA, while the expression levels of other members were reduced significantly. Under JA treatment, the expression of *ZmEPF10, L2,* and *L3* was significantly reduced, while *ZmEPF8* was induced obviously. Under IAA treatment, the expression of *ZmEPF6, 10, 14, L2,* and *L3* was significantly reduced, while *ZmEPF8* was induced obviously. These results suggested that *ZmEPF/EPFL* members respond to different hormone treatments.

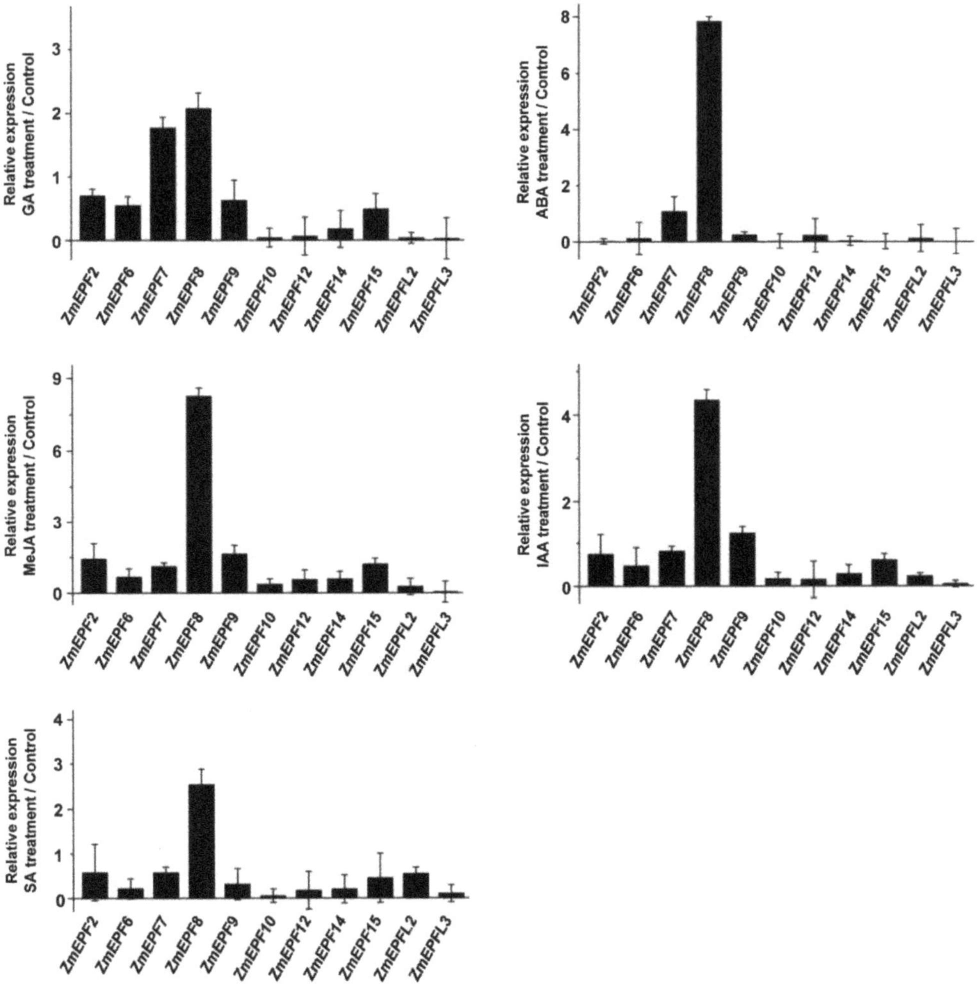

Figure 9. Expression level of *ZmEPF/EPFL* members under different hormone treatments. Expression levels of *ZmEPF2, 6, 7, 8, 9, 10, 12, 14, 15, L2,* and *L3* were detected by RT-qPCR under different hormone treatments including ABA, GA, IAA, MeJA, and SA treatments. Data are means (±SE) of three biological replicates.

3. Discussion

Previous studies reported that EPF/EPFL proteins participate in various biological processes in several plant species, including the regulation of stomatal density and aggregation, plant growth, fertility, and yield. However, there is still a lack of systematic identification and comprehensive analyses of the ZmEPF/EPFL family in evolution. In this work, we performed bioinformatics analyses of the EPF/EPFL family members in maize. A total of 18 ZmEPF/EPFL proteins were identified. Their phylogenetic analysis, physicochemical properties, *cis*-acting element prediction in promoters, and expression pattern analysis were carried out. These results contribute to a better understanding of the evolutionary history and functional mechanisms of the EPF/EPFL family in maize.

Phylogenetic tree analysis showed some EPF/EPFL members from monocotyledons (rice and maize) and dicotyledons (Arabidopsis) were distributed in the same branch (Figure 2). For example, ZmEPF3 clusters with AtEPF1 and Os04g0457700 (OsEPF2) and ZmEPF9 and ZmEPF11 clusters with AtEPF2 and Os04g0637300 (OsEPF1), all belonging to Class IV (Figure 2), suggest that the EPF/EPFL family emerged before the differentiation of dicotyledons and monocotyledons. EPF/EPFL members were divided into six subgroups (Figure 2), suggesting that they were highly conserved during the evolution of Arabidopsis, rice, and maize. Moreover, AtEPF1 and AtEPF2 act as negative regulatory factors to regulate stomatal development [10,16], *OsEPF1* and *OsEPF2* regulate stomatal density [19], and overexpression of OsEPF1 or OsEPF2 significantly reduces the frequency of stomatal density in Arabidopsis *epf2* and *epf1epf2* mutants, demonstrating that EPF1 and EPF2 are indeed functionally conserved between rice and Arabidopsis. This evidence implies that ZmEPF3, 9, and 11 may function in stomatal development and/or density regulation in maize due to their co-cluster with OsEPF1 and OsEPF2 in Class IV. Similarly, ZmEPFL1, L2, and L3 clustered in Class V with AtEPFL9 (Figure 2), and AtEPFL9 functions as a positive regulatory factor in regulating stomatal density and degree of stomatal aggregation [15,21], suggesting that ZmEPFL1, L2, and L3 also may function in stomatal development in maize. These results indicated that some EPF/EPFL homologies were conserved during evolution.

It was previously reported that the EPF/EPFL gene can appropriately reduce stomatal density in the early stages of plant development, reducing transpiration and water loss while maintaining photosynthesis, thereby improving plant drought resistance and water use efficiency [26], suggesting that the *EPF/EPFL* gene family may play an important role in defense. *Cis*-acting element analysis showed that ZmEPF/EPFL members contained defense and stress-responsive elements, low temperature-responsive elements, and drought-inducible elements (Figure 5), speculating that ZmEPF/EPFL members may respond to stress signals. For example, the ZmEPF11 promoter region contains a defense and stress-responsive element (Figure 5), and it is expressed and induced under salt and water deficit, suggesting that it may play a role in response to flooding, salt, UV, and drought (Figure 7). The ZmEPF9 promoter region contains a number of elements, including a defense and stress-responsive element and a drought-inducible element (Figure 5), and it is expressed as induced under water deficit (Figure 7B), suggesting that it may play a role in response to waterlogging and drought. Additionally, RT-qPCR results showed that the expression of *ZmEPF7*, *8*, and *15* was induced under heat stress and *ZmEPF7*, *10*, *12*, *14*, and *L2* were reduced under cold stress (Figure 8). These data indicated that *ZmEPF/EPFL* members respond to abiotic stresses.

Plant hormones play an important role in the formation and development of stomata [27–30]. For example, auxin (IAA) regulates stomatal development in guard cells. The higher the concentration of auxin, the greater the degree of stomatal opening [31]. Several EPF/EPFL members were reported to function in stomatal development and regulation [10,16,21]. How these members and hormone signals function in the regulation of stomata together is unknown in maize. *Cis*-acting element analysis of the promoter region of ZmEPF/EPFL was performed (Figure 5). Results showed that *ZmEPF3*, *12*, and *13* contain auxin-responsive regulatory elements, speculating that these three members may be involved in stomatal development through responding to auxin signals. All ZmEPF/EPFL members contained methyl jasmonate-responsive elements (Figure 5), which is consistent with previous studies on the inhibition of stomatal development by low concentrations of jasmine [32]. Additionally, RT-qPCR results showed that the expression of some *ZmEPF/EPFL* was inhibited or induced under hormone treatments (Figure 9). These data implied that ZmEPF/EPFL members may participate in stomatal development by responding to hormone signals.

In this study, we identified maize EPF/EPFL family members and their potential functions in stomatal development and stress responses. As to what the specific biological functions of each family member are, they need to be analyzed through further in-depth studies. The above exploration results provide a theoretical basis for facilitating further research on the function of *ZmEPF/EPFL* genes.

4. Materials and Methods

4.1. Materials

All materials used in the experiment were sourced from the maize B73 inbred line preserved in the laboratory. Plump seeds were selected to be planted in a tissue culture room in a paper cup filled with nutrient soil. The samples were harvested at the three leaf and one heart stages, frozen in liquid nitrogen, and then stored in a −70 °C refrigerator for subsequent experiments.

Seedlings at the three leaves and one heart stage were treated with low temperature (placed in a transparent refrigerator at 4 °C) for 24 h, methyl jasmonate (MeJA, 0.1 mmol/L), abscisic acid (ABA, 0.1 mmol/L), salicylic acid (SA, 0.1 mmol/L), gibberellin (GA, 0.1 mg/L), and auxin (IAA, 0.1 mg/L) for 6 h, respectively. Mock treatment was used as the corresponding control (CK), with three biological replicates for each treatment. Samples from each treatment were harvested for subsequent RT-qPCR analysis.

4.2. Identification of ZmEPF/EPFL Family Members and Analysis of Protein Physicochemical Properties

The protein sequences of 11 Arabidopsis EPF/EPFL family members were obtained from the TAIR website. Maize genome sequences, proteome sequences, and annotation files were downloaded from the ensembl plant (https://plants.ensembl.org/index.html (accessed on 7 January 2024)) database. The proteins containing the same domain were selected via protein bidirectional BLAST using TBtools-II software, and the maize EPF/EPFL family members were verified by EPF (Pfam17181), stomagen-like superfamily, and PLN03207 domain [33,34]. The protein physicochemical properties of maize EPF/EPFL family members were analyzed online using ExPASy (https://www.expasy.org/ (accessed on 7 January 2024)) website, including relative molecular mass, theoretical isoelectric point, instability index, and the grand average of hydropathicity [35]. Subcellular localizations of ZmEPF/EPFL family members were predicted using Cell-Ploc (http://www.csbio.sjtu.edu.cn/bioinf/Cell-PLoc-2/ (accessed on 8 January 2024)) combined with CELLO (http://cello.life.nctu.edu.tw/ (accessed on 8 January 2024)) online website.

4.3. Distribution of ZmEPF/EPFL Family Members on the Chromosome

Based on the GTF file, the chromosome location information was obtained from gene IDs of the identified maize EPF/EPFL family members. Distribution map on chromosomes of maize EPF/EPFL family members using TBtools-II software [33,34].

4.4. Gene Structure Analysis of ZmEPF/EPFL Family Members

The genomic DNA sequences of maize EPF/EPFL family members were obtained by NCBI (https://www.ncbi.nlm.nih.gov/ (accessed on 1 January 2024)) and Maize GDB (http://www.maizegdb.org/ (accessed on 1 January 2024)), and the gene structure maps were drawn using GSDS (http://gsds.cbi.pku.edu.cn/ (accessed on 9 January 2024)) online software.

4.5. Conserved Motif, Domain Prediction, and Phylogenetic Tree Construction of ZmEPF/EPFL Proteins

Conserved motifs of maize EPF/EPFL proteins were predicted using the online MEME (http://meme-suite.org/tools/meme (accessed on 9 January 2024)) software. Conserved domains of maize EPF/EPFL proteins were predicted using CD search tool on the NCBI official website. The conserved motif and domain of maize EPF/EPFL proteins were visualized using the Gene Structure View (Advanced) function of TBtools-II software. All the EPF/EPFL protein sequences of maize, rice, and Arabidopsis were aligned via ClustalW method, and the phylogenetic tree was constructed by MEGA7 (V7.0.26) [36]. The phylogenetic tree was further beautified using the iTol online website (https://itol.embl.de/ (accessed on 25 January 2024)).

4.6. Analysis of cis-Acting Elements in the Promoters of ZmEPF/EPFL Family Members

The upstream 2000 bp promoter sequences of the ZmEPF/EPFL transcription start site were obtained according to the official maize database MaizeGDB (https://maizegdb.org/ (accessed on 18 January 2024)). The *cis*-acting elements were predicted using the online tool PlantCARE (https://bioinformatics.psb.ugent.be/webtools/plantcare/html/ (accessed on 12 February 2024)), and figures of *cis*-acting elements were drawn using TBtools-II software [33,34].

4.7. Expression Patterns of ZmEPF/EPFL Family Members

The transcriptome data of maize EPF/EPFL family members in multiple tissues and under abiotic stresses were downloaded from qTeller (https://qteller.maizegdb.org/genes_by_name_B73v5.php (accessed on 27 February 2024)). The heatmaps were drawn using TBtools-II software [33,34].

4.8. Collinearity Analysis

The genome files and gene annotation files of *Zea mays*, *Oryza sativa*, and *Arabidopsis thaliana* were obtained from the Ensembl Plants database, and the intra-species covariate analysis of *Zea mays* and inter-species covariate analysis of *Zea mays*, *Oryza sativa*, and *Arabidopsis thaliana* were plotted by the MCScanX function in the TBtools-II software [33,34].

4.9. RNA Extraction and Real-Time Fluorescence Quantitative PCR

The total RNAs were extracted from the leaves of each treated plant using an RNA-easy isolation reagent (Vazyme Biotech Co., Ltd., Nanjing, China). The first strands of cDNA were synthesized using the HiFiScript cDNA Synthesis Kit (Vazyme Biotech Co., Ltd.) and then used as the templates of the real-time fluorescence quantitative PCR (qPCR) (Vazyme Biotech Co., Ltd.) for further hormone treatments and cold stress-responsive analysis. The RT-qPCR was performed on the CFX96 real-time PCR instrument (Bio-Rad, Hercules, CA, USA), and *ZmActin* (*GRMZM2G126010*) was used as the reference gene. All the primers are listed in Supplementary Table S1. Three biological and technical replicates are performed, respectively.

4.10. Subcellular Localization

For subcellular localization, the full-length ZmEPF6, ZmEPF12, and ZmEPFL2 were cloned into binary vector pGWB6, respectively. Agrobacterium tumefaciens strain (*GV3101*) harboring GFP-ZmEPF6, GFP-ZmEPF12, and GFP-ZmEPFL2 constructs were transiently expressed in tobacco epidermal cells, as previously described [37]. The GFP signals were observed and imaged using ZEISS LSM 880 confocal microscope (Carl-Zeiss, Jena, Germany).

Supplementary Materials: The following supporting information can be downloaded at: https://www.mdpi.com/article/10.3390/ijms25137196/s1.

Author Contributions: R.L. and S.-K.C. designed the experiments. R.L., K.X., Y.L., W.Z., H.J., X.L., T.M., J.Y., and S.-K.C. performed the experiments and bioinformatic analysis. R.L., J.Z., H.D., and S.-K.C. analyzed the data and wrote the manuscript. All authors have read and agreed to the published version of the manuscript.

Funding: This research was funded by the Hubei Natural Science Foundation (2022CFB183).

Institutional Review Board Statement: Not applicable.

Informed Consent Statement: Not applicable.

Data Availability Statement: Data are contained within the article and supplementary materials.

Acknowledgments: We thank Tsuyoshi Nakagawa (Shimane University, Japan) for the pGWB vectors.

Conflicts of Interest: The authors declare no conflicts of interest.

References

1. Zhang, H.M.; Zhu, J.H.; Gong, Z.Z.; Zhu, J.K. Abiotic stress responses in plants. *Nat. Rev. Genet.* **2022**, *23*, 104–119. [CrossRef] [PubMed]
2. Bertolino, L.T.; Caine, R.S.; Gray, J.E. Impact of stomatal density and morphology on water-use efficiency in a changing world. *Front. Plant Sci.* **2019**, *10*, 225. [CrossRef] [PubMed]
3. Bawa, G.; Yu, X.; Liu, Z.; Zhou, Y.; Sun, X. Surviving the enemies: Regulatory mechanisms of stomatal function in response to drought and salt stress. *Environ. Exp. Bot.* **2023**, *209*, 105291. [CrossRef]
4. Naithani, S.; Mohanty, B.; Elser, J.; D'Eustachio, P.; Jaiswal, P. Biocuration of a transcription factors network involved in submergence tolerance during seed germination and coleoptile elongation in rice (*Oryza sativa*). *Plants* **2023**, *12*, 2146. [CrossRef] [PubMed]
5. Qi, J.S.; Song, C.P.; Wang, B.S.; Zhou, J.M.; Kangasjärvi, J.; Zhu, J.K.; Gong, Z.Z. Reactive oxygen species signaling and stomatal movement in plant responses to drought stress and pathogen attack. *J. Integr. Plant Biol.* **2018**, *60*, 805–826. [CrossRef] [PubMed]
6. Zheng, W.N.; Jiang, Y.X.; Wang, X.F.; Huang, S.J.; Yuan, M.; Guo, Y. AP3M harbors actin filament binding activity that is crucial for vacuole morphology and stomatal closure in. *Proc. Natl. Acad. Sci. USA* **2019**, *116*, 18132–18141. [CrossRef] [PubMed]
7. Chater, C.C.C.; Caine, R.S.; Fleming, A.J.; Gray, J.E. Origins and evolution of stomatal development. *Plant Physiol.* **2017**, *174*, 624–638. [CrossRef] [PubMed]
8. Bowles, A.M.C.; Williamson, C.J.; Williams, T.A.; Lenton, T.M.; Donoghue, P.C.J. The origin and early evolution of plants. *Trends Plant Sci.* **2023**, *28*, 312–329. [CrossRef] [PubMed]
9. Dong, J.; Bergmann, D.C. Stomatal patterning and development. *Curr. Top. Dev. Biol.* **2010**, *91*, 267–297.
10. Hara, K.; Kajita, R.; Torii, K.U.; Bergmann, D.C.; Kakimoto, T. The secretory peptide gene *EPF1* enforces the stomatal one-cell-spacing rule. *Genes Dev.* **2007**, *2*, 1720–1725. [CrossRef]
11. Zhao, L.; Sack, F.D. Ultrastructure of stomatal development in Arabidopsis (*Brassicaceae*) leaves. *Am. J. Bot.* **1999**, *86*, 929–939. [CrossRef] [PubMed]
12. Shimada, T.; Sugano, S.S.; Hara-Nishimura, I. Positive and negative peptide signals control stomatal density. *Cell. Mol. Life Sci.* **2011**, *68*, 2081–2088. [CrossRef] [PubMed]
13. Caine, R.S.; Chater, C.C.C.; Fleming, A.J.; Gray, J.E. Stomata and sporophytes of the model moss *Physcomitrium patens*. *Front. Plant Sci.* **2020**, *11*, 643. [CrossRef] [PubMed]
14. Hughes, J.; Hepworth, C.; Dutton, C.; Dunn, J.A.; Hunt, L.; Stephens, J.; Waugh, R.; Cameron, D.D.; Gray, J.E. Reducing stomatal density in barley improves drought tolerance without impacting on yield. *Plant Physiol.* **2017**, *174*, 776–787. [CrossRef] [PubMed]
15. Lu, J.; He, J.; Zhou, X.; Zhong, J.; Li, J.; Liang, Y.K. Homologous genes of epidermal patterning factor regulate stomatal development in rice. *J. Plant Physiol.* **2019**, *234*, 18–27. [CrossRef] [PubMed]
16. Hara, K.; Yokoo, T.; Kajita, R.; Onishi, T.; Yahata, S.; Peterson, K.M.; Torii, K.U.; Kakimoto, T. Epidermal cell density is autoregulated via a secretory peptide, EPIDERMAL PATTERNING FACTOR 2 in Arabidopsis leaves. *Plant Cell Physiol.* **2009**, *50*, 1019–1031. [CrossRef] [PubMed]
17. Wei, D.; Chang, P.; Liu, J.; Yang, Y.; Zhang, X.; Chen, L.; Hu, Y. Genome-wide identification of *EPF/EPFL* gene family in wheat (*Triticum aestivum*) and analysis of *TaEPF1-2B* associated with stomatal traits. *J. Trit. Crop.* **2021**, *41*, 1317–1329.
18. Caine, R.S.; Chater, C.C.; Kamisugi, Y.; Cuming, A.C.; Beerling, D.J.; Gray, J.E.; Fleming, A.J. An ancestral stomatal patterning module revealed in the non-vascular land plant *Physcomitrella patens*. *Development* **2016**, *143*, 3306–3314. [CrossRef] [PubMed]
19. Mohammed, U.; Caine, R.S.; Atkinson, J.A.; Harrison, E.L.; Wells, D.; Chater, C.C.; Gray, J.E.; Swarup, R.; Murchie, E.H. Rice plants overexpressing *OsEPF1* show reduced stomatal density and increased root cortical aerenchyma formation. *Sci. Rep.* **2019**, *9*, 5584. [CrossRef]
20. Hunt, L.; Gray, J.E. The signaling peptide EPF2 controls asymmetric cell divisions during stomatal development. *Curr. Biol.* **2009**, *19*, 864–869. [CrossRef]
21. Takata, N.; Yokota, K.; Ohki, S.; Mori, M.; Taniguchi, T.; Kurita, M. Evolutionary relationship and structural characterization of the *EPF/EPFL* gene family. *PLoS ONE* **2013**, *8*, e65183. [CrossRef]
22. Hunt, L.; Bailey, K.J.; Gray, J.E. The signalling peptide EPFL9 is a positive regulator of stomatal development. *New Phytol.* **2010**, *186*, 609–614. [CrossRef]
23. Bessho-Uehara, K.; Wang, D.R.; Furuta, T.; Minami, A.; Nagai, K.; Gamuyao, R.; Asano, K.; Angeles-Shim, R.B.; Shimizu, Y.; Ayano, M.; et al. Loss of function at RAE2, a previously unidentified EPFL, is required for awnlessness in cultivated Asian rice. *Proc. Natl. Acad. Sci. USA* **2016**, *113*, 8969–8974. [CrossRef]
24. Xiong, L.; Huang, Y.; Liu, Z.; Li, C.; Yu, H.; Shahid, M.Q.; Lin, Y.; Qiao, X.; Xiao, J.; Gray, J.E.; et al. Small EPIDERMAL PATTERNING FACTOR-LIKE2 peptides regulate awn development in rice. *Plant Physiol.* **2022**, *190*, 516–531. [CrossRef]
25. Guo, T.; Lu, Z.Q.; Xiong, Y.; Shan, J.X.; Ye, W.W.; Dong, N.Q.; Kan, Y.; Yang, Y.B.; Zhao, H.Y.; Yu, H.X.; et al. Optimization of rice panicle architecture by specifically suppressing ligand-receptor pairs. *Nat. Commun.* **2023**, *14*, 1640. [CrossRef]

26. Liu, S.; Jia, F.L.; Jiao, Z.Y.; Wang, J.J.; Xia, X.L.; Yin, W.L. Ectopic expression of secretory peptide PdEPF3 in confers drought tolerance with reduced stomatal density. *Acta Soc. Bot. Pol.* **2019**, *88*, 12. [CrossRef]
27. Zhao, X.; Zhang, K.; Zhang, H.; Bi, M.; He, Y.; Cui, Y.; Tan, C.; Ma, J.; Qi, M. Tomato short internodes and pedicels encode an LRR receptor-like serine/threonine-protein kinase ERECTA regulating stem elongation through modulating gibberellin metabolism. *Front. Plant Sci.* **2023**, *14*, 1283489. [CrossRef]
28. Xia, Y.; Du, K.; Ling, A.; Wu, W.; Li, J.; Kang, X. Overexpression of *PagSTOMAGEN*, a positive regulator of stomatal density, promotes vegetative growth in poplar. *Int. J. Mol. Sci.* **2022**, *23*, 10165. [CrossRef]
29. Franco, J. BZR1, you have an invite: EPFL-ERECTA wants to join your female germline specification network. *Plant Cell* **2023**, *35*, 1298–1299. [CrossRef]
30. Vaghela, P.; Gandhi, G.; Trivedi, K.; Anand, K.G.V.; Chavda, D.; Manna, M.; Seth, T.; Seth, A.; Shanmugam, M.; Ghosh, A. Underpinning beneficial maize response to application of minimally processed homogenates of red and brown seaweeds. *Front. Plant Sci.* **2023**, *14*, 1273355. [CrossRef]
31. Tameshige, T.; Ikematsu, S.; Torii, K.U.; Uchida, N. Stem development through vascular tissues: EPFL-ERECTA family signaling that bounces in and out of phloem. *J. Exp. Bot.* **2017**, *68*, 45–53. [CrossRef] [PubMed]
32. Hu, Z.; Zhang, H.; Shi, K. Plant peptides in plant defense responses. *Plant Signal. Behav.* **2018**, *13*, e1475175. [CrossRef]
33. Chen, C.; Chen, H.; Zhang, Y.; Thomas, H.R.; Frank, M.H.; He, Y.; Xia, R. TBtools: An integrative toolkit developed for interactive analyses of big biological data. *Mol. Plant* **2020**, *13*, 1194–1202. [CrossRef] [PubMed]
34. Chen, C.; Wu, Y.; Li, J.; Wang, X.; Zeng, Z.; Xu, J.; Liu, Y.; Feng, J.; Chen, H.; He, Y.; et al. TBtools-II: A "one for all, all for one" bioinformatics platform for biological big-data mining. *Mol. Plant* **2023**, *16*, 1733–1742. [CrossRef] [PubMed]
35. Wilkins, M.R.; Gasteiger, E.; Bairoch, A.; Sanchez, J.C.; Williams, K.L.; Appel, R.D.; Hochstrasser, D.F. Protein identification and analysis tools in the ExPASy server. *Methods Mol. Biol.* **1999**, *112*, 531–552. [PubMed]
36. Kumar, S.; Stecher, G.; Tamura, K. MEGA7: Molecular evolutionary genetics analysis version 7.0 for bigger datasets. *Mol. Biol. Evol.* **2016**, *33*, 1870–1874. [CrossRef]
37. Luan, Q.L.; Zhu, Y.X.; Ma, S.; Sun, Y.; Liu, X.Y.; Liu, M.; Balint-Kurti, P.J.; Wang, G.F. Maize metacaspases modulate the defense response mediated by the NLR protein Rp1-D21 likely by affecting its subcellular localization. *Plant J.* **2021**, *105*, 151–166. [CrossRef]

International Journal of
Molecular Sciences

MDPI

Article

Transcriptomic and Hormonal Changes in Wheat Roots Enhance Growth under Moderate Soil Drying

Ying Li [1,2,3], Shuqiu Jiang [1], Yonghui Hong [1], Zixuan Yao [1], Yadi Chen [4], Min Zhu [1,3], Jinfeng Ding [1,3], Chunyan Li [1,3], Xinkai Zhu [1,3], Weifeng Xu [5], Wenshan Guo [1,3], Nanyan Zhu [1,*] and Jianhua Zhang [2,6,*]

1 Jiangsu Key Laboratory of Crop Genomics and Physiology/Jiangsu Key Laboratory of Crop Cultivation and Physiology, College of Agriculture, Yangzhou University, Yangzhou 225009, China; yingli@yzu.edu.cn (Y.L.); dx120230131@stu.yzu.edu.cn (S.J.); mz120211245@stu.yzu.edu.cn (Y.H.); 006514@yzu.edu.cn (M.Z.); jfdin@yzu.edu.cn (J.D.); licy@yzu.edu.cn (C.L.); xkzhu@yzu.edu.cn (X.Z.); guows@yzu.edu.cn (W.G.)
2 Department of Biology, Hong Kong Baptist University, Hong Kong 999077, China
3 Jiangsu Co-Innovation Center for Modern Production Technology of Grain Crops, Yangzhou University, Yangzhou 225009, China
4 College of Horticulture and Landscape Architecture, Yangzhou University, Yangzhou 225009, China; yadichen@yzu.edu.cn
5 Joint International Research Laboratory of Water and Nutrient in Crop, College of JunCao Science and Ecology, Fujian Agriculture and Forestry University, Fuzhou 350002, China; wfxu@fafu.edu.cn
6 School of Life Sciences and State Key Laboratory of Agrobiotechnology, The Chinese University of Hong Kong, Hong Kong 999077, China
* Correspondence: dx120200148@stu.yzu.edu.cn (N.Z.); jzhang@hkbu.edu.hk (J.Z.)

Abstract: Understanding the mechanisms that regulate plant root growth under soil drying is an important challenge in root biology. We observed that moderate soil drying promotes wheat root growth. To understand whether metabolic and hormonic changes are involved in this regulation, we performed transcriptome sequencing on wheat roots under well-watered and moderate soil drying conditions. The genes upregulated in wheat roots under soil drying were mainly involved in starch and sucrose metabolism and benzoxazinoid biosynthesis. Various plant hormone-related genes were differentially expressed during soil drying. Quantification of the plant hormones under these conditions showed that the concentrations of abscisic acid (ABA), cis-zeatin (CZ), and indole-3-acetic acid (IAA) significantly increased during soil drying, whereas the concentrations of salicylic (SA), jasmonic (JA), and glycosylated salicylic (SAG) acids significantly decreased. Correlation analysis of total root length and phytohormones indicated that CZ, ABA, and IAA are positively associated with wheat root length. These results suggest that changes in metabolic pathways and plant hormones caused by moderate soil drying help wheat roots grow into deeper soil layers.

Keywords: plant hormone; metabolism; soil drying; wheat; root

1. Introduction

Drought limits crop growth and yield worldwide [1–3]. Modulation of root systems is a key feature of plant responses to soil drying [4]. Rhizosphere soils form a complex environment, and roots subjected to soil drying face multiple challenges that can alter their growth and development [5]. Maintenance of root growth is essential for plant adaptation to soil drying. However, the mechanisms by which soil drying mediates developmental changes are unclear.

Wheat (*Triticum aestivum* L.) is important for global food security, but its production is severely reduced by drought in many regions [3]. Researchers have found that a 40% reduction in water can cause a 20.6% yield loss in wheat, a threat that is exacerbated by climatic warming and population expansion [6–8]. Beneficial plant root traits and irrigation management techniques that can enhance water-use capacities and increase crop yields under reduced water input need to be identified urgently [9]. Moderate soil drying is a promising approach for promoting root growth, reducing water inputs, and maintaining or increasing crop yields.

Plant hormones such as abscisic acid (ABA), auxin, brassinosteroid, cytokinin, ethylene, gibberellic acid (GA), jasmonic acid (JA), salicylic acid (SA), and strigolactone are important plant growth regulators that play pivotal roles in responses to various environmental stresses [10–13]. Classic studies have shown that ABA biosynthesis and concentrations increase under drought conditions, and low levels of applied ABA or osmotic stress can increase root growth [4]. Drought stress causes ABA to accumulate in plant roots, enter the xylem, and be transported to the guard cells of leaves via transpiration, inducing stomatal closure, which reduces water transpiration under drought stress [14,15]. Drought stress can induce the expression of genes involved in the water deficit response in roots, such as pathogenesis-related protein 4, ABA deficient protein 4, ABA insensitive protein 5, protein phosphatase 2C, and ABA-8′ hydroxylase [10,16]. The first plant hormone discovered, auxin, regulates virtually all aspects of plant growth and development, as well as adaptation to drought, high temperature, and salinity [1,5,17]. The auxin-responsive *Gretchen Hagen 3* (*GH3*) family gene *WES1* (for WESO 1, meaning a dwarfed stature in Korean), which encodes indole-3-acetic acid (IAA) amino acid synthase, is up-regulated under stress conditions, catalyzing the binding to IAA of amino acids, causing inactivation and thereby reducing endogenous auxin levels, and activating the expression of the stress-related genes *pathogenesis-related protein 1* (*PR-1*) and *C-repeat binding factor* (*CBF*) to adapt to stress [18]. Other plant hormones also play important roles under drought conditions; for instance, perturbation of ethylene or cytokinin pathways can affect growth and survival under drought conditions [11–13]. In addition to plant hormones, several metabolic pathways regulate the adaptive responses of plant roots to soil drying [19–26].

In nature, plant roots grow in a complex soil environment with extremely diverse soil microbial communities, and studies suggest that rhizosphere microorganisms in the soil significantly affect root growth and resistance to various stresses [5,27]. For example, auxin-producing bacterial strains isolated from soil (*Chryseobacterium culicis* and *Paenibacillus polymyxa*) produce IAA and enhance barley root growth [27]. However, it is not clear how plant hormone actions and metabolic pathways are integrated into plant root growth and stress adaptation under dry soil conditions.

We hypothesized that metabolic pathways and phytohormones play important roles in mediating root growth during soil drying. We analyzed differentially expressed genes (DEGs) in wheat roots between well-watered (WW) and soil-drying (SD) conditions using transcriptome sequencing and examined the concentrations of various phytohormones under soil drying. Our results suggest that starch and sucrose metabolism and benzoxazinoid biosynthesis promote wheat root elongation under moderate soil drying, allowing roots to obtain more water from the soil.

2. Results

2.1. Moderate Soil Drying Promotes Wheat Root Growth

Our previous study indicated that moderate drought can promote root growth in *Arabidopsis* and tomatoes, allowing the plants to obtain more water [28]. To investigate the effect of soil drying on wheat root growth, wheat was grown under well-watered (WW) and soil-drying (SD) conditions. Under WW, soil water was maintained at 80% of the field capacity for the first 4 days and then dried for 2 days, whereas in SD, soil water was maintained at 45% of the field capacity for the first 4 days before drying for 2 days. We found that moderate soil drying significantly promoted wheat root growth (Figure 1A). Compared to WW, the root length and fresh and dry weights per plant under SD were 19.1%, 29.4%, and 53.3% higher, respectively (Figure 1B–D). In addition, there was no significant difference in shoot fresh weight (Figure S1A) or dry weight (Figure S1B) between WW and SD. Taken together, moderate soil drying can significantly promote root growth.

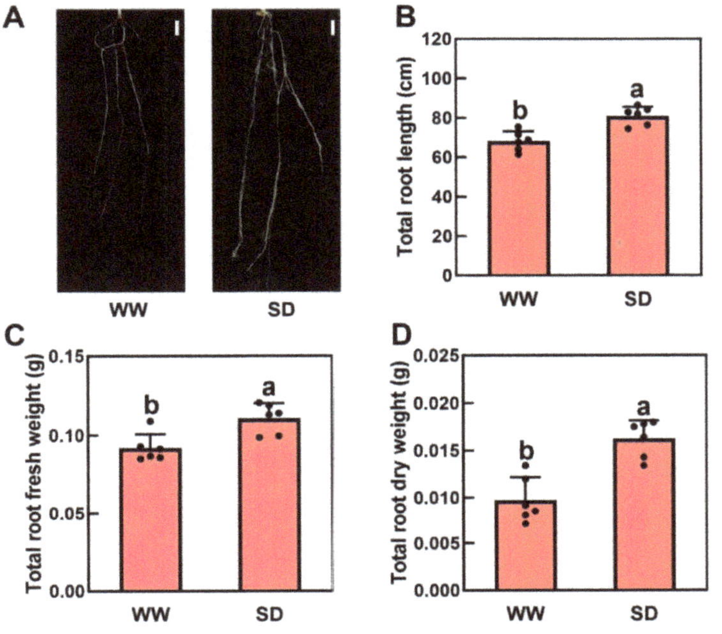

Figure 1. Effects of soil drying on wheat roots. (**A**) Representative images of wheat roots in well-watered (WW) and soil drying (SD) conditions. Scale bar = 1 cm. (**B–D**) Impact of soil drying on the total root length (**B**), root fresh weight (**C**), and root dry weight (**D**). The data in (**B–D**) are the mean ± standard error (n = 6). The same letter indicates no significant difference, whereas different letters indicate significant differences ($p < 0.05$, Duncan's test).

2.2. Starch and Sucrose as Primary Metabolites and Benzoxazinoids as Secondary Metabolites Are Involved in Wheat Root Growth Under Soil Drying

To investigate the mechanism by which moderate soil drying promotes wheat root growth, we collected roots under WW and SD conditions for RNA-Seq analysis. After filtering the raw data, checking the sequencing error rate, and checking the GC content distribution, clean reads (Table S1) for subsequent RNA-Seq analysis were obtained. The expression of all genes in each sample was calculated, and the distribution of gene expression levels in different samples was evaluated (Figure S2). The violin plot for each sample displays five statistics (from top to bottom, the maximum value, upper quartile, median, lower quartile, and minimum value); this shows that the data distribution in each sample is concentrated, which ensures the accuracy and reliability of the RNA-Seq analysis in this experiment. Pearson correlation analysis showed that the square of the correlation coefficient (R2) between each pair of samples exceeded 0.82 (Figure 2A), and principal component analysis (PCA) revealed that the six samples could be clearly assigned to the WW and SD groups (Figure 2B), indicating that the data from each independent biological replicate in this experiment had good reproducibility and could be used for further analysis. The gene expression of each sample was analyzed; 49,567 genes were detected in the WW group (3078 unique genes) versus 49,378 genes in the SD group (2889 unique genes) (Figure 2C). Differential gene clustering analysis showed that many genes were differentially expressed in the different conditions (Figure S3). Compared to WW, SD treatment significantly upregulated (more than 2-fold) 1958 genes and downregulated (less than 2-fold) 2529 genes (Figures 2D and S3), indicating that the expression of many genes in wheat roots changed after soil drying.

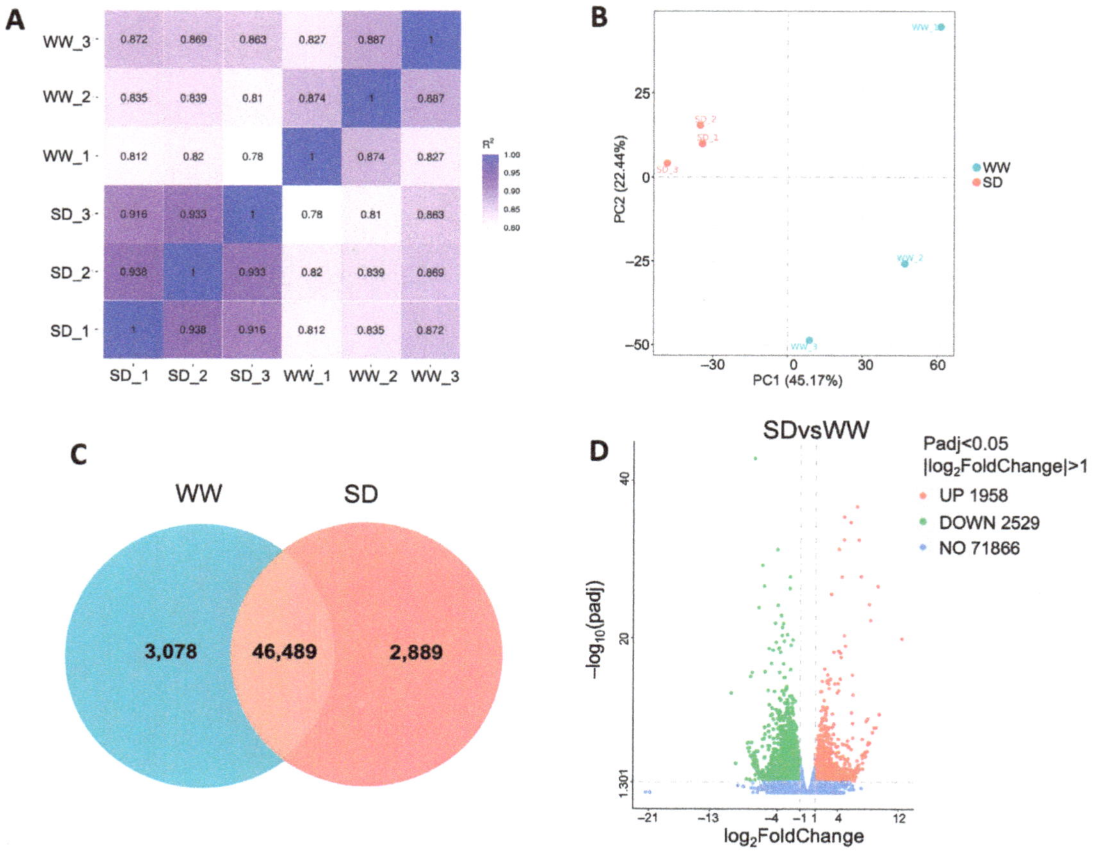

Figure 2. DEGs in wheat roots between well-watered (WW) and soil-drying (SD) conditions. (**A**) Pearson correlation coefficient of gene expression; (**B**) Principal components analysis of the transcriptome; (**C**) Venn diagram of DEGs; and (**D**) number of DEGs in wheat roots between the WW and SD conditions. Each point in D represents a gene, with red representing upregulated (greater than 2-fold), green representing downregulated (less than 2-fold), and blue representing genes whose expression did not change between WW and SD.

To identify which functional classes or biological processes the DEGs are involved in, we performed Gene Ontology (GO) functional enrichment analysis. GO is a comprehensive database that describes gene functions, including biological processes (BP), cellular components (CC), and molecular functions (MF). We selected the ten processes with the greatest significant differences from each functional category for analysis, totaling 30 biological processes involving response to water deprivation, response to water, structural consistency of chromatin, vacuole, and other processes (Figure 3). These results indicated that many genes responsive to water deficiency are involved in wheat root elongation under soil drying.

Next, we performed a Kyoto Encyclopaedia of Genes and Genomes (KEGG) pathway enrichment analysis on the DEGs. Figure 4 shows the ten KEGG pathways with the greatest differences; the DEGs were mainly enriched in metabolic pathways. Compared to WW, the KEGG pathways with more upregulated genes under SD treatment included biotin, linoleic acid, cyanoamino acid, starch, and sucrose metabolism, as well as the biosynthesis of benzoxazinoids and various plant secondary metabolites. Of these, SD treatment had the most upregulated genes for the biosynthesis of various plant secondary metabolites and starch and sucrose metabolism. The upregulated genes in the secondary metabolite pathway were mainly enriched in benzoxazinoid biosynthesis (Figure 5). Collectively, these results suggest that starch, sucrose, and benzoxazinoids contribute to wheat root elongation under soil drying.

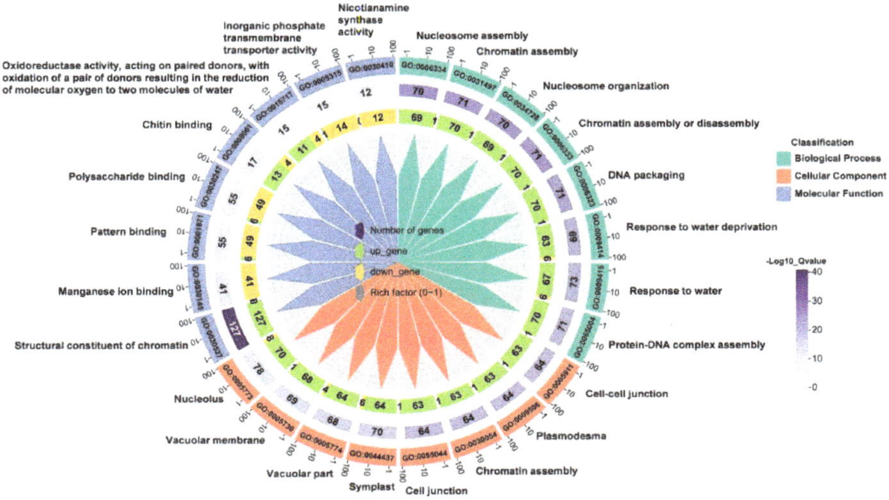

Figure 3. GO biological process enrichment chart of DEGs between well-watered (WW) and soil-drying (SD) conditions. The outermost circle of the image represents the term name enriched with DEGs. From outer to inner, the first layer is the GO term number, layer 2 is the number of genes in the GO term, and layer 3 shows the numbers of up- and down-regulated genes in the GO term. The color range from white to purple represents the magnitude of significance, and the gray label indicates the proportion of differentially expressed genes in the enriched pathway to the total enriched genes.

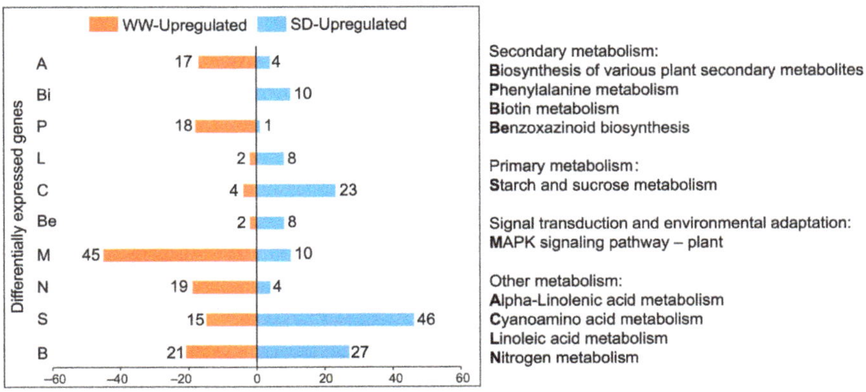

Figure 4. KEGG pathway enrichment chart of DEGs between well-watered (WW) and soil-drying (SD) conditions. Orange (blue) represents the number of upregulated genes under WW (SD) conditions.

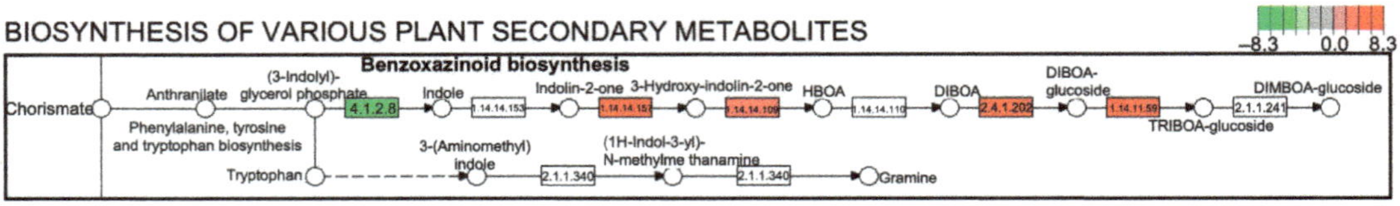

Figure 5. Benzoxazinoid biosynthesis in the KEGG pathway between well-watered (WW) and soil-drying (SD) conditions. The color and number in the upper right box of the metabolic pathway map represent the up- and down-regulated gene expression levels in the pathway.

2.3. Soil Drying Increases the CZ, ABA, and IAA Levels in Wheat Roots

Studies have indicated that phytohormones are involved in root growth under drought, so we next analyzed plant hormone-regulated DEGs. Compared with WW, 29 auxin-related genes were differentially expressed in wheat roots under SD treatment (19 significantly upregulated, ten significantly downregulated) (Figure 6A); 13 ABA-related genes showed differential expression (five significantly upregulated, eight significantly downregulated) (Figure 6B); four jasmonic acid (JA)-related genes were significantly downregulated (Figure 6C); three cytokinin-related genes showed differential expression (two significantly upregulated, one significantly downregulated) (Figure 6D); and three salicylic acid (SA)-related genes were significantly downregulated (Figure 6E). To verify the reliability of the transcriptome data, we randomly selected plant hormone-related genes for qPCR detection. The results (Figure 7) showed that qPCR data had a high correlation with the data obtained from the RNA-Seq (Figure 6), with a correlation ranging from 0.82 to 0.91, indicating that the RNA-Seq results were reliable ($p < 0.05$). To further examine the role of plant hormones in wheat roots under different water treatments, we measured the concentrations of 34 phytohormones (Figure S4) in wheat roots in eight major categories: auxin, abscisic acid, cytokinins, melatonin, jasmonic acid, ethylene, salicylic acid, and gibberellins. Only 16 plant hormones were detected and quantified (Figure 8), and the other 18 plant hormones were not detected due to their low content in samples. Compared to WW, concentrations of cis-zeatin (CZ) (Figure 8C), abscisic acid (ABA) (Figure 8J), and indole-3-acetic acid (IAA) (Figure 8M) were significantly increased under SD treatment. Four plant hormones were significantly decreased: JA (Figure 8F), SA (Figure 8K), glycosylated salicylic acid (SAG) (Figure 8L), and indole-3-formaldehyde (ICAld) (Figure 8O). Correlation analysis of total root length and plant hormones showed that CZ, ABA, and IAA are positively correlated with root length (Figure 9). These results indicate that CZ, ABA, and IAA are significantly increased under moderate soil drying, which may contribute to wheat root elongation under soil drying.

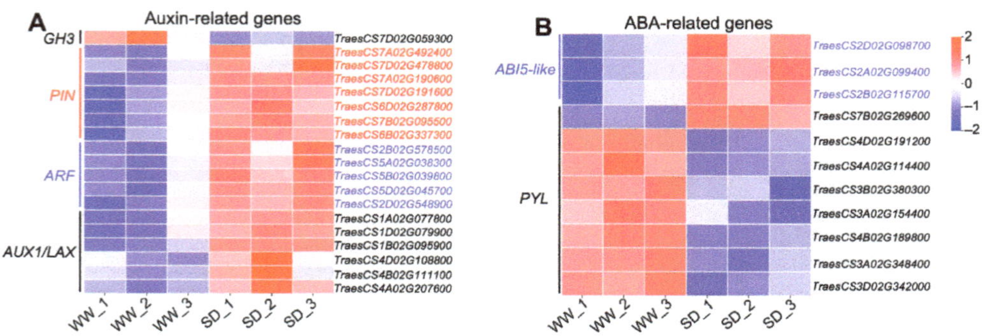

Figure 6. *Cont.*

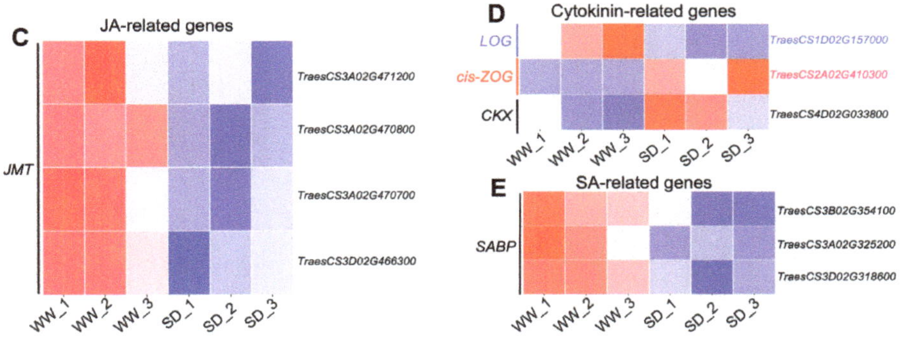

Figure 6. Heatmap of DEGs related to phytohormones between well-watered (WW) and soil-drying (SD) conditions. Heat map of auxin-related (**A**), ABA-related (**B**), JA-related (**C**), cytokinin-related (**D**), and SA-related (**E**) DEGs between WW and SD. The color of the heat map represents the expression of genes, with blue (red) indicating low (high) gene expression.

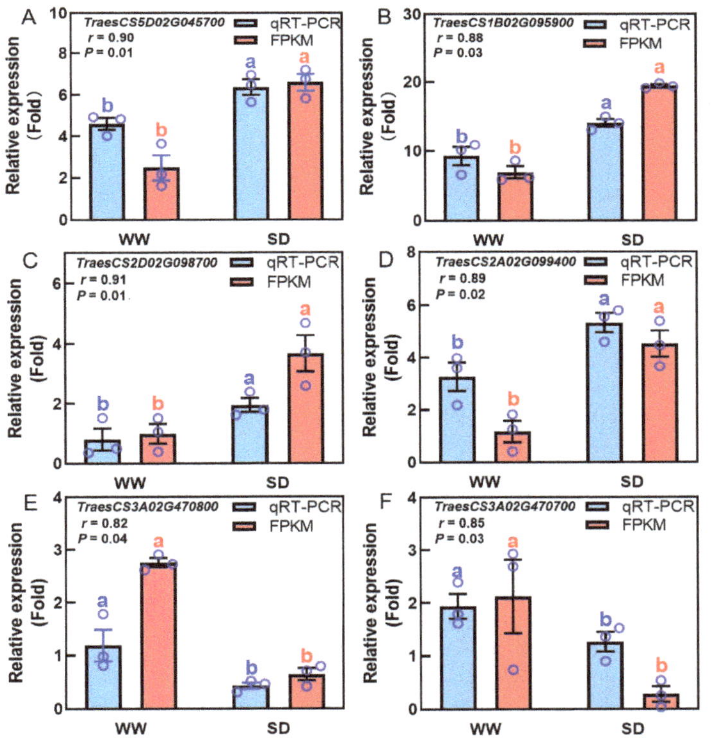

Figure 7. Correlation analysis between transcriptome profiling and qRT-PCR verification results of phytohormone-related genes in wheat roots under well-watered (WW) and soil-drying (SD) conditions. Auxin—(**A**,**B**), ABA—(**C**,**D**), and JA-related (**E**,**F**) gene expression in wheat roots under WW and SD conditions. The data in (**A**–**F**) are presented using mean ± standard error (n = 3). FPKM: the number of transcripts per million mapped reads per thousand bases of transcription. The same letter indicates no significant difference, while different letters indicate significant differences ($p < 0.05$, Student's t-test).

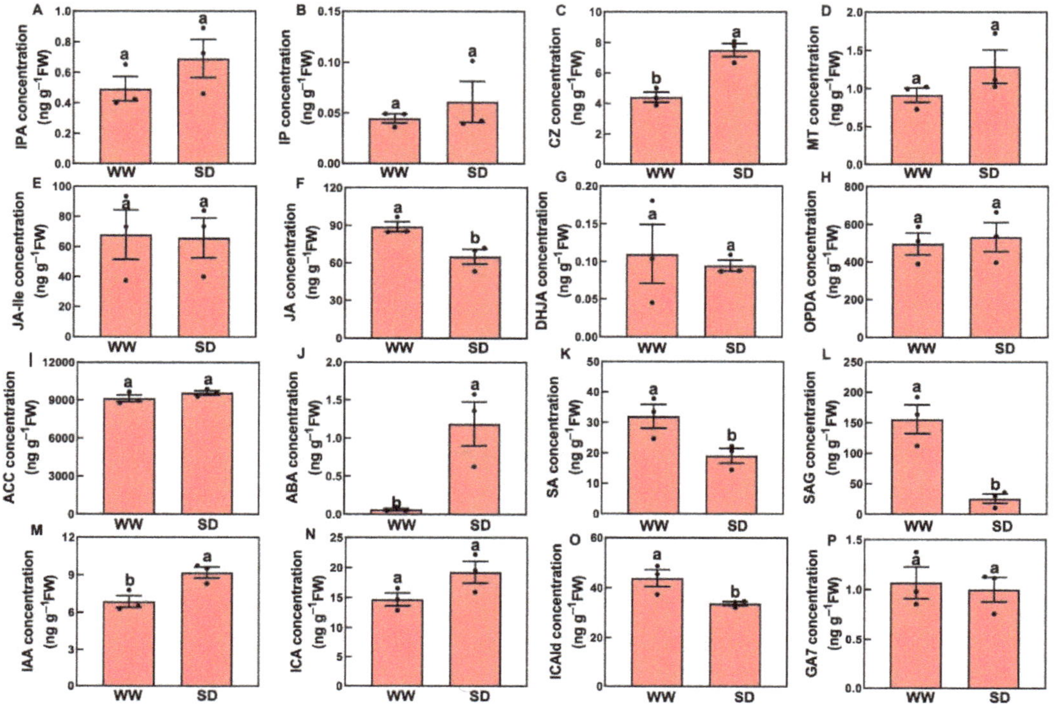

Figure 8. Changes in phytohormones in roots between well-watered (WW) and soil-drying (SD) conditions. (**A–C**) Changes in cytokinins, including isopentenyladenine nucleoside (IPA), isopentenyladenine (IP), and cis-zeatin (CZ), (**D**) Changes in melatonin. (**E–H**) Changes in jasmonic acids, including N-jasmonic acid isoleucine (JA-Ile), jasmonic acid (JA), dihydrojasmonic acid (DHJA), and 12-oxo plant dienoic acid (OPDA), (**I–J**) Changes in 1-aminocyclopropane-1-carboxylic acid (ACC, ethylene) and abscisic acid. (**K–L**) Changes in salicylic acids, including salicylic acid (SA) and glycosylated salicylic acid (SAG), (**M–O**). Changes in auxins, including indole-3-acetic acid (IAA), 3-indole-3-carboxylic acid (ICA), and indole-3-formaldehyde (ICAld), (**P**) Changes in gibberellin 7 (gibberellins). The data in the figure is presented using mean ± standard error (n = 3). The same letter indicates no significant difference, while different letters indicate significant differences ($p < 0.05$, Student's *t*-test).

2.4. Analysis of the Interactions of Plant Hormone Levels and Metabolic Pathways

To analyze the relationships between hormone-coupled metabolic pathways, we performed correlation analysis of the KEGG metabolic pathway analysis on the plant hormones. As shown in Figure 10, ABA and JA affected the biosynthesis of plant secondary metabolites; SA, IAA, IP, ABA, JA, and Ile affected the plant hormone signal transduction pathway; and CZ, SA, IAA, ABA, JA, and ethylene affected the biosynthesis of plant hormones pathway. Among these plant hormones, the log2Fold Change value of ABA is the largest and affects almost all metabolic pathways. These results suggest that moderate soil drying affects plant hormone signaling transduction and plant hormone synthesis pathways, especially altering the ABA and IAA concentrations, ultimately promoting wheat root growth to obtain more water from the soil.

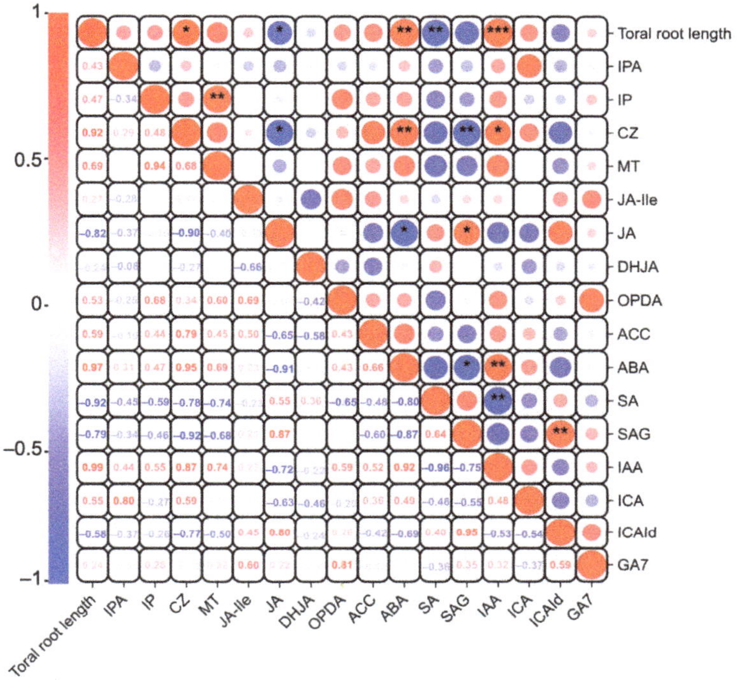

Figure 9. Correlation analysis of total root length and plant hormones in wheat roots between well-watered (WW) and soil-drying (SD) conditions. Asterisk indicates a significant correlation: * represents $p < 0.05$, ** represent $p < 0.01$, and *** represent $p < 0.001$. Red indicates positive correlations, while blue indicates negative ones. The size of the circle is positively correlated with the Pearson correlation coefficient.

Figure 10. Metabolite molecular network of differentially expressed plant hormones in wheat roots between well-watered (WW) and soil-drying (SD) conditions. Green dots represent metabolic pathways,

and other dots represent metabolite molecules. The size of each metabolic pathway point represents the number of metabolite molecules connected to it; the greater the number, the larger the point. Metabolite molecular points indicate the size of $\log_2$Fold Change values through gradual color changes, and multiple groups with no metabolite $\log_2$Fold Change information are compared.

3. Discussion

Under mild soil drying, root elongation is maintained or even promoted [29,30]. In this study, our results indicate that moderate soil drying promotes wheat root growth (Figure 1), which is consistent with results in *Arabidopsis* and tomato [10,28]. Wheat root transcriptome analysis indicates that the genes upregulated under soil drying were mainly enriched in starch and sucrose metabolism and benzoxazinoid biosynthesis (Figures 2–5). Root growth is closely related to soil and rhizosphere microorganisms. Studies have found that sugar concentrations in maize root exudates affect the rhizosphere bacterial community [25]. Sugars are derived from plant photosynthesis, are transported by the phloem to various places, and also play an important role in the regulation of plant growth and development. Sugars have been found in abundance in root exudate compounds and have been suggested to affect the root-associated microbiomes [25]. Chaparro et al. found that the abundances of *Acidobacteria*, *Actinobacteria*, *Bacteroidetes*, and *Cyanobacteria* were correlated with the levels of root exudated amino acids, phenols, sugars, and alcohols [22]. Cordovez et al. found that sugars in tomato root exudates attract *Pseudomonas* to colonize them and help them resist fungi [26]. In addition, roots can shape a specific rhizosphere microbial community structure by secreting various secondary metabolites [24]. Benzoxazinoids have a screening effect on rhizosphere microorganisms, and the tolerance of bacteria to benzoxazinoids shapes the maize rhizosphere microbiome [23]. Therefore, soil drying may affect the secretion of these metabolites by altering the expression of genes related to sucrose and benzoxazinoids, reshaping the microbial structure of the rhizosphere soil, thereby affecting wheat root growth.

Several plant hormones regulate root growth and stress adaptation responses [15,17,18]. Auxin accumulation and gradient modulation by multiple auxin transporters play crucial roles in the regulation of plant growth and development [18,27,31–33]. ABA is one of the main plant hormones involved in water deficiency and the induction of stomatal closure [14,15,17]. Under moderate water stress, ABA establishes and maintains root meristem function and stimulates root elongation [34–36]. ABA response and auxin transport play a key role in root elongation under drought stress; for example, auxin transporter mutants *aux1-7* and *eir1-4* (*pin2*) and ABA synthesis mutant *aba3-1* exhibit significantly lower root elongation rates than the wild type under osmotic stress conditions [1]. We found that IAA- and ABA-related genes are involved in wheat root growth (Figures 6 and 7). The hormone quantification results showed that moderate drought stress increases the IAA and ABA levels (Figure 8), consistent with our previous finding that moderate drought increases the IAA and ABA concentrations in tomato and *Arabidopsis* roots [10,28]. The correlation analysis of the metabolic pathway analysis of plant hormones showed that ABA also affects the biosynthesis of secondary metabolites (Figure 10). Previous studies have shown that lower auxin concentrations lead to a reduction in the meristem size of the root tip and reduced root growth [17]. ABA accumulation restricts ethylene synthesis in roots under water stress conditions, and ethylene can also interact with auxin in plant roots [1,17,30]. In addition, we found that moderate drought increases plant CZ levels [21], which is consistent with our results (Figures 8 and 9). Therefore, an increase in CZ concentrations may benefit wheat root growth during soil drying.

Plant hormones not only affect the root structure [19,20] but are also secreted outside the roots, affecting the soil rhizosphere microbial community to regulate plant growth [25,37]. Some previous studies characterized changes in the rhizosphere microbiome due to added plant hormones or changes in phytohormone signal transduction in roots. For instance, immune-signaling phytohormone SA mutants and added SA were employed to indicate that SA dramatically regulates colonization of the root microbiome by specific bacterial taxa [38]. JA is another important phytohormone involved in defense signaling that has been suggested to shape the rhizosphere microbiome. JA and sugars are important root exudates that affect the composition of the maize rhizobacterial communities, and the soil within the bacterial community within rows is modulated by different JA concentrations [25]. Changes in SA signals can alter the relative abundance of specific bacterial communities, such as actinomycetes in the rhizosphere microbial community [37]. In addition, hormonal crosstalk networks suggest various layers of complexity in the regulation of root growth by water stress conditions; for example, the responses of phytohormone synthesis, transport, and signal transduction components to water stress are complex and nonlinear, and understanding the effects of one phytohormone requires comprehensive consideration of how this phytohormone affects all other components [17]. Our results indicate that moderate soil drying reduced the JA and SA concentrations (Figures 8 and 9), perhaps due to the soil conditions. In summary, all of these results suggest that moderate soil drying affects plant metabolic and hormone pathways or the rhizosphere microbial community, ultimately promoting wheat root growth to obtain more water from deep soil. Understanding the mechanisms regulating root elongation under soil drying is pivotal to our understanding of the rooting patterns of plants under natural conditions, as efforts intensify to improve the water-use efficiency traits of crops in order to meet the demand for food under increasingly climate-challenged growth conditions.

4. Materials and Methods

4.1. Plant Materials and Growth Conditions

This study used the wheat (*Triticum aestivum* L.) variety 'Fielder'. Wheat seeds with full grains and uniform size were selected before treatment. Wheat seeds were surface-sterilized using 10% (v/v) sodium hypochlorite for 15 min and rinsed five times with double-distilled water [39]. The seeds were subsequently placed on moistened filter paper in darkness for 36 h at 28 °C to synchronize germination. Then, the device was placed in a greenhouse with a 14 h light (under 100 mmol photons m^{-2} s^{-1} illumination) and a 10 h dark cycle, a 26 °C light and a 22 °C dark temperature cycle, and a relative humidity of 60%.

4.2. Soil Drying Treatment

The soil used in this study was collected from a wheat field in Yangzhou, Jiangsu Province, China (119°53′ E, 32°31′ N), from a depth of 0–20 cm, and air dried for 7 days. Soil chemical factors are shown in Table S2. The air-dried soil was sieved through a 4 mm mesh to remove any coarse material and vegetative matter. Then, 20% sterile distilled water by weight was added to the soil, and the mixture was mixed thoroughly and passed through a 4 mm mesh again to ensure uniform soil particle size. For subsequent water addition from the bottom to help the water absorb naturally and prevent the soil from hardening, nine small holes of uniform size were created in the bottoms of the black pots (9 cm upper diameter, 6 cm lower diameter, 17 cm height). Then, the sieved soil (500 g) was placed in the black pots. Subsequently, uniform seedlings whose roots had grown to a length of about 2 cm at 36 h after germination were cultivated. The soil-water treatment was performed as previously described by Xu et al. [27], with slight modifications. Different amounts of water were added to the bottom trays of the pots, and the wheat plants were subjected to two water treatments: well-watered conditions (WW, maintaining 80% of field capacity for the first 4 days, and then drying for 2 days) and moderate soil drying (SD, maintaining 45% of field capacity for the first 4 days, and then drying for 2 days).

4.3. Root Sample Collection and Analysis

Wheat roots were carefully collected and shaken after the pots were disassembled. The roots sampled from the washed plants were weighed to obtain the fresh weight and placed together in an oven (40 °C for 7 days) to obtain the dry weight. The total root length was measured using ImageJ software (Version 1.54d).

4.4. Transcriptome Sequencing Process and Quantitative Real-Time PCR (qRT-PCR) Analysis

The surfaces of wheat roots were washed clean, and then the roots were quickly frozen in liquid nitrogen. The root samples were sent via cold chain to Nuohe Zhiyuan Technology Service (Beijing, China) for total RNA extraction using a RNAprep Pure Plant Plus Kit (DP441; TIANGEN, Beijing, China) according to the manufacturer's instructions and sequencing using an Agilent 2100 bioanalyzer to evaluate the RNA quality and quantity. After passing inspection, different libraries were pooled according to the effective concentration and target offline data volume requirements for Illumina sequencing. Reads with adapters, poly-N, or low quality were removed to obtain clean data. All subsequent analyses are based on high-quality analysis using clean data. Aegilops tauschii (Ats) was used as the reference genome. The high-quality reads were annotated using HISAT2 (ver. 2.0.5). Gene expression, DEGs, and gene function were analyzed on the Nuohe Zhiyuan Cloud platform, and charts were drawn on the website https://www.chiplot.online/ (accesed on 24 January 2024).

Total RNA was extracted from roots using a RNAprep Pure Plant Plus Kit (DP441; TIANGEN, China) according to the manufacturer's instructions. First-strand cDNA was synthesized using a First-Strand cDNA Synthesis Kit (TransGen Biotech, Beijing, China) according to the manufacturer's instructions. Quantitative real-time PCR (qRT-PCR) assays were performed as described by Li et al. [39]. The expression levels were normalized to the expression of the wheat *Actin-6* gene. The primers specific to each gene are listed in Table S3.

4.5. Root Plant Hormone Determination

After soil drying treatment, wash the surface of the wheat roots with 1 × PBS (130 mM NaCl, 7 mM $Na_2HPO_4 \cdot 7H_2O$, 3 mM $NaH_2PO_4 \cdot 4H_2O$, pH 7.0), dry it with sterile filter paper, cut the wheat roots into sterile foil with sterile scissors, and quickly place it in liquid nitrogen for freezing. At least six seedlings were pooled for each biological replicate for hormonomic analysis, with three biological replicates for each sample. The root samples were sent via cold chain to PANOMIX Biomedical Tech (Suzhou, China) for plant hormone detection using liquid chromatography-mass spectrometry (LC-MS/MS). UPLC separation was performed on an ExionLC UPLC system (AB Sciex, Boston, MA, USA) equipped with an Acquity UPLC® CSH C18 (1.7 μm, 2.1 mm × 150 mm, Waters, Milford, MA, USA) column. The temperature of the column was set at 40 °C. The sample injection volume was 2 μL. The eluents consisted of 0.05% formic acid with 2 mM ammonium formate water (eluent A) and 0.05% formic acid methanol (eluent B). The MS analysis was performed using an AB Sciex Triple Quadrupole 6500 Plus mass spectrometer (AB Sciex, USA) in the multiple reaction monitoring (MRM) mode. Figure S4 shows the fitting of the standard curves of plant hormones in wheat roots under WW and SD conditions.

4.6. Statistical Analysis

All experimental data were analyzed using SPSS version 17.0. Each treatment was repeated at least three times, and the differences between the two groups were examined using the Student's *t*-test; Duncan's test was used to analyze the differences between two or more datasets [1,39]. $p < 0.05$ was considered a significant difference.

5. Conclusions

This study comprehensively elucidated the responding mechanisms of soil drying on wheat root elongation based on physiological-biochemical performance and transcriptional analysis. RNA-Seq analysis indicated that the upregulated genes in the wheat roots under SD were mainly enriched in starch and sucrose metabolism pathways and the biosynthesis pathway of benzoxazinoids. Various plant hormone-related genes were also differentially expressed under SD. The concentrations of CZ, ABA, and IAA increased significantly under SD, whereas the SA, JA, SAG, and indole-3-formaldehyde (ICADd) concentrations significantly decreased under SD. ABA, JA, and ethylene affect the biosynthesis of plant secondary metabolites under SD. These results suggest that changes in these metabolic pathways and plant hormones contribute to wheat root elongation to obtain water from deeper layers. In conclusion, this study investigated transcriptomes and hormones in wheat roots under soil drying stress and will be useful for clarifying the mechanism of drought stress in wheat.

Supplementary Materials: The following supporting information can be downloaded at: https://www.mdpi.com/article/10.3390/ijms25179157/s1.

Author Contributions: Y.L. planned and designed this research. Y.H., S.J., Y.C., Z.Y., N.Z., M.Z., J.D. and C.L. performed experiments and analyzed data. W.G., X.Z., N.Z., J.Z. and W.X. wrote the manuscript. N.Z. and J.Z. agree to serve as the authors responsible for scientific communication. All authors have read and agreed to the published version of the manuscript.

Funding: This research is funded by the China Postdoctoral Science Foundation (2023M733003), the Hong Kong Scholars Program (XJ2023052), the National Natural Science Foundation of China (32301761), the Natural Science Foundation of Jiangsu Province (BK20220566), and the Postgraduate Research & Practice Innovation Program of Jiangsu Province (SJCX24_2278). We are also grateful for a Project Funded by the Priority Academic Program Development of Jiangsu Higher Education Institutions (PAPD).

Institutional Review Board Statement: Not applicable.

Informed Consent Statement: Not applicable.

Data Availability Statement: All data that support the findings of this study are available from the authors upon reasonable request.

Acknowledgments: We thank Hongxiang Ma (Yangzhou University) for providing the wheat 'Fielder' seeds.

Conflicts of Interest: The authors declare no conflicts of interest.

References

1. Xu, W.; Jia, L.; Shi, W.; Liang, J.; Zhou, F.; Li, Q.; Zhang, J. Abscisic acid accumulation modulates auxin transport in the root tip to enhance proton secretion for maintaining root growth under moderate water stress. *New Phytol.* **2012**, *197*, 139–150. [CrossRef]
2. Uga, Y.; Sugimoto, K.; Ogawa, S.; Rane, J.; Ishitani, M.; Hara, N.; Kitomi, Y.; Inukai, Y.; Ono, K.; Kanno, N.; et al. Control of root system architecture by DEEPER ROOTING 1 increases rice yield under drought conditions. *Nat. Genet.* **2013**, *45*, 1097–1102. [CrossRef]
3. Mao, H.; Li, S.; Chen, B.; Jian, C.; Mei, F.; Zhang, Y.; Li, F.; Chen, N.; Li, T.; Du, L.; et al. Variation in cis-regulation of a NAC transcription factor contributes to drought tolerance in wheat. *Mol. Plant* **2021**, *15*, 276–292. [CrossRef]
4. Zhang, J.; Davies, W. Increased Synthesis of ABA in Partially Dehydrated Root Tips and ABA Transport from Roots to Leaves. *J. Exp. Bot.* **1987**, *38*, 2015–2023. [CrossRef]
5. Li, Y.; Zeng, H.; Xu, F.; Yan, F.; Xu, W. H+-ATPases in Plant Growth and Stress Responses. *Annu. Rev. Plant Biol.* **2022**, *73*, 495–521. [CrossRef]
6. Hickey, L.T.; Hafeez, A.N.; Robinson, H.; Jackson, S.A.; Leal-Bertioli, S.C.M.; Tester, M.; Gao, C.; Godwin, I.D.; Hayes, B.J.; Wulff, B.B.H. Breeding crops to feed 10 billion. *Nat. Biotechnol.* **2019**, *37*, 744–754. [CrossRef]
7. Lesk, C.; Rowhani, P.; Ramankutty, N. Influence of extreme weather disasters on global crop production. *Nature* **2016**, *529*, 84–87. [CrossRef]
8. Daryanto, S.; Wang, L.; Jacinthe, P.-A. Global Synthesis of Drought Effects on Maize and Wheat Production. *PLoS ONE* **2016**, *11*, e0156362. [CrossRef]
9. Thorup-Kristensen, K.; Kirkegaard, J. Root system-based limits to agricultural productivity and efficiency: The farming systems context. *Ann. Bot.* **2016**, *118*, 573–592. [CrossRef]
10. Li, Y.; Chen, Y.; Jiang, S.; Dai, H.; Xu, W.; Zhang, Q.; Zhang, J.; Dodd, I.C.; Yuan, W. ABA is required for differential cell wall acidification associated with root hydrotropic bending in tomato. *Plant Cell Environ.* **2023**, *47*, 38–48. [CrossRef] [PubMed]
11. Tran, L.S.; Urao, T.; Qin, F.; Maruyama, K.; Kakimoto, T.; Shinozaki, K.; Yamaguchi-Shinozaki, K. Functional analysis of AHK1/ATHK1 and cytokinin receptor histidine kinases in response to abscisic acid, drought, and salt stress in *Ara-bidopsis*. *Proc. Natl. Acad. Sci. USA* **2007**, *104*, 20623–20628. [CrossRef] [PubMed]

12. Nishiyama, R.; Watanabe, Y.; Fujita, Y.; Le, D.T.; Kojima, M.; Werner, T.; Vankova, R.; Yamaguchi-Shinozaki, K.; Shinozaki, K.; Kakimoto, T.; et al. Analysis of Cytokinin Mutants and Regulation of Cytokinin Metabolic Genes Reveals Important Regulatory Roles of Cytokinins in Drought, Salt and Abscisic Acid Responses, and Abscisic Acid Biosynthesis. *Plant Cell* **2011**, *23*, 2169–2183. [CrossRef] [PubMed]
13. Cheng, M.-C.; Liao, P.-M.; Kuo, W.-W.; Lin, T.-P. The Arabidopsis ETHYLENE RESPONSE FACTOR1 Regulates Abiotic Stress-Responsive Gene Expression by Binding to Different cis-Acting Elements in Response to Different Stress Signals. *Plant Physiol.* **2013**, *162*, 1566–1582. [CrossRef] [PubMed]
14. Ohashi, Y.; Nakayama, N.; Saneoka, H.; Fujita, K. Effects of drought stress on photosynthetic gas exchange, chlorophyll fluorescence and stem diameter of soybean plants. *Biol. Plant.* **2006**, *50*, 138–141. [CrossRef]
15. Takahashi, F.; Kuromori, T.; Urano, K.; Yamaguchi-Shinozaki, K.; Shinozaki, K. Drought Stress Responses and Resistance in Plants: From Cellular Responses to Long-Distance Intercellular Communication. *Front. Plant Sci.* **2020**, *11*, 556972. [CrossRef]
16. Karanja, J.K.; Aslam, M.M.; Qian, Z.; Yankey, R.; Dodd, I.C.; Weifeng, X. Abscisic Acid Mediates Drought-Enhanced Rhizosheath Formation in Tomato. *Front. Plant Sci.* **2021**, *12*, 658787. [CrossRef]
17. Rowe, J.H.; Topping, J.F.; Liu, J.; Lindsey, K. Abscisic acid regulates root growth under osmotic stress conditions via an interacting hormonal network with cytokinin, ethylene and auxin. *New Phytol.* **2016**, *211*, 225–239. [CrossRef]
18. Park, J.-E.; Park, J.-Y.; Kim, Y.-S.; Staswick, P.E.; Jeon, J.; Yun, J.; Kim, S.-Y.; Kim, J.; Lee, Y.-H.; Park, C.-M. GH3-mediated Auxin Homeostasis Links Growth Regulation with Stress Adaptation Response in Arabidopsis. *J. Biol. Chem.* **2007**, *282*, 10036–10046. [CrossRef]
19. Verma, P.; Yadav, A.N.; Khannam, K.S.; Mishra, S.; Kumar, S.; Saxena, A.K.; Suman, A. Appraisal of diversity and func-tional attributes of thermotolerant wheat associated bacteria from the peninsular zone of India. Saudi. *J. Biol. Sci.* **2019**, *26*, 1882–1895.
20. Yadav, A.N.; Kumar, V.; Dhaliwal, H.S.; Prasad, R.; Saxena, A.K. *Chapter 15—Microbiome in Crops: Diversity, Distribution, and Potential Role in Crop Improvement. Crop Improvement through Microbial Biotechnology;* Elsevier: Amsterdam, The Netherlands, 2018; pp. 305–332.
21. Alharby, H.F.; Al-Zahrani, H.S.; Alzahrani, Y.M.; Alsamadany, H.; Hakeem, K.R.; Rady, M.M. Maize Grain Extract *En-riched* with Polyamines Alleviates Drought Stress in *Triticum aestivum* through Up-Regulation of the Ascorbate–Glutathione Cycle, *Glyox-alase* System, and Polyamine Gene Expression. *Agronomy* **2021**, *11*, 949. [CrossRef]
22. Chaparro, J.M.; Badri, D.V.; Vivanco, J.M. Rhizosphere microbiome assemblage is affected by plant development. *ISME J.* **2013**, *8*, 790–803. [CrossRef]
23. Thoenen, L.; Giroud, C.; Kreuzer, M.; Waelchli, J.; Gfeller, V.; Deslandes-Hérold, G.; Mateo, P.; Robert, C.A.M.; Ahrens, C.H.; Rubio-Somoza, I.; et al. Bacterial tolerance to host-exuded specialized metabolites structures the maize root microbiome. *Proc. Natl. Acad. Sci. USA* **2023**, *120*, e2310134120. [CrossRef] [PubMed]
24. Moe, L.A. Amino acids in the rhizosphere: From plants to microbes. *Am. J. Bot.* **2013**, *100*, 1692–1705. [CrossRef]
25. Lopes, L.D.; Wang, P.; Futrell, S.L.; Schachtman, D.P. Sugars and Jasmonic Acid Concentration in Root Exudates Affect Maize Rhizosphere Bacterial Communities. *Appl. Environ. Microbiol.* **2022**, *88*, e0097122. [CrossRef]
26. Cordovez, V.; Rotoni, C.; Dini-Andreote, F.; Oyserman, B.; Carrión, V.J.; Raaijmakers, J.M. Successive plant growth amplifies genotype-specific assembly of the tomato rhizosphere microbiome. *Sci. Total Environ.* **2021**, *772*, 144825. [CrossRef]
27. Xu, F.; Liao, H.; Yang, J.; Zhang, Y.; Yu, P.; Cao, Y.; Fang, J.; Chen, S.; Li, L.; Sun, L.; et al. Auxin-producing bacteria promote barley rhizosheath formation. *Nat. Commun.* **2023**, *14*, 5800. [CrossRef]
28. Li, Y.; Yuan, W.; Li, L.; Dai, H.; Dang, X.; Miao, R.; Baluška, F.; Kronzucker, H.J.; Lu, C.; Zhang, J.; et al. Comparative analysis reveals gravity is involved in the MIZ1-regulated root hydrotropism. *J. Exp. Bot.* **2020**, *71*, 7316–7330. [CrossRef]
29. van der Weele, C.M.; Spollen, W.G.; Sharp, R.E.; Baskin, T.I. Growth of *Arabidopsis thaliana* seedlings under water deficit studied by control of water potential in nutrient-agar media. *J. Exp. Bot.* **2000**, *51*, 1555–1562. [CrossRef]
30. Sharp, R.E.; LeNoble, M.E. ABA, ethylene and the control of shoot and root growth under water stress. *J. Exp. Bot.* **2002**, *53*, 33–37. [CrossRef]
31. Pitts, R.J.; Cernac, A.; Estelle, M. Auxin and ethylene promote root hair elongation in Arabidopsis. *Plant J.* **1998**, *16*, 553–560. [CrossRef]
32. Blilou, I.; Xu, J.; Wildwater, M.; Willemsen, V.; Paponov, I.; Friml, J.; Heidstra, R.; Aida, M.; Palme, K.; Scheres, B. The PIN auxin efflux facilitator network controls growth and patterning in Arabidopsis roots. *Nature* **2005**, *433*, 39–44. [CrossRef]
33. Jones, A.R.; Kramer, E.M.; Knox, K.; Swarup, R.; Bennett, M.J.; Lazarus, C.M.; Leyser, H.M.O.; Grierson, C.S. Auxin transport through non-hair cells sustains root-hair development. *Nat. Cell Biol.* **2008**, *11*, 78–84. [CrossRef]
34. Liang, Y.; Mitchell, D.M.; Harris, J.M. Abscisic acid rescues the root meristem defects of the *Medicago truncatula latd* mutant. *Dev. Biol.* **2006**, *304*, 297–307. [CrossRef]
35. Rosales, M.A.; Maurel, C.; Nacry, P. Abscisic Acid Coordinates Dose-Dependent Developmental and Hydraulic Responses of Roots to Water Deficit. *Plant Physiol.* **2019**, *180*, 2198–2211. [CrossRef]
36. Yang, L.; Zhang, J.; He, J.; Qin, Y.; Hua, D.; Duan, Y.; Chen, Z.; Gong, Z. ABA-Mediated ROS in Mitochondria Regulate Root Meristem Activity by Controlling PLETHORA Expression in Arabidopsis. *PLoS Genet.* **2014**, *10*, e1004791. [CrossRef]
37. Li, X.; Jousset, A.; de Boer, W.; Carrión, V.J.; Zhang, T.; Wang, X.; Kuramae, E.E. Legacy of land use history determines reprogramming of plant physiology by soil microbiome. *ISME J.* **2018**, *13*, 738–751. [CrossRef]

38. Lebeis, S.L.; Paredes, S.H.; Lundberg, D.S.; Breakfield, N.; Gehring, J.; McDonald, M.; Malfatti, S.; Del Rio, T.G.; Jones, C.D.; Tringe, S.G.; et al. Salicylic acid modulates colonization of the root microbiome by specific bacterial taxa. *Science* **2015**, *349*, 860–864. [CrossRef] [PubMed]
39. Li, Y.; Yuan, W.; Li, L.; Miao, R.; Dai, H.; Zhang, J.; Xu, W. Light-Dark Modulates Root Hydrotropism Associated with Gravitropism by Involving Amyloplast Response in Arabidopsis. *Cell Rep.* **2020**, *32*, 108198. [CrossRef] [PubMed]

International Journal of
Molecular Sciences

MDPI

Article

Transcriptomic Insights into Salt Stress Response in Two Pepper Species: The Role of MAPK and Plant Hormone Signaling Pathways

Muhammad Aamir Farooq [1,2], Muhammad Zeeshan Ul Haq [1], Liping Zhang [1,2], Shuhua Wu [1,2], Naveed Mushtaq [1,2], Hassam Tahir [1,2] and Zhiwei Wang [1,2,*]

1 Key Laboratory for Quality Regulation of Tropical Horticultural Crops of Hainan Province, School of Breeding and Multiplication (Sanya Institute of Breeding and Multiplication), Center of Nanfan and High-Efficiency Tropical Agriculture, Hainan University, Sanya 572025, China
2 Key Laboratory for Quality Regulation of Tropical Horticultural Crops of Hainan Province, School of Tropical Agriculture and Forestry, Hainan University, Haikou 570228, China
* Correspondence: wangzhiwei@hainanu.edu.cn

Abstract: Salt stress imposes significant plant limitations, altering their molecular, physiological, and biochemical functions. Pepper, a valuable herbaceous plant species of the *Solanaceae* family, is particularly susceptible to salt stress. This study aimed to elucidate the physiological and molecular mechanisms that contribute to the development of salt tolerance in two pepper species (*Capsicum baccatum* (moderate salt tolerant) and *Capsicum chinense* (salt sensitive)) through a transcriptome and weighted gene co-expression network analysis (WGCNA) approach to provide detailed insights. A continuous increase in malondialdehyde (MDA) and hydrogen peroxide (H_2O_2) levels in *C. chinense* and higher activities of catalase (CAT), superoxide dismutase (SOD), and peroxidase (POD) in *C. baccatum* indicated more tissue damage in *C. chinense* than in *C. baccatum*. In transcriptome analysis, we identified 39 DEGs related to salt stress. Meanwhile, KEGG pathway analysis revealed enrichment of MAPK and hormone signaling pathways, with six DEGs each. Through WGCNA, the ME.red module was identified as positively correlated. Moreover, 10 genes, *A-ARR* (CQW23_24856), *CHIb* (CQW23_04881), *ERF1b* (CQW23_08898), *PP2C* (CQW23_15893), *ABI5* (CQW23_29948), *P450* (CQW23_16085), *Aldedh1* (CQW23_06433), *GDA* (CQW23_12764), *Aldedh2* (CQW23_14182), and *Aldedh3* (CQW23_11481), were validated by qRT-PCR. This study provides valuable insights into the genetic mechanisms underlying salt stress tolerance in pepper. It offers potential targets for future breeding efforts to enhance salt stress resilience in this crop.

Keywords: pepper; salt stress; transcriptome; WGCNA; enzymatic antioxidants

1. Introduction

Abiotic stresses pose significant limitations on crop yield [1]. These stresses, such as salinity, temperature fluctuations, drought, flooding, and heavy metal, profoundly impact the growth and yield of crops [2,3]. Alarmingly, around 90% of arable lands are susceptible to one or more of these stressors [4], resulting in yield losses of up to 70% in major food crops [5]. Abiotic stress triggers molecular, physiological, and biochemical alterations in plants, leading to changes in morphology and disruptions in plant functions [6,7]. Soil salinity significantly reduces crop yield across various global regions [8]. Currently, cultivated and irrigated lands afflicted by salt constitute about 20–30% of the worldwide soil area, with projections indicating a potential doubling of salinized regions by 2050 [9]. This increase is attributed to high evapotranspiration rates, inadequate rainfall, unsustainable agricultural practices, and ineffective irrigation and drainage methods [10]. Consequently, conventional high-yielding crop varieties often struggle to tolerate these elevated salt concentrations [11].

Salinity triggers complex mechanisms within plants, impacting various physiological, biochemical, and molecular pathways [12]. Initially, it diminishes soil osmotic potential, imposing osmotic stress that reduces root water content and nutrient uptake [13]. Furthermore, salinity stress induces the overproduction of reactive oxygen species (ROS), leading to oxidative damage in cellular components such as the plasma membrane, mitochondria, and chloroplasts [14]. This oxidative stress disrupts protein biosynthesis, ion homeostasis, and photosynthesis while downregulating antioxidant activities within the plant cell [15]. To avoid oxidative stress damage, plants activate antioxidant enzymes like superoxide dismutase (SOD), peroxidase (POD), and catalase (CAT) and accumulate nonenzymatic antioxidant compounds, including primary and secondary metabolites like proline, phenolic acids, carotenoids, or alkaloids [16]. Similarly, ROS initiation during salt stress triggers Mitogen-activated protein (MAP) cascades and hormonal and other metabolic pathways aiding ionic balance and osmotic equilibrium [17]. MAPK responds to osmotic stress by elevating osmolyte synthesis [18]. MAPKs are supposed to maintain ROS balance [19]. Phytohormones, like ABA (abscisic acid), also regulate osmotic stress caused by salinity and ROS-mediated pathways [20]. ABA also influences ion balance, elevating K^+ and Ca^{2+} and hindering Na^+ and Cl^- absorption [21]. It controls Na^+/K^+ ions and H_2O_2 concentrations during salt stress [22], activating MAPK via H_2O_2 from ROS scavenging [23].

Pepper is an important member of the Solanaceae family, with a broad diversity in fruit morphology and flavor [24]. It originated from Central/South America and Mexico and later spread globally to Europe, Asia, and Africa. Its production has increased over the last 20 years from 17 to 36 million tons, and the cultivated area has expanded by about 35% worldwide [25]. It is cultivated in most regions of China, particularly in the east, west, and south, with a share of 7.76% of the total national vegetable output [26]. It is a tropical and subtropical annual crop that requires a warm (18–30 °C) and humid climate for optimal growth and fruit production [27]. Still, it can also be cultivated in temperate climates in greenhouses [28]. Several stresses (biotic and abiotic) significantly affect peppers, which decrease yields and fruit quality [29]. In the previous study, salt stress is considered most relevant for pepper plants and has moderately sensitive, sensitive, or highly susceptible effects [30]. A high salt concentration source that affects plants may be soil or irrigation water [31]. In pepper plants, the dry weight and marketable yield diminish by 46% and 25%, respectively, when rinsed with water at 4.4 dS m^{-1} [32].

Previous research has delved into salt-related genes and pathways in model plants, overlooking pepper [33]. Pepper is particularly susceptible to salt stress, so it is essential to understand the molecular mechanisms governing its response to this environmental challenge. We examined two pepper species, *Capsicum baccatum* (moderately sensitive) and *Capsicum chinense* (sensitive), in response to salt stress. This study aims to compare the physiological responses, identify enriched pathways and DEGs, and identify potential genetic targets. These findings will contribute to a deeper understanding of pepper's intricate molecular mechanisms underlying salt stress tolerance. It offers potential targets for future breeding efforts to enhance salt stress resilience in this economically important crop species.

2. Results

2.1. Physiological and Phenotypical Alterations

Following 24 h of exposure to salt stress, observable alterations began to occur in the *C. chinense* plants, with the lower leaves exhibiting a gradual yellowing and a decline in turgor pressure compared to the *C. baccatum* plants, which became more pronounced by the 144 h mark (Figure 1).

Proline levels exhibited a more significant increase in *C. chinense* compared to *C. baccatum* at all time points. The highest proline level was observed in *C. chinense* at 144 h (Figure 2a). In *C. chinense*, the highest level of SOD was observed at 24 h and 48 h compared to its control (CK; 0 h). However, in *C. baccatum*, there was an initial decrease in SOD levels at 24 h and 48 h compared to its respective CK. Interestingly, in *C. baccatum*, the SOD levels peaked at the 144 h time point compared to CK, whereas in *C. chinense*, a decrease in SOD levels was observed at 144 h when compared to the other time points and CK (Figure 2b). The highest H_2O_2 level was documented at 144 h in *C. chinense* in contrast to *C. baccatum*, which remained consistently higher at all time points and compared to CK. Conversely, the maximum H_2O_2 activity in *C. baccatum* was recorded after 24 h of salt stress across all time points and compared to CK (Figure 2c). Similarly, CAT activity peaked after 24 h of stress in *C. baccatum* and compared to CK, while it displayed a declining trend, particularly pronounced in *C. chinense* over time and compared to CK (Figure 2d). Nonetheless, the peak level of MDA was documented in *C. chinense* at 96 h compared to CK, surpassing that of *C. baccatum* at all time points and CK (Figure 2e). Conversely, POD activity exhibited its highest levels in *C. baccatum* compared to *C. chinense*, particularly evident after 48 h of salt stress when compared across all time points and compared to CK (Figure 2f). Furthermore, compared to the CK of each species, it was apparent that total chlorophyll contents, Chl a and Chl b contents, exhibited a more pronounced decrease in *C. chinense* than in *C. baccatum* (Figure 2g–i). In contrast, carotenoid levels in *C. baccatum* showed an increase up to 24 h and 48 h after salt stress, surpassing the levels in *C. chinense* at these time points, but displayed a more substantial decrease in carotenoids in *C. baccatum* than in *C. chinense* at the 144 h mark following salt stress (Figure 2j). Overall, the physiological indicators suggest that *C. chinense* exhibits a higher sensitivity to salt stress when compared to *C. baccatum*.

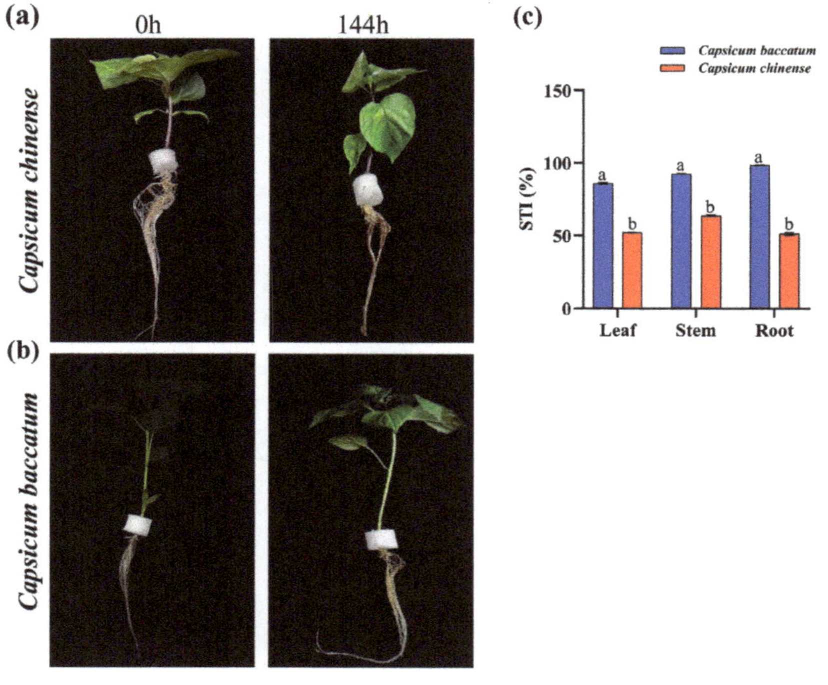

Figure 1. (**a**) The external appearance of the leaves and roots from *C. chinense* exposed to salt (NaCl) stress was observed at various time intervals (0 h and 144 h). (**b**) The external appearance of the leaves and roots from *C. baccatum* exposed to NaCl stress was observed at various time intervals (0 h and 144 h); (**c**) stress tolerance index percentage (STI %) of leaf, stem, and root dried samples of *C. baccatum* and *C. chinense* at 100 mM NaCl stress. The lowercase alphabets indicating significant differences at $p < 0.05$.

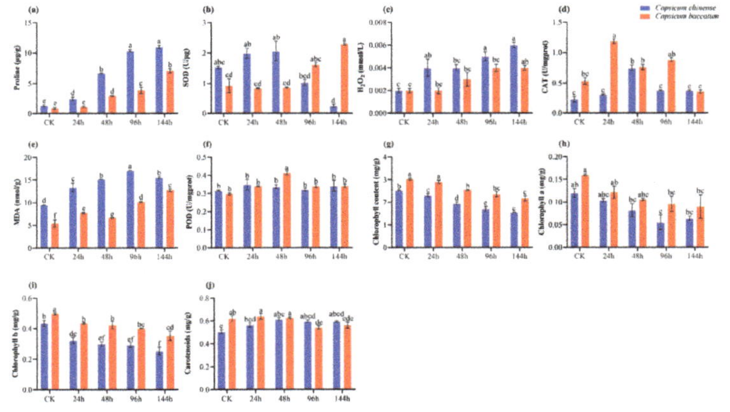

Figure 2. The biochemical parameters are (**a**) Proline, (**b**) SOD, (**c**) H_2O_2, (**d**) Catalase, (**e**) MDA, (**f**) POD, (**g**) Chlorophyll contents, (**h**) Chl. a and (**i**) Chl. b, and (**j**) Carotenoids of *C. chinense* and *C. baccatum* leaves under salt stress at various intervals. The presented data is the average ± standard deviation of three replicates, with lowercase alphabets indicating significant differences at $p < 0.05$.

2.2. Stress Tolerance Index Percentage (STI%)

The STI% results showed that the *C. baccatum* root showed more salt stress tolerance than the stem and roots as compared to the *C. chinense* root, stem, and leaf (Figure 1c). The STI% showed *C. baccatum* tolerance against salt stress.

2.3. Microscopic Scanning of Leaf and Root Tissues

We evaluated the leaf stomata, root structure, and cross-section in both species treated with salt stress (Figure 3). SEM results showed that compared to their respective controls, *C. chinense* salt-treated samples stomata showed more closing as compared to *C. baccatum*; the root scan and its cross-section analysis also showed weak health due to the change in the root tip appearance, degradation of vascular tissues, and increased cell wall thickening in the cortex region as compared to *C. chinense* control and *C. baccatum* control and treated samples (Figure 3a,b). Moreover, a noticeable decline in leaf stomata and cross-section in length, width, pore length, and pore width of *C. chinense* was recorded compared to its control and *C. baccatum* control and treatment samples (Supplementary Table S1). These results showed *C. chinense* sensitivity against salt stress.

2.4. Analysis of RNA Sequencing

Transcriptome sequencing of pepper cultivars *C. chinense* and *C. baccatum* yielded valuable and distinct data about their responses under salt stress conditions. The essential RNA-seq characteristics are summarized in Supplementary Table S2. Upon scrutinizing the raw sequencing data, it was determined that *C. chinense* generated a range of 38–56 million clean reads, while *C. baccatum* produced 38–45 million clean reads. The data exhibited a consistent overall error rate of 0.03%, with quality scores (Q20 $\leq$ 95–97% and Q30 $\leq$ 92–94%) and a GC content ranging from 40% to 43%, as summarized in Supplementary Table S2. Among all the different comparisons, 33,022 genes were detected across and within both species. Moreover, we constructed Venn diagrams based on the control with all-time points within the species and between the species; 675 genes were detected among the controls and all-time points in *C. chinense* (Figure 4a), 75 in *C. baccatum* (Figure 4b), and 4866 genes among control groups of both species with all-time points group (Figure 4c), respectively. Moreover, upregulated and downregulated genes were shown among all-time points within and between species (Figure 4d), where maximum upregulated and downregulated genes were detected after 24 h in the salt stress group of both species. Similarly, the heat map of the genes is given in Figure 4e.

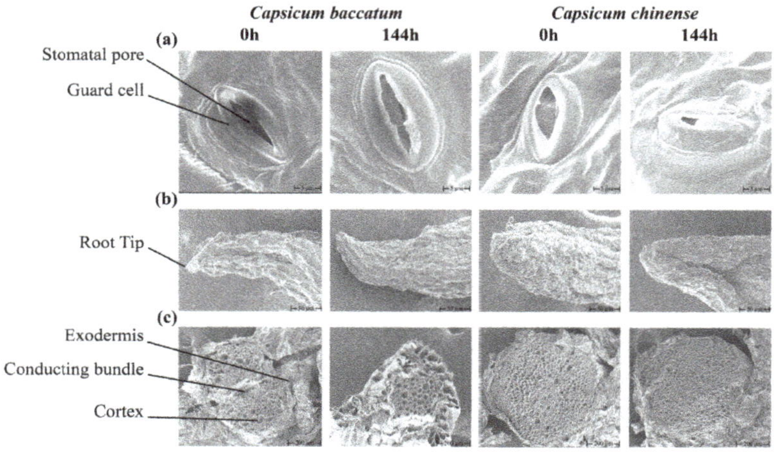

Figure 3. The SEM images of the leaf stomata (**a**), root tips (**b**), and root cross-section (**c**) of *C. baccatum* and *C. chinense* exposed to salt stress were observed at 0 h and 144 h.

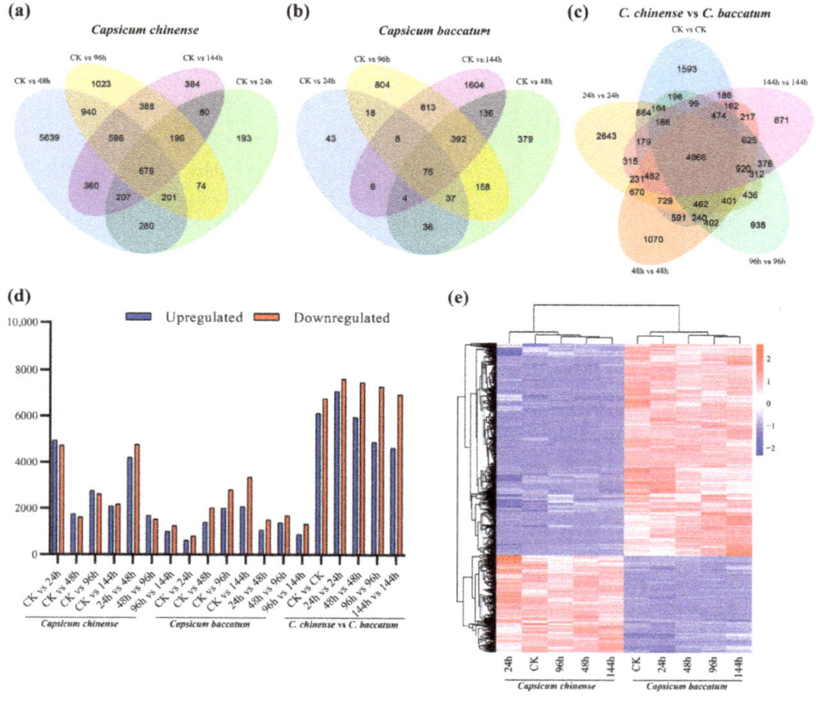

Figure 4. Venn diagrams were constructed to represent commonly expressed genes in different comparisons: (**a**) all treatments paired with control in *C. chinensis*, (**b**) all treatments paired with control in *C. baccatum*, and (**c**) amongst and between control pairs and treatment pairs of both *C. chinensis* and *C. baccatum*, under salt stress. Meanwhile, (**d**) shows the overall number of upregulated and downregulated genes within and between the species at all-time points in salt stress response; moreover, the gene heat map (**e**) shows their respective pattern under salt stress.

2.5. Identification of the Module Using WGCNA

To evaluate the expression patterns of pepper genes linked to salt stress, we developed a gene co-expression network using the WGCNA method. A comprehensive set of 16,117 genes was classified into 17 co-expression modules, with each module being visually distinguished by a unique shade. These modules were subsequently utilized for further research, as seen in Figure 5a. Significantly, our study revealed the presence of two modules strongly associated with salt resistance, as evidenced by a p-value ≤ 0.05. The ME.red module had a favorable correlation with resistance to salt stress (Figure 5b), including transcription factors like *ERF1b2*, *A-RR9*, ABA receptor *PYL4/6*, *P450*, *pKinase*, low kDa *HSP*, etc. Conversely, the ME.yellow module displayed a favorable association with physicochemical properties (Figure 5c). On initial findings, we identified 1059 genes within the ME.red module, and pathway enrichment analysis of the genes within the ME.red module revealed that the primary enriched pathways were related to plant MAPK and hormone signal transduction.

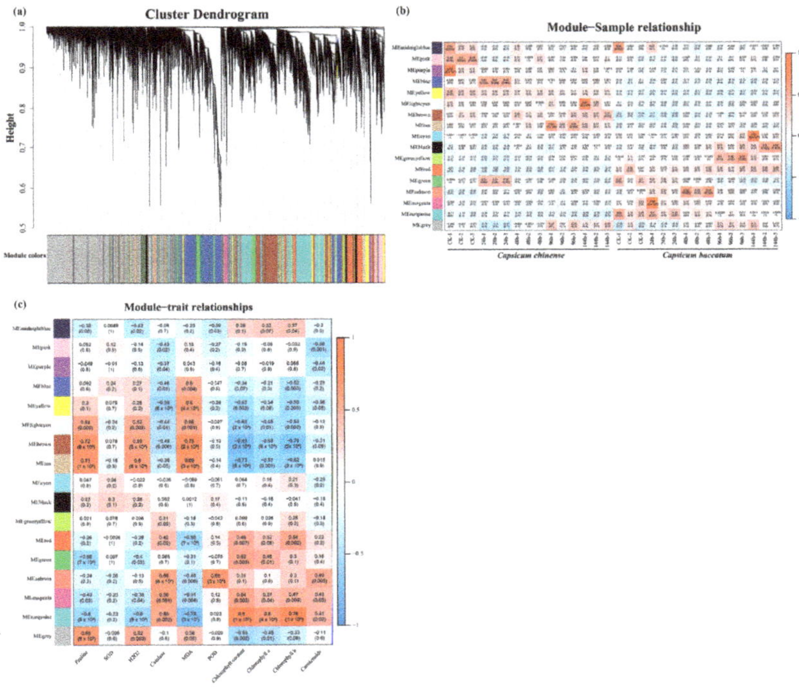

Figure 5. (**a**) The genes have been classified into 17 discrete modules, as seen in the sequence diagram in the figure's uppermost section. The lower portion of the diagram depicts the gene modules using a range of colors, along with the corresponding count of bases (**b**,**c**). The abscissa is the sample, the ordinate is the module, and the number of each grid represents the correlation between the module and the sample. The closer the value is to 1, the stronger the positive correlation between the module and the sample. The closer it is to −1, the stronger the negative correlation between the module and the sample. The number in parentheses represents the p-value of significance, and the lower the number, the stronger the significance (denoted as **b**,**c**). However, Figure (c) depicts the physiological traits of the abscissa. The presented heat map visually represents the eigengene adjacency for each module, indicating the degree of correlation between these modules. The coefficient correlation between the gene module and the sample/trait is determined within each cell, and subsequently, the corresponding p-value is presented directly below.

2.6. GO and KEGG Enrichment Analysis

Enrichment analyses like GO and KEGG were applied to find significant pathways under 100 mM salt stress. We employed GO functional enrichment analysis to categorize the functions of the identified DEGs in response to salt stress. For Biological Process (BP), the most prominent processes included RNA processing (GO:0006396), cell communication (GO:0007154), and cellular response to stimulus (GO:0051716) processes, featuring 179, 273, and 307 genes, respectively. Similarly, for Cellular Components (CC), the top three enriched components were the organelle part (GO:0044422), intracellular organelle part (GO:0044446), and bounding membrane of organelle (GO:0098588), with 385, 385, and 66 genes. However, for the Molecular Function (MF), the three most important categories were RNA binding (GO:0003723), calcium ion binding (GO:0005509), and transferase activity, transferring glycosyl groups (GO:0016757) with 322, 191, and 357 genes, respectively (Figure 6a).

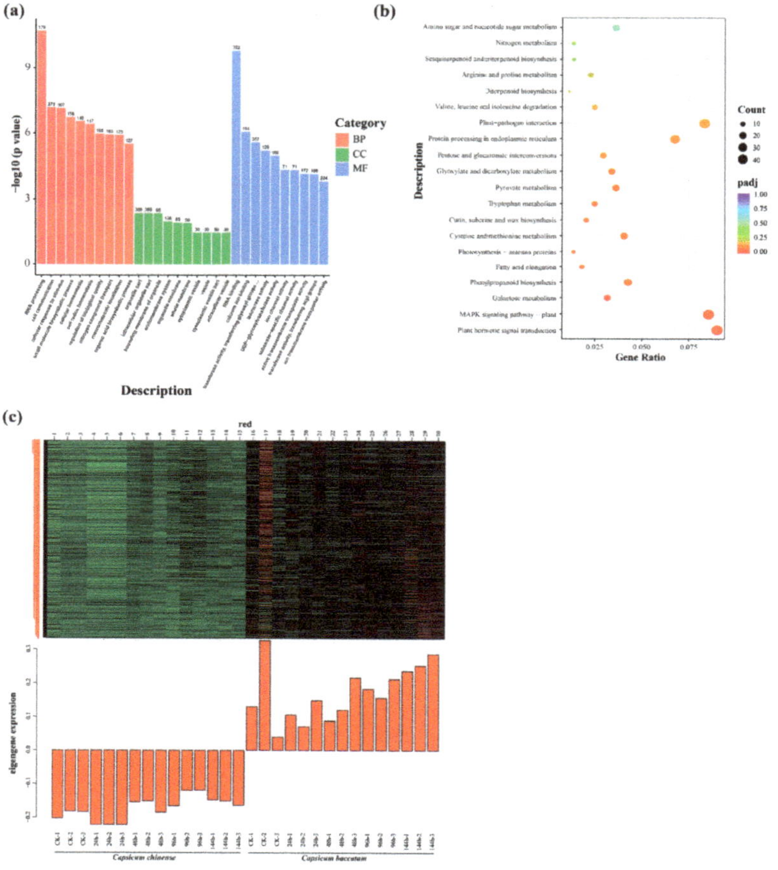

Figure 6. GO (**a**) and KEGG (**b**) enriched DEGs functional analyses; (**c**) expression pattern of genes in the ME.red module selecting using the log10 FPKM values

KEGG pathway enrichment analysis was performed for both cultivars exposed to salt stress. Among the top 20 pathways, the plants MAPK signaling (cann04016) and plant hormone signal transduction (cann04075) pathways, phenylpropanoid biosynthesis (cann00940), galactose metabolism (cann00052), protein processing in endoplasmic reticulum (cann04141), and cysteine and methionine metabolism (cann00270) pathways were mainly enriched pathways (Figure 6b). Hence, considering the outcomes from the KEGG analysis, the plant MAPK signaling and hormone signal transduction pathways were singled out as the most crucial pathways, with pivotal roles in regulating the salt stress response in pepper. The DEGs of these pathways were also positively associated with the ME.red module in WGCNA. Moreover, the epigene expression heat map of the ME.red module is given below (Figure 6c).

2.7. MAPK Signaling Pathway

This work identified six DEGs highly enriched in the MAPK signaling pathway in pepper. These DEGs are known to have a role in the defensive response and adaptation to various abiotic stimuli, including salt, drought, and cold stress. The dynamic transcriptome analysis showed that *ERF1b* transcription factor (TF) upregulation in both species interacts with acidic endochitinase *CHIb* in the defense response of the phytohormone ethylene pathway. However, *ERFb* showed more expression in *C. baccatum* than in *C. chinense*, while acidic *CHIb* showed more expression in *C. chinense*, indicating a possible role in the pepper response to salinity (Figure 7). Similarly, in stress adaptation, we found four TFs, where an abscisic acid receptor *PYL4* downregulation inhibits two *PP2C* (protein phosphatase 2C 51 and probable protein phosphatase 2C 24) that interact with the *MAPKKK18*. However, both *PP2C* TFs showed upregulation in both species cultivars, but more upregulation was seen in *C. chinense* than *C. baccatum*, while *MAPKK18* showed more expression in *C. baccatum* than *C. chinense*, which shows their probable role in the pepper response to salt stress (Figure 7).

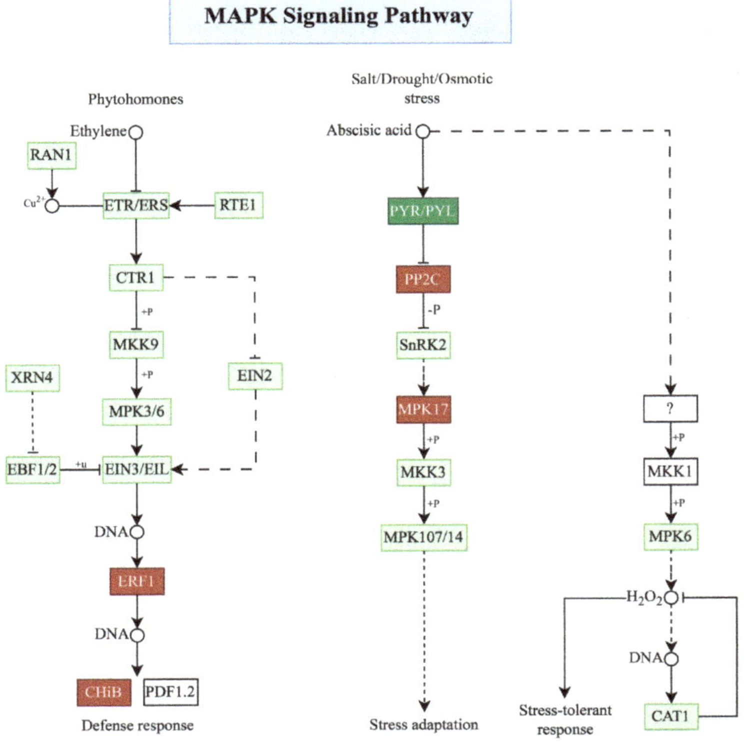

Figure 7. Plant MAPK signaling pathway; upregulated genes are represented in dark red, while the downregulated gene is expressed in dark green color.

2.8. Plant Hormone Signal Transduction Pathway

Under salt stress, cytokinins (Ck) are involved in cell division and shoot initiation during the zeatin biosynthesis process. In this process, Ck negative regulator *A-AAR* downregulation irreversibly inhibits *B-ARR* in *C. chinense* while it was absent in *C. baccatum*, which shows it might have a role in sensitivity against salinity. Similarly, in carotenoid synthesis under plant hormonal signal transduction, ABA plays a crucial role in stomatal closure and seed dormancy. In this process, ABA receptor *PYL4* downregulation inhibits *PP2C* (protein phosphatase 2C 51 and probable protein phosphatase 2C 24), which upregulates and interacts with *ABF* TF, *BZIP_1* (Abscisic acid–insensitive 5-like protein 5). Interestingly. *ABI5* was found to be downregulated in *C. chinense* and upregulated in *C. baccatum*, indicating its possible role in pepper sensitivity against salinity stress (Figure 8). Moreover, in cysteine and methionine metabolism, ethylene, as an important phytohormone, functions as fruit ripening and senescence under stress conditions; in our study, *ERF1b* was found to be upregulated more in *C. baccatum* than *C. chinense*, which predicts it may have a role in pepper response towards salt stress.

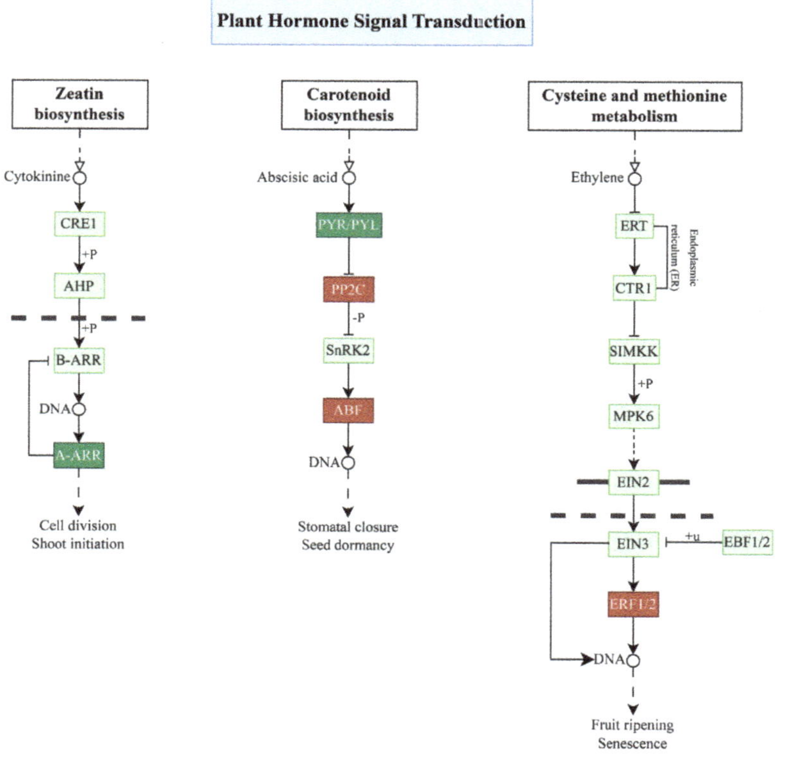

Figure 8. In the Plant hormone signal transduction pathway; upregulated genes are represented in dark red, while the downregulated genes are expressed in dark green.

2.9. qRT-PCR Validation

The qRT-PCR analysis validated RNA-seq expression patterns of 10 specifically selected DEGs. This study was performed across five specific time periods when salt stress treatment was applied in both *C. baccatum* and *C. chinense*, as depicted in Figure 9. Consistent outcomes were reported across multiple genes, encompassing those implicated in MAPK and plant hormone signal transduction pathways (*A-ARR*, *CHIb*, *ERF1b*, *PP2C*, and

ABI5), stilbenoid, diarylheptanoid, and gingerol biosynthesis (*P450*), alanine, aspartate, and glutamate metabolism (*Aldedh1*), phenylpropanoid biosynthesis (*GDA*), and arginine and proline metabolism (*Aldedh2* and *Aldedh3*) (Figure 9a,b). The expression patterns obtained from RNA-Seq data and quantitative real-time polymerase chain reaction (qRT-PCR) demonstrated a significantly strong positive correlation across various time points.

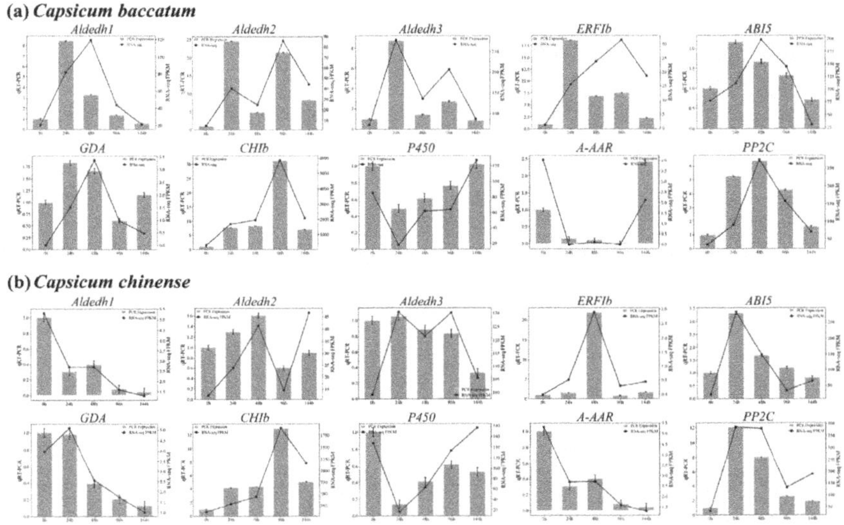

Figure 9. RNA-Seq transcriptional data validation of randomly selected DEGs through qRT-PCR between *C. baccatum* (**a**) and *C. chinense* (**b**) against salt stress; represents the same DEGs comparison between these two species.

3. Discussion

3.1. Phenotypical and Physiological Changes

Salinity poses a significant challenge to plant growth and development, affecting crucial processes such as water absorption [34]. This environmental stressor alters plants' biochemical, physiological, and morphological reactions, ultimately reducing growth, yield, and quality. To cope with salt stress, plants have evolved mechanisms to manage water relations and accumulate ions, thereby developing tolerance to osmotic stress through the accumulation of organic osmolytes or ions [35]. Salt-tolerant plants also regulate their levels of sodium (Na^+) and chloride (Cl^-) ions by transporting or storing them in vacuoles to prevent their toxicity [36]. A high salt level triggers the production of ROS, which can cause damage to proteins, lipids, and DNA [37]. To respond to ROS, plants activate their defense system, which includes POD, SOD, CAT, and other enzymatic antioxidant enzymes and some nonenzymatic compounds like proline, carotenoids, and other such compounds [38]. These enzymatic antioxidants and compounds are crucial for cell growth and survival to scavenge ROS in maintaining the ROS and antioxidant defense balance [39]. In the present study, under salt stress, we investigated the response of two pepper species, *C. baccatum* and *C. chinense*. We observed a distinct pattern of enzymatic and no enzymatic activities in both species. At 24 h and 48 h after salt stress treatment, *C. chinense* showed an early and sustained increase in SOD activity, as compared to *C. baccatum* with a delayed response initially followed by a peak at 144 h. The CAT activity peaked after 24 h in *C. baccatum*, whereas it continuously declined over time in *C. chinense*. Similarly, POD activity was higher in *C. baccatum* compared to *C. chinense* in a continuous manner, particularly after 48 h of treatment. These outcomes highlight the importance of enzymatic antioxidant activities in scavenging the ROS to avoid membrane damage [40], which indicates a plant's response to stress defense and resistance [41].

Moreover, our investigation also revealed the proline, H_2O_2, and MDA role between both species under salt stress. We noticed a greater increase in proline levels in *C. chinense* at all time points compared to *C. baccatum*. Similarly, H_2O_2 remained consistent in *C. baccatum*, but it peaked at 144 h in *C. chinense*, at all-time points showing consistency with earlier findings that high salt stress can cause H_2O_2 rapid bursts in plant cells, which could be an important signal for salinity response [42]. We recorded a maximum level of MDA in *C. chinense* at 96 h after salt stress compared to *C. baccatum* at all time intervals, which suggests greater membrane damage in salt-sensitive species, consistent with the earlier research [43,44]. Additionally, this study also found a more consistent decline in chlorophyll in *C. chineses* under salt stress as compared to *C. baccatum*, and these findings are in agreement with studies showing that salt-tolerant plants normally experience less chlorophyll decrease or degradation under abiotic stresses [45–47].

3.2. Role of MAPK Signaling Pathway

Salinity adversely affects the crop plants' biochemical and physiological pathways [48]. We investigated the molecular response of *C. baccatum* and *C. chinense* under NaCl stress, focusing on the MAPK signaling pathway. Under salt stress, the ethylene-responsive factor (*ERF1b*) showed upregulation in both species but with more expression in *C. baccatum* and more interaction with *CHIb*, which can activate defense-related genes like *PDF1.2*, potentially indicating its robust defense response as compared to *C. chinense*, consistent with the earlier findings [49]. Similarly, we found that *PYL4* (abscisic acid receptor) downregulation and inhibition of *PP2C* genes (protein phosphate 51C and 24C), which interact with mitogen-activated protein kinases (*MAPKKK18*). For *PP2C*, both genes showed more expression in *C. chinense*, whereas *MAPKKK18* showed more expression in *C. baccatum*. These findings are consistent with prior studies suggesting that plants can adapt to stress conditions through ABA signaling, which involves ABA receptors, pyrabactin-resistant regulatory components (*PYR1/PYL/RCAR* proteins), and *PP2C* genes [50–52]. MAPK cascades are also implicated in ABA signaling processes, which include stomatal regulation and antioxidant defense [53]. For example, in stressful conditions, ABA regulates MAPK signaling components [54].

3.3. Role Hormone Signal Transduction Pathway

Phytohormones are important players and play a key role in plant growth and development by involving various plant defense mechanisms under salt stress [55]. For instance, ABA, gibberellic acid (GA), and jasmonic acid (JA) improve alfalfa's salt stress response [56]. Similarly, in regulating the effects of salinity-related factors, ethylene works with a range of other phytohormones and stress signaling molecules to finely orchestrate the plant's responses, whether in the context of salinity or normal growth [57,58]. Our study delves into the key regulatory roles of phytohormones like Cks, ABA, and ethylene in shaping pepper crop transcriptional mechanisms under salt stress. Cks are important indicators of plant sensitivity to salinity, as they influence different physiological processes [52,59]. In our study, we observed the downregulation of *ARR 4/9* signaling genes in *C. chinense*, which might negatively impact the Ck pathway, indicating their potential involvement in salt response; these results are consistent with prior findings [60,61]. Previous research suggests ethylene is a significant regulator of NaCl stress response at various cellular levels and transcription factors [62]. Our findings revealed a higher upregulation of *ERF1b* expression in *C. baccatum* than in *C. chinense* under stress conditions, indicating their importance in pepper's tolerance response. Moreover, ABA's vital role in salt stress responses indicates its influence on various physiological mechanisms [63]. In our study, the downregulation of the ABA receptor *PYL4* inhibits *PP2C* genes, leading to the upregulation of *bZIP_1* (ABI5), which positively impacts the ABA signaling pathway in pepper under salt stress. *ABI5* showed a higher expression in *C. baccatum* than in *C.chinense*, which indicates that *ABI5* might have a role in tolerance against salt stress. Prior findings support our findings by emphasizing the significance of ABA, its receptors, *PP2C* phosphatases, and ABF transcription factors in plant responses under salt stress [64,65]. This study underscores the importance of the intricate interplay of phytohormone signaling pathways in regulating pepper's response to salt stress, providing valuable insights into potential targets for improving salt tolerance in crops. Further research must elucidate the specific mechanisms underlying hormone-mediated responses to salt stress in pepper plants.

4. Materials and Methods

4.1. Plant Material and NaCl Treatments

Two pepper-cultivated species seeds were germinated on MS media and screened against 100 mM salt stress. The *C. baccatum* germplasm (HNUCB226), moderately salt-sensitive, and *C. chinense* germplasm (HNUCC275), salt-sensitive (based on screening), were selected for this study [66]. At the 3–4 leaf growth stage (BBCH 13–14), seedlings were transferred into 800 mL boxes filled with Hoagland nutritional solution and grown in a growth chamber (PQX-330B-30HM, Ningbo Life Technology Co. Ltd., Ningbo, China) under conditions of 26 °C temperature, 70% humidity, and a light cycle of 16 h followed by 8 h of darkness. After 4–5 weeks of growth, the plants were treated with 100 mM NaCl for 144 h [67]. A total of 72 plants were selected, with 36 plants in each species; each species contained 3 replicates, and each replicate contained 12 plants. After stress treatment, samples were collected at 0 h, 24 h, 48 h, 96 h, and 144 h, respectively, with three biological repeats. Subsequently, the collected samples were stored at −80 °C for future analysis.

4.2. Response of Biochemical Indexes

To assess the enzyme (SOD, POD, and CAT) activities, 0.1 g of leaf samples were homogenized in 900 μL of phosphate buffer (pH 7.8). Samples were centrifuged (TGL-1850 Micro refrigerated centrifuge, Sichuan Shuke Instrument Co., Ltd., Chengdu, China) at 10,000× *g* for 10 min, and the supernatants were used to determine the antioxidant enzyme activities according to their respective kits (A001-1-2, A084-3-1, and A007-1-1) from Nanjing Jiancheng Biotechnology. Additionally, proline and MDA content and H_2O_2 concentration content in pepper leaves were quantified using assay kits (G0111F96, G0110F, and A064-1-1) from Suzhou Geruisi Biotechnology Co. Ltd., (Suzhou, China) and Nanjing Jiancheng Biotechnology (Nanjing, China), adhering to the instructions provided by the manufacturers, respectively. Moreover, for the determination of the total chlorophyll contents, chlorophyll a, chlorophyll b, and carotenoids, 0.1 g of fresh leaf samples were ground with 10 mL of extraction solution. Samples were centrifuged at 10,000× *g* for 10 min, and the supernatant was used to observe the values of Chl a, Chl b, and carotenoid according to the commercially available kit (G0163F) from Suzhou Geruisi Biotechnology Co. Ltd., Suzhou, China.

4.3. Stress Tolerance Index Percentage (STI%)

The leaf, stem, and root samples from the control and treatment of both species were dried in an oven (DGG-9053A, Shanghai Enxin Instrument Co., Ltd., Shanghai, China) at 65 °C for 72 h. Samples were weighed via a weighing apparatus, and stress tolerance index percentage (STI%) was calculated using the following formula previously described [68].

STI (%) = (Dry weight of leaf/stem/root of treated samples/Dry weight of leaf/stem/root of control samples) × 100.

4.4. Scanning of Leaf and Root

Root and leaf samples were washed with 75% ethanol and were subsequently dried. After drying, samples were attached to metallic stubs with scotch tape and sputter coated with gold and aluminum in a J20 Ion Sputter Coater for 90 s before scanning. Finally, images were taken using a scanning electron microscope (Thermoscientific Verios G4 UC) [69].

4.5. RNA Extraction, Library Construction, and RNA Sequencing

RNA extraction was performed on all obtained samples using the NEB extraction kit. In a subsequent step, the cDNA library was constructed using the RNA Library Prep kit for Illumina® (NEB, Ipswich, Massachusetts, USA). Clusters were generated, and sequencing was performed using the Illumina Novaseq platform, producing paired-end reads with a length of 150 base pairs. The HISAT2 alignment tool mapped paired-end clean reads of both species with the reference genome *C. baccatum* (https://ftp.ncbi.nlm.nih.gov/genomes/all/GCA/002/271/885/GCA_002271885.2_ASM227188v2/GCA_002271885.2_ASM227188v2_genomic.gtf.gz: accessed on 15 May 2023). FeatureCounts (v1.5.0-p3) was used to count the reads numbers mapped to each gene, then the FPKM of each gene was calculated based on the length of the gene and read counts mapped to this gene. Differential expression analysis (DEA) was conducted with DESeq2 software (version 1.20.0). DESeq2 provides statistical routines for determining differential expression in digital gene expression data using a model based on the negative binomial distribution. The resulting *p*-values were adjusted using Benjamini and Hochberg's approach to control the false discovery rate. Genes with an adjusted $p \leq 0.05$ found by DESeq2 were assigned as differentially expressed [70]. The clusterProfiler R package (3.8.1) was used to analyze the Gene Ontology (GO) of differentially expressed genes and the statistical enrichment of DEGs in the KEGG pathway [71,72].

4.6. Co-Expression Modules Identification

The R WGCNA package (a set of functions used to calculate various weighted association analyses, which can be used for network construction, gene screening, gene cluster identification, topological feature calculation, data simulation, and visualization) was employed to establish a gene co-expression network. The networks were visualized using Cytoscape version 3.6.048. Using the heat map function in the R package, high-throughput heat maps were generated and visually displayed to showcase the gene expression levels for specific DEGs [73].

4.7. qRT-PCR Validation

The RNA-seq samples created the quantitative real-time PCR analysis library. Ten randomly selected DEG expression patterns were validated via qRT-PCR. Primers for these genes were designed using PRIMER5 software ((v5.0) Supplementary Table S3). The HiScript III (+gDNA wiper) kit was used for reverse transcription. SYBR qPCR-Mix was used for qPCR analysis, and QuantStudio™ Design and Analysis Software 1.3 calculated the expression levels of each sample. Moreover, the $2^{-\Delta\Delta CT}$ method was then used to determine the relative expression levels of the chosen DEGs.

4.8. Statistics

Data were analyzed using the statistical software SAS 9.4 (SAS Institute Inc., Cary, NC, USA). A two-way ANOVA was employed to analyze the salt stress changes at different intervals. The graphs were generated using GraphPad Prism 8.0.1, and the significant differences were examined using Tukey's HSD tests following $p \leq 0.05$.

5. Conclusions

This study investigated the physiological and transcriptomic responses of two pepper species (*C. chinense* and *C. baccatum*) to salt stress. Our findings illuminate the complex mechanisms, particularly emphasizing the roles of MAPK and hormone signal transduction pathways. Physiologically, significant differences between the two species were observed, with *C. chinense* exhibiting higher sensitivity, as evidenced by pronounced alterations such as enzymatic antioxidants and nonenzymatic compounds, whereas *C. baccatum* responded more resilient to salt stress. The DEGs involved in plant signaling pathways, including ethylene-responsive factors (*ERFs*), ABA receptors (*PYL*), protein phosphatases (*PP2C*), and basic leucine zipper protein (*BZIP_1*), further highlighted the species-specific responses to salt stress. Overall, this study provides valuable insights into salt stress tolerance's molecular and physiological mechanisms in pepper plants, identifying key genes and pathways for potential use in breeding salt-tolerant cultivars. These findings also enhance the understanding of plant stress responses, with implications for improving crop resilience to environmental challenges.

Supplementary Materials: The following supporting information can be downloaded at: https://www.mdpi.com/article/10.3390/ijms25179355/s1.

Author Contributions: Conceptualization, Z.W. and M.A.F.; methodology, M.A.F., Z.W. and L.Z.; software, M.A.F., M.Z.U.H., H.T., S.W. and Z.W.; validation, M.A.F., N.M., S.W. and M.Z.U.H.; formal analysis, M.A.F., H.T., L.Z. and M.Z.U.H.; investigation, M.A.F. and Z.W.; resources, M.A.F., Z.W., H.T., S.W., M.Z.U.H. and L.Z.; data curation, M.A.F., H.T., M.Z.U.H. and N.M.; writing—original draft preparation, M.A.F., Z.W. and M.Z.U.H.; writing—review and editing, M.A.F., N.M., L.Z., S.W., H.T., M.Z.U.H. and Z.W.; visualization, M.A.F. and Z.W.; supervision, Z.W.; project administration, Z.W.; funding acquisition, Z.W. All authors have read and agreed to the published version of the manuscript.

Funding: This research was funded by the Hainan Province Science and Technology Special Fund (ZDYF2023XDNY028) and the Collaborative Innovation Center of Nanfan and High-Efficiency Tropical Agriculture (XTCX2022NYB03), Hainan University.

Institutional Review Board Statement: Not applicable.

Informed Consent Statement: Not applicable.

Data Availability Statement: The original data can be found in NCBI BioProject: PRJNA1056267 (https://dataview.ncbi.nlm.nih.gov/object/PRJNA1056267?reviewer=4jvkg624h1schnmnovkjinudoq: accessed on 25 December 2023).

Conflicts of Interest: The authors declare no conflicts of interest.

References

1. Islam, M.R.; Mia, M.B.; Islam, T. Role of abiotic stresses on photosynthesis and yield of crop plants, with special reference to wheat. In *Abiotic Stresses in Wheat*; Elsevier: Amsterdam, The Netherlands, 2023; pp. 179–193.
2. Rathod, A.; Verma, N.S. Impact of Abiotic Stress on Agronomical Crops. In *Frontiers of Agronomy*; Elite Publishing House: Rohini, ND, USA, 2023; p. 27.
3. Gosai, H.G.; Sharma, A.; Mankodi, P. Climate Change's Impact on Agricultural Food Production. In *Food Security in a Developing World: Status, Challenges, and Opportunities*; Springer: Berlin/Heidelberg, Germany, 2024; pp. 117–132.
4. Datta, D.; Ghosh, S.; Das, K.; Singh, S.V.; Mazumdar, S.P.; Mandal, S.; Singh, Y. Biochar imparting abiotic stress resilience. In *Biochar Production for Green Economy*; Elsevier: Amsterdam, The Netherlands, 2024; pp. 227–248.
5. Mantri, N.; Patade, V.; Penna, S.; Ford, R.; Pang, E. Abiotic stress responses in plants: Present and future. In *Abiotic Stress Responses in Plants: Metabolism, Productivity and Sustainability*; Springer: New York, NY, USA, 2012; pp. 1–19.
6. Abd El-Mageed, T.A.; Rady, M.O.; Semida, W.M.; Shaaban, A.; Mekdad, A.A. Exogenous micronutrients modulate morpho-physiological attributes, yield, and sugar quality in two salt-stressed sugar beet cultivars. *J. Soil Sci. Plant Nutr.* **2021**, *21*, 1421–1436. [CrossRef]
7. Semida, W.M.; Abd El-Mageed, T.A.; Abdelkhalik, A.; Hemida, K.A.; Abdurrahman, H.A.; Howladar, S.M.; Leilah, A.A.; Rady, M.O. Selenium modulates antioxidant activity, osmoprotectants, and photosynthetic efficiency of onion under saline soil conditions. *Agronomy* **2021**, *11*, 855. [CrossRef]
8. Devkota, K.P.; Devkota, M.; Rezaei, M.; Oosterbaan, R. Managing salinity for sustainable agricultural production in salt-affected soils of irrigated drylands. *Agric. Syst.* **2022**, *198*, 103390. [CrossRef]
9. Machado, R.M.A.; Serralheiro, R.P. Soil salinity: Effect on vegetable crop growth. Management practices to prevent and mitigate soil salinization. *Horticulturae* **2017**, *3*, 30. [CrossRef]
10. Kaya, C.; Ashraf, M. The endogenous L-cysteine desulfhydrase and hydrogen sulfide participate in supplemented phosphorus-induced tolerance to salinity stress in maize (*Zea mays*) plants. *Turk. J. Bot.* **2020**, *44*, 36–46. [CrossRef]
11. Kamran, M.; Parveen, A.; Ahmar, S.; Malik, Z.; Hussain, S.; Chattha, M.S.; Saleem, M.H.; Adil, M.; Heidari, P.; Chen, J.-T. An overview of hazardous impacts of soil salinity in crops, tolerance mechanisms, and amelioration through selenium supplementation. *Int. J. Mol. Sci.* **2019**, *21*, 148. [CrossRef]
12. Zhonghua, T.; Yanju, L.; Xiaorui, G.; Yuangang, Z. The combined effects of salinity and nitrogen forms on Catharanthus roseus: The role of internal ammonium and free amino acids during salt stress. *J. Plant Nutr. Soil Sci.* **2011**, *174*, 135–144. [CrossRef]

13. Gupta, B.; Huang, B. Mechanism of salinity tolerance in plants: Physiological, biochemical, and molecular characterization. *Int. J. Gen.* **2014**, *2014*, 701596. [CrossRef]
14. Shelke, D.B.; Nikalje, G.C.; Chambhare, M.R.; Zaware, B.N.; Penna, S.; Nikam, T.D. Na^+ and Cl^- induce differential physiological, biochemical responses and metabolite modulations in vitro in contrasting salt-tolerant soybean genotypes. *3 Biotech* **2019**, *9*, 91. [CrossRef]
15. Tahir, H.; Sajjad, M.; Qian, M.; Zeeshan Ul Haq, M.; Tahir, A.; Chen, T.; Shaopu, S.; Farooq, M.A.; Ling, W.; Zhou, K. Transcriptomic Analysis Reveals Dynamic Changes in Glutathione and Ascorbic Acid Content in Mango Pulp across Growth and Development Stages. *Horticulturae* **2024**, *10*, 694. [CrossRef]
16. Hasanuzzaman, M.; Raihan, M.R.H.; Masud, A.A.C.; Rahman, K.; Nowroz, F.; Rahman, M.; Nahar, K.; Fujita, M. Regulation of reactive oxygen species and antioxidant defense in plants under salinity. *Int. J. Mol. Sci.* **2021**, *22*, 9326. [CrossRef]
17. Mansoor, S.; Ali, A.; Kour, N.; Bornhorst, J.; AlHarbi, K.; Rinklebe, J.; Abd El Moneim, D.; Ahmad, P.; Chung, Y.S. Heavy metal induced oxidative stress mitigation and ROS scavenging in plants. *Plants* **2023**, *12*, 3003. [CrossRef] [PubMed]
18. Zhou, X.; Naguro, I.; Ichijo, H.; Watanabe, K. Mitogen-activated protein kinases as key players in osmotic stress signaling. *Biochim. Biophys. Acta (BBA)—Gen. Subj.* **2016**, *1860*, 2037–2052. [CrossRef]
19. Singh, A.; Kumar, A.; Yadav, S.; Singh, I.K. Reactive oxygen species-mediated signaling during abiotic stress. *Plant Gene* **2019**, *18*, 100173. [CrossRef]
20. Liu, Y.; Wang, L.; Liu, C.; Yin, H.; Liu, H.; Luo, H.; He, M.; Zhou, Y. Cgbzip1: A bzip transcription factor from chrysanthemum grandiflora confers plant tolerance to salinity and drought stress. *Agronomy* **2022**, *12*, 556. [CrossRef]
21. Gurmani, A.; Bano, A.; Khan, S.; Din, J.; Zhang, J. Alleviation of salt stress by seed treatment with abscisic acid (ABA), 6-benzylaminopurine (BA) and chlormequat chloride (CCC) optimizes ion and organic matter accumulation and increases yield of rice (*Oryza sativa* L.). *Aust. J. Crop Sci.* **2011**, *5*, 1278–1285.
22. Ma, L.; Zhang, H.; Sun, L.; Jiao, Y.; Zhang, G.; Miao, C.; Hao, F. NADPH oxidase AtrbohD and AtrbohF function in ROS-dependent regulation of Na^+/K^+ homeostasis in Arabidopsis under salt stress. *J. Exp. Bot.* **2012**, *63*, 305–317. [CrossRef]
23. Yang, Y.; Guo, Y. Unraveling salt stress signaling in plants. *J. Integr. Plant Biol.* **2018**, *60*, 796–804. [CrossRef]
24. Khoury, C.K.; Carver, D.; Barchenger, D.W.; Barboza, G.E.; van Zonneveld, M.; Jarret, R.; Bohs, L.; Kantar, M.; Uchanski, M.; Mercer, K. Modelled distributions and conservation status of the wild relatives of chile peppers (*Capsicum* L.). *Divers. Distrib.* **2020**, *26*, 209–225. [CrossRef]
25. Tripodi, P.; Kumar, S. *The Capsicum Genome*; Springer International Publishing: Cham, Switzerland, 2019; pp. 1–8.
26. Zou, Z.; Zou, X. Geographical and ecological differences in pepper cultivation and consumption in China. *Front. Nutri.* **2021**, *8*, 718517. [CrossRef]
27. Olatunji, T.L.; Afolayan, A.J. The suitability of chili pepper (*Capsicum annuum* L.) for alleviating human micronutrient dietary deficiencies: A review. *Food Sci. Nutri.* **2018**, *6*, 2239–2251. [CrossRef]
28. Wahyuni, Y.; Ballester, A.-R.; Sudarmonowati, E.; Bino, R.J.; Bovy, A.G. Secondary metabolites of Capsicum species and their importance in the human diet. *J. Nat. Prod.* **2013**, *76*, 783–793. [CrossRef] [PubMed]
29. Bacelar, E.; Pinto, T.; Anjos, R.; Morais, M.C.; Oliveira, I.; Vilela, A.; Cosme, F. Impacts of climate change and mitigation strategies for some abiotic and biotic constraints influencing fruit growth and quality. *Plants* **2024**, *13*, 1942. [CrossRef] [PubMed]
30. Bojórquez-Quintal, E.; Velarde-Buendía, A.; Ku-González, Á.; Carillo-Pech, M.; Ortega-Camacho, D.; Echevarría-Machado, I.; Pottosin, I.; Martínez-Estévez, M. Mechanisms of salt tolerance in habanero pepper plants (*Capsicum chinense* Jacq.): Proline accumulation, ions dynamics and sodium root-shoot partition and compartmentation. *Front. Plant Sci.* **2014**, *5*, 605.
31. Zaman, M.; Shahid, S.A.; Heng, L. *Guideline for Salinity Assessment, Mitigation and Adaptation Using Nuclear and Related Techniques*; Springer Nature: Berlin, Germany, 2018.
32. De Pascale, S.; Ruggiero, C.; Barbieri, G.; Maggio, A. Physiological responses of pepper to salinity and drought. *J. Am. Soc. Hortic. Ssc.* **2003**, *128*, 48–54. [CrossRef]
33. Rolly, N.K.; Imran, Q.M.; Shahid, M.; Imran, M.; Khan, M.; Lee, S.-U.; Hussain, A.; Lee, I.-J.; Yun, B.-W. Drought-induced AtbZIP62 transcription factor regulates drought stress response in Arabidopsis. *Plant Physiol. Biochem.* **2020**, *156*, 384–395. [CrossRef]
34. Khan, M.A.H.; Baset Mia, M.A.; Quddus, M.A.; Sarker, K.K.; Rahman, M.; Skalicky, M.; Brestic, M.; Gaber, A.; Alsuhaibani, A.M.; Hossain, A. Salinity-induced physiological changes in pea (*Pisum sativum* L.): Germination rate, biomass accumulation, relative water content, seedling vigor and salt tolerance index. *Plants* **2022**, *11*, 3493. [CrossRef]
35. Mansour, M.M.F.; Salama, K.H.A. Cellular mechanisms of plant salt tolerance. In *Microorganisms in Saline Environments: Strategies and Functions*; Springer International Publishing: Cham, Switzerland, 2019; pp. 169–210.
36. Rady, M.M.; Taha, S.S.; Kusvuran, S. Integrative application of cyanobacteria and antioxidants improves common bean performance under saline conditions. *Sci. Hortic.* **2018**, *233*, 61–69. [CrossRef]
37. Parihar, P.; Singh, S.; Singh, R.; Singh, V.P.; Prasad, S.M. Effect of salinity stress on plants and its tolerance strategies: A review. *Environ. Sci. Pollut. Res.* **2015**, *22*, 4056–4075. [CrossRef]
38. Kesawat, M.S.; Satheesh, N.; Kherawat, B.S.; Kumar, A.; Kim, H.-U.; Chung, S.-M.; Kumar, M. Regulation of reactive oxygen species during salt stress in plants and their crosstalk with other signaling molecules—Current perspectives and future directions. *Plants* **2023**, *12*, 864. [CrossRef]
39. Jomova, K.; Raptova, R.; Alomar, S.Y.; Alwasel, S.H.; Nepovimova, E.; Kuca, K.; Valko, M. Reactive oxygen species, toxicity, oxidative stress, and antioxidants: Chronic diseases and aging. *Arch. Toxicol.* **2023**, *97*, 2499–2574. [PubMed]

40. Guan, Q.; Wang, Z.; Wang, X.; Takano, T.; Liu, S. A peroxisomal APX from *Puccinellia tenuiflora* improves the abiotic stress tolerance of transgenic Arabidopsis thaliana through decreasing of H_2O_2 accumulation. *J. Plant Physiol.* **2015**, *175*, 183–191. [CrossRef] [PubMed]
41. Meng, F.; Duan, Y.; Yang, Y.; Liu, G.; Liu, H. Effects of mixed saline stress on photosynthetic characteristics and antioxidant enzymes activity of cassava seedlings. *Chin. Agric. Sci. Bull.* **2019**, *35*, 34–39.
42. Wang, Y.; Diao, P.; Kong, L.; Yu, R.; Zhang, M.; Zuo, T.; Fan, Y.; Niu, Y.; Yan, F.; Wuriyanghan, H. Ethylene enhances seed germination and seedling growth under salinity by reducing oxidative stress and promoting chlorophyll content via ETR2 pathway. *Front. Plant Sci.* **2020**, *11*, 1066. [CrossRef] [PubMed]
43. Ashraf, M.; Harris, P.J. Potential biochemical indicators of salinity tolerance in plants. *Plant Sci.* **2004**, *166*, 3–16. [CrossRef]
44. Szabados, L.; Savouré, A. Proline: A multifunctional amino acid. *Trends Plant Sci.* **2010**, *15*, 89–97. [CrossRef]
45. Rao, A.; Ahmad, S.D.; Sabir, S.M.; Awan, S.I.; Shah, A.H.; Abbas, S.R.; Shafique, S.; Khan, F.; Chaudhary, A. Potential antioxidant activities improve salt tolerance in ten varieties of wheat (*Triticum aestivum* L.). *Am. J. Plant Sci.* **2013**, *4*, 69–76. [CrossRef]
46. Santos, C.V. Regulation of chlorophyll biosynthesis and degradation by salt stress in sunflower leaves. *Sci. Hortic.* **2004**, *103*, 93–99. [CrossRef]
47. Tania, S.S.; Rhaman, M.S.; Rauf, F.; Rahaman, M.M.; Kabir, M.H.; Hoque, M.A.; Murata, Y. Alleviation of Salt-Inhibited germination and seedling growth of kidney bean by seed priming and exogenous application of salicylic acid (SA) and hydrogen peroxide (H_2O_2). *Seeds* **2022**, *1*, 87–98. [CrossRef]
48. Arzani, A.; Kumar, S.; Mansour, M.M.F. Salt tolerance in plants: Molecular and functional adaptations. *Front. Plant Sci.* **2023**, *14*, 1280788. [CrossRef]
49. Solano, R.; Stepanova, A.; Chao, Q.; Ecker, J.R. Nuclear events in ethylene signaling: A transcriptional cascade mediated by ETHYLENE-INSENSITIVE3 and ETHYLENE-RESPONSE-FACTOR1. *Genes Dev.* **1998**, *12*, 3703–3714. [CrossRef] [PubMed]
50. Lee, S.C.; Luan, S. ABA signal transduction at the crossroad of biotic and abiotic stress responses. *Plant Cell Environ.* **2012**, *35*, 53–60. [CrossRef]
51. Nakashima, K.; Fujita, Y.; Kanamori, N.; Katagiri, T.; Umezawa, T.; Kidokoro, S.; Maruyama, K.; Yoshida, T.; Ishiyama, K.; Kobayashi, M. Three Arabidopsis SnRK2 protein kinases, SRK2D/SnRK2.2, SRK2E/SnRK2.6/OST1 and SRK2I/SnRK2.3, involved in ABA signaling are essential for the control of seed development and dormancy. *Plant Cell Physiol.* **2009**, *50*, 1345–1363. [CrossRef]
52. Umezawa, T.; Sugiyama, N.; Takahashi, F.; Anderson, J.C.; Ishihama, Y.; Peck, S.C.; Shinozaki, K. Genetics and phosphoproteomics reveal a protein phosphorylation network in the abscisic acid signaling pathway in *Arabidopsis thaliana*. *Sci. Signal.* **2013**, *6*, rs8. [CrossRef]
53. Danquah, A.; De Zélicourt, A.; Colcombet, J.; Hirt, H. The role of ABA and MAPK signaling pathways in plant abiotic stress responses. *Biotechnol. Adv.* **2014**, *32*, 40–52. [CrossRef]
54. Liu, Y. Roles of mitogen-activated protein kinase cascades in ABA signaling. *Plant Cell Rep.* **2012**, *31*, 1–12. [CrossRef]
55. Mangal, V.; Lal, M.K.; Tiwari, R.K.; Altaf, M.A.; Sood, S.; Kumar, D.; Bharadwaj, V.; Singh, B.; Singh, R.K.; Aftab, T. Molecular insights into the role of reactive oxygen, nitrogen and sulphur species in conferring salinity stress tolerance in plants. *J. Plant Growth Regul.* **2023**, *42*, 554–574. [CrossRef]
56. Wang, X.; Yin, J.; Wang, J.; Li, J. Integrative analysis of transcriptome and metabolome revealed the mechanisms by which flavonoids and phytohormones regulated the adaptation of alfalfa roots to NaCl stress. *Front. Plant Sci.* **2023**, *14*, 1117868. [CrossRef] [PubMed]
57. Atkinson, N.J.; Urwin, P.E. The interaction of plant biotic and abiotic stresses: From genes to the field. *J. Exp. Bot.* **2012**, *63*, 3523–3543. [CrossRef] [PubMed]
58. Kissoudis, C.; van de Wiel, C.; Visser, R.G.; van der Linden, G. Enhancing crop resilience to combined abiotic and biotic stress through the dissection of physiological and molecular crosstalk. *Front. Plant Sci.* **2014**, *5*, 207. [CrossRef]
59. Pavlů, J.; Novák, J.; Koukalová, V.; Luklová, M.; Brzobohatý, B.; Černý, M. Cytokinin at the crossroads of abiotic stress signalling pathways. *Int. J. Mol. Sci.* **2018**, *19*, 2450. [CrossRef]
60. Waidmann, S.; Ruiz Rosquete, M.; Schöller, M.; Sarkel, E.; Lindner, H.; LaRue, T.; Petřík, I.; Dünser, K.; Martopawiro, S.; Sasidharan, R. Cytokinin functions as an asymmetric and anti-gravitropic signal in lateral roots. *Nat. Commun.* **2019**, *10*, 3540. [CrossRef]
61. Zubo, Y.O.; Schaller, G.E. Role of the cytokinin-activated type-B response regulators in hormone crosstalk. *Plants* **2020**, *9*, 166. [CrossRef] [PubMed]
62. Wang, J.; An, C.; Guo, H.; Yang, X.; Chen, J.; Zong, J.; Li, J.; Liu, J. Physiological and transcriptomic analyses reveal the mechanisms underlying the salt tolerance of *Zoysia japonica* Steud. *BMC Plant Biol.* **2020**, *20*, 114. [CrossRef]
63. Ruggiero, A.; Landi, S.; Punzo, P.; Possenti, M.; Van Oosten, M.J.; Costa, A.; Morelli, G.; Maggio, A.; Grillo, S.; Batelli, G. Salinity and ABA seed responses in pepper: Expression and interaction of ABA core signaling components. *Front. Plant Sci.* **2019**, *10*, 304. [CrossRef] [PubMed]
64. Zhao, B.-Y.; Hu, Y.-F.; Li, J.-J.; Yao, X.; Liu, K.-D. BnaABF2, a bZIP transcription factor from rapeseed (*Brassica napus* L.), enhances drought and salt tolerance in transgenic Arabidopsis. *Bot. Stud.* **2016**, *57*, 12. [CrossRef] [PubMed]
65. Di, F.; Jian, H.; Wang, T.; Chen, X.; Ding, Y.; Du, H.; Lu, K.; Li, J.; Liu, L. Genome-wide analysis of the PYL gene family and identification of PYL genes that respond to abiotic stress in *Brassica napus*. *Genes* **2018**, *9*, 156. [CrossRef]

66. Niu, M.; Wei, L.; Peng, Y.; Huang, Y.; Bie, Z. Mechanisms of increasing salt resistance of vegetables by grafting. *Veg. Res.* **2022**, *2*, 8. [CrossRef]
67. Ma, J.; Wang, Y.; Wang, L.-Y.; Lin, D.; Yang, Y. Transcriptomic analysis reveals the mechanism of the alleviation of salt stress by salicylic acid in pepper (*Capsicum annuum* L.). *Mol. Biol. Rep.* **2023**, *50*, 3593–3606. [CrossRef]
68. Kotaś, J.; Stasicka, Z. Chromium occurrence in the environment and methods of its speciation. *Environ. Pollut.* **2000**, *107*, 263–283. [CrossRef]
69. Shu, H.; Altaf, M.A.; Mushtaq, N.; Fu, H.; Lu, X.; Zhu, G.; Cheng, S.; Wang, Z. Physiological and transcriptome analysis of the effects of exogenous strigolactones on drought responses of pepper seedlings. *Antioxidants* **2023**, *12*, 2019. [CrossRef] [PubMed]
70. Anders, S.; Huber, W. Differential expression analysis for sequence count data. *J. Genome Boil.* **2010**, *11*, R106. [CrossRef] [PubMed]
71. Young, M.D.; Wakefield, M.J.; Smyth, G.K.; Oshlack, A. Gene ontology analysis for RNA-seq: Accounting for selection bias. *Genome Biol.* **2010**, *11*, R14. [CrossRef]
72. Kanehisa, M.; Goto, S. KEGG: Kyoto encyclopedia of genes and genomes. *Nucleic Acids Res.* **2000**, *28*, 27–30. [CrossRef] [PubMed]
73. Mumtaz, M.A.; Zhou, Y.; Gao, C.; Kamran, H.M.; Altaf, M.A.; Hao, Y.; Shu, H.; Zhang, Y.; Lu, X.; Abbas, H.M.K.; et al. Interaction between transcriptional activator BRI1-EMS-SUPPRESSOR 1 and HSPs regulates heat stress tolerance in pepper. *Environ. Exp. Bot.* **2023**, *211*, 105341. [CrossRef]

International Journal of
Molecular Sciences

MDPI

Article

Regulating Leaf Photosynthesis and Soil Microorganisms through Controlled-Release Nitrogen Fertilizer Can Effectively Alleviate the Stress of Elevated Ambient Ozone on Winter Wheat

Nanyan Zhu [1,2,3], Yinsen Qian [3], Lingqi Song [4], Qiaoqiao Yu [3], Haijun Sheng [4], Ying Li [3,5,6,*] and Xinkai Zhu [3,6,7,*]

1 College of Animal Science and Technology, Yangzhou University, Yangzhou 225000, China; dx120200148@stu.yzu.edu.cn
2 College of JunCao Science and Ecology (College of Carbon Neutrality), Fujian Agriculture and Forestry University, Fuzhou 350002, China
3 Jiangsu Key Laboratory of Crop Genomics and Physiology/Jiangsu Key Laboratory of Crop Cultivation and Physiology, College of Agricultural, Yangzhou University, Yangzhou 225000, China; dx120220114@stu.yzu.edu.cn (Y.Q.); 15046086520@163.com (Q.Y.)
4 College of Environmental Science and Engineering, Yangzhou University, Yangzhou 225009, China; slq231010@163.com (L.S.); hjsheng@yzu.edu.cn (H.S.)
5 Department of Biology, Hong Kong Baptist University, Hong Kong 999077, China
6 Jiangsu Co-Innovation Center for Modern Production Technology of Grain Crops, Yangzhou University, Yangzhou 225009, China
7 Joint International Research Laboratory of Agriculture and Agri-Product Safety, The Ministry of Education of China, Yangzhou University, Yangzhou 225009, China
* Correspondence: yingli@yzu.edu.cn (Y.L.); wheat@yzu.edu.cn (X.Z.)

Abstract: The mitigation mechanisms of a kind of controlled-release nitrogen fertilizer (sulfur-coated controlled-release nitrogen fertilizer, SCNF) in response to O_3 stress on a winter wheat (*Triticum aestivum* L.) variety (Nongmai-88) were studied in crop physiology and soil biology through the ozone-free-air controlled enrichment (O_3-FACE) simulation platform and soil microbial metagenomics. The results showed that SCNF could not delay the O_3-induced leaf senescence of winter wheat but could enhance the leaf size and photosynthetic function of flag leaves, increase the accumulation of nutrient elements, and lay the foundation for yield by regulating the release rate of nitrogen (N). By regulating the soil environment, SCNF could maintain the diversity and stability of soil bacterial and archaeal communities, but there was no obvious interaction with the soil fungal community. By alleviating the inhibition effects of O_3 on N-cycling-related genes (*ko00910*) of soil microorganisms, SCNF improved the activities of related enzymes and might have great potential in improving soil N retention. The results demonstrated the ability of SCNF to improve leaf photosynthetic function and increase crop yield under O_3-polluted conditions in the farmland ecosystem, which may become an effective nitrogen fertilizer management measure to cope with the elevated ambient O_3 and achieve sustainable production.

Keywords: ozone; controlled-release nitrogen fertilizer; flowering period (Z60); flag leaf; soil microorganisms; nitrogen-cycling

1. Introduction

The atmospheric photochemical pollutant ozone (O_3) possesses potent oxidation and toxic effects on plant growth and production [1]. In recent years, the widespread utilization of fossil fuels has resulted in a global rise in ambient O_3 concentration [2]. Currently, we are confronted with a critical situation characterized by heightened levels of ambient O_3 [3]. It has been reported that the ambient concentration of O_3 in China has reached 41 ppb and is rising at a rate of 3 ppb per year. Particularly, in economically developed regions like

Jiangsu Province, it has escalated to levels between 60 and 70 ppb, posing a significant threat to agricultural production and food security [4].

Up to now, the stress effects of elevated ambient O_3 on plant apparent traits [5], soil nutrient cycling processes [6], and soil microbial communities [7] have been extensively investigated. Relevant studies have demonstrated that O_3 inhibits crop growth and biomass accumulation, affecting crop quality and yield [8,9], and ultimately reducing the carbon sequestration capacity of farmland ecosystems [10]. O_3 directly acts on plant leaves and inhibits photosynthesis, which will reduce the chlorophyll content and stomatal conductance of leaves [11]; and indirectly affects the underground process of plants by reducing the carbon content in root exudates and changing the types of root exudates [12]. In addition, O_3 can also change the soil's physical and chemical properties, affect the circulation of nutrients and the activity of related enzymes, and disrupt the stability of microbial community structure by inhibiting the nutrient and energy supply from the environment to microorganisms [13].

Soil microorganisms provide power and guarantee nutrient cycling and energy flow between plants and soil by participating in the decomposition and transformation of soil nutrients, which is an important part of the farmland ecosystem [14]. More importantly, due to the sensitivity of soil microorganisms to environmental changes, changes in microbial community structure and metabolic processes occur before plants exhibit observable O_3 stress symptoms [15], which is of great value for exploring O_3 mitigation options for crop production stress. However, previous studies on soil microbial communities under O_3 stress have mostly focused on soil bacterial and fungal communities [7,16], and the research on soil archaea is still limited. Compared with the soil bacterial community, although the abundance of soil archaeal community soil is quite low, it is involved in important processes and plays a vital role in the soil nutrient cycling processes [17–19]. Therefore, it is a potential soil nutrient pool [20] and also an important topic of soil microbial research. The emergence of metagenomics provides a new perspective for analyzing the responses of soil microbial communities to various environmental stresses. The metagenomic sequencing process can omit the separation and purification process of microorganisms and simultaneously perform the sequencing and assembly of soil bacteria, fungi, and archaea [21], intuitively showing the differences in soil microbial community structure under different field management modes. Moreover, the sequences obtained by metagenomic sequencing can be compared with the Kyoto Encyclopedia of Genes and Genomes (KEGG) and enzyme databases to further explore the changes in metabolic functions of microorganisms from the perspective of functional genes and enzymes [22].

The current research on how to alleviate the impact of ambient O_3 on agricultural production primarily focuses on optimizing crop varieties to improve resistance and tolerance toward ambient O_3 [23], regulating water and fertilizer management to optimize the growth environment [24], and using exogenous chemical protectants to enhance the antioxidant capacity [25,26]. The research on the impacts of O_3 stress on crop production and soil microbial function, as well as the mechanisms underlying various mitigation measures, remains insufficiently comprehensive. O_3 is mainly absorbed into the crops through the stomata on the leaves, so the leaves are the initial sensor of crops under O_3 stress and the entry point to alleviate its effects [11]. The feasibility of controlled-release nitrogen fertilizer alleviating O_3 stress is that controlling the release rate of nitrogen fertilizer to coordinate the supply of soil nitrogen may alleviate the decrease in chlorophyll and leaf area and reduce the inhibition of O_3 on photosynthesis of crop leaves [27]. Sulfur-coated controlled-release nitrogen fertilizer (SCNF) is a long-acting nitrogen fertilizer that delays the release rate of N elements. It is considered to be an effective product to reduce N loss and field management pressure by controlling the balance between the supply rate of soil inorganic-N and the absorption demand of crops [28,29]. Previous studies have shown that the twice-split application of SCNF can improve crop yield and N use efficiency while reducing the fertilization times compared to the multiple application of urea in conventional planting management [30]. Therefore, we hypothesized that substituting SCNF for urea, which is commonly used today, may alleviate the inhibitory effect of elevated ambient O_3 on farmland ecosystems.

Wheat (*Triticum aestivum* L.) is a typical O_3-sensitive crop in the world's important food crops [31]. According to statistics, from 2014 to 2019, the loss of wheat yield caused by ambient O_3 stress in the Yangtze River Delta region accounted for about 20% of the actual production [32]. The reduction in wheat yield caused by elevated ambient O_3 is one of the important problems affecting world food security that needs to be solved urgently [33]. Some reports have shown that photosynthesis is the physiological basis of high yield of crops, and the contribution of photosynthetic products of functional leaves to yield can reach 70–80% in the late growth period of wheat [34]. The flag leaf is the first leaf below a wheat ear. Although the shortest leaf age and smallest leaf area, the flag leaf contains more chlorophyll than others, which directly affects the accumulation of carbohydrates in wheat grains and plays a decisive role in yield [35]. Therefore, it is of great significance to alleviate the effect of O_3 on the morphology and physiology of flag leaves and maintain photosynthetic function to stabilize wheat yield.

In summary, it is of great theoretical and practical significance to carry out research on the responses and mitigation technologies to the elevated ambient O_3 in typical farmland ecosystems in Jiangsu Province and to clarify the effects of SCNF on wheat production and soil biological characteristics under O_3 stress. We hypothesized that, by regulating the N supply to the soil, SCNF may regulate the morphological and physiological characteristics of wheat as well as the physical and chemical properties of soil, thereby mitigating the wheat yield loss caused by the O_3-stressed environment. In this study, ozone-free-air controlled enrichment (O_3-FACE) facilities and meteorological monitoring devices were used to simulate the scenario of elevated ambient O_3, and field experiments were conducted to study the effects of SCNF instead of urea on the photosynthetic function of wheat flag leaves, wheat yield composition, and soil biological characteristics under O_3 stress. Therefore, our aims were to (1) investigate the response of soil chemical properties, soil enzyme activities, and soil microbial communities to different N sources under O_3 stress and (2) to evaluate the effects of different fertilizer management models on wheat physiological indices and yield and clarify the potential mechanisms of influence between above-ground plant physiology and subsurface biological processes.

2. Results

2.1. Effect of O_3 and SCNF on Agronomic Parameters

The photosynthetic functions of wheat flag leaves at the flowering period (Z60), accumulation of nutrient elements, and yield components at the maturity stage (Z92) were measured under different treatments (Figure 1). The results showed that leaf size (LS), net photosynthetic rate (*Pn*), and SPAD value decreased significantly ($p < 0.05$) under the O_3-polluted conditions (E_CK and E_S); while leaf mass per area (LMA) was significantly increased. Compared with E_CK, the LS, *Pn*, and SPAD values of E_S were significantly improved (Figure 1a–d).

The premise of wheat yield is the photosynthetic capacity and nutrient accumulation of functional leaves. The accumulation of TN, TP, and TK (Figure 1e–g) at the maturity stage (Z92) was significantly positively correlated with yield, respectively (Spearman's *r* value = 0.97, 0.58, 0.89, respectively). Under O_3 stress, the nutrient accumulation and wheat yield components (Figure 1h–k) decreased significantly. However, the O_3 stress was alleviated by the fertilizer management mode of SCNF replacing urea, which was reflected in the agronomic parameters of E_S being significantly higher than those of E_CK.

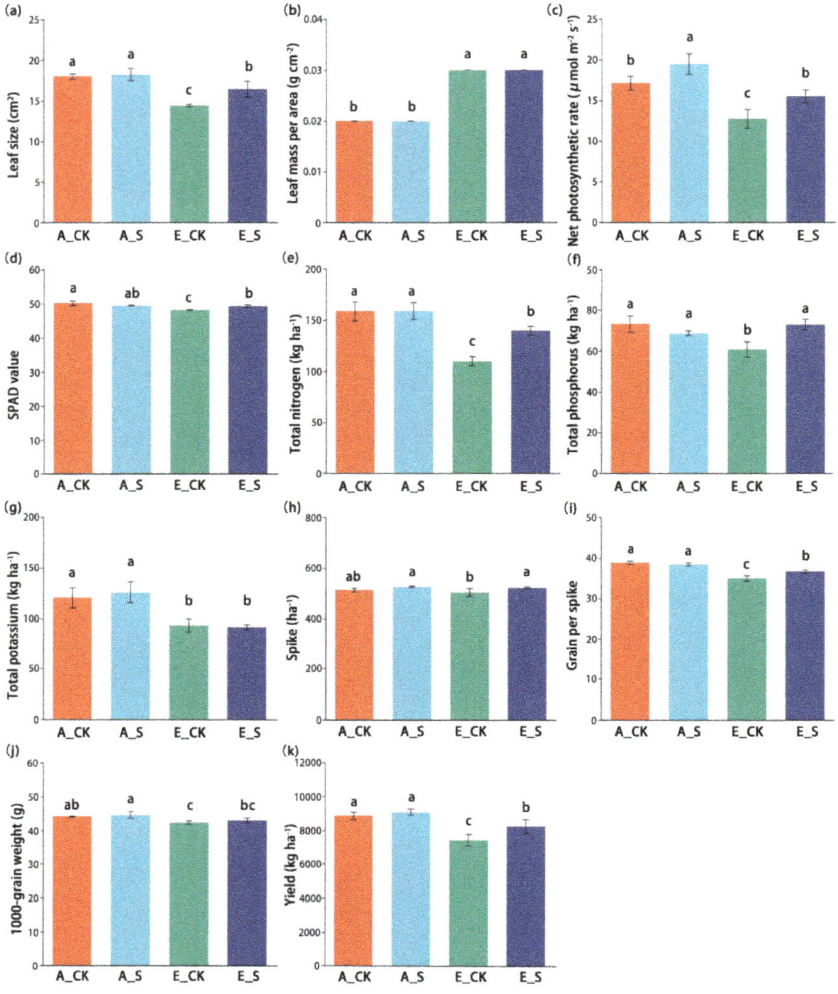

Figure 1. (**a–d**) Photosynthetic functions of wheat flag leaves in the flowering period (Z60), (**e–g**) nutrient accumulation of functional leaves, and (**h–k**) yield components at the maturity stage (Z92). A_CK, normal atmospheric environment + urea; A_S, normal atmospheric environment + sulfur-coated controlled release nitrogen fertilizer (SCNF); E_CK, ozone-free-air controlled enrichment (O_3-FACE) + urea; E_S, O_3-FACE + SCNF. Error bars mean the standard error and the different lowercase letters indicate significant differences between various treatments based on a one-way ANOVA followed by Duncan's multiple-range tests ($p < 0.05$).

2.2. Effect of O_3 and SCNF on Soil Chemical Properties

The soil chemical properties of each treatment at flowering (Z60) and maturity (Z92) stages are shown in Figure 2a,b. In these two growth stages, soil pH and content of OM were relatively stable, while the content of inorganic-N (NO_3^-–N and NH_4^+–N) fluctuated significantly. Both O_3 and SCNF significantly promoted the activity of SUE (soil urease) at the wheat flowering period (Z60), but had no significant effect on the activity of SNR (soil nitrate reductase; Figure 2c).

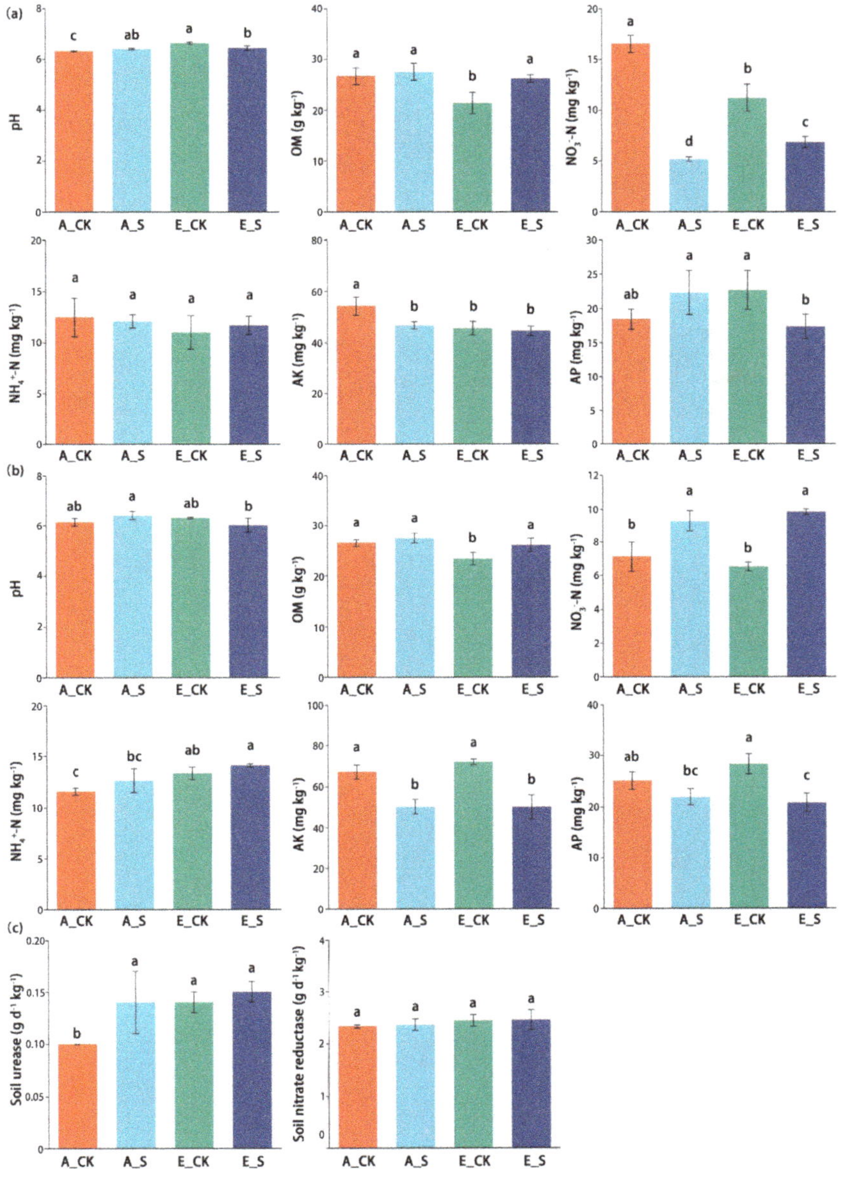

Figure 2. Soil chemical properties at the (**a**) flowering (Z60) and (**b**) mature period in different treatments and (**c**) activities of soil urease and soil nitrate reductase at the flowering period (Z60). OM, organic matter; $NO_3{}^{-}$–N, nitrate nitrogen; $NH_4{}^{+}$–N, ammonium nitrogen; AK, available potassium; AP, available phosphorus. Error bars mean the standard error and the different lowercase letters indicate significant differences between various treatments based on a one-way ANOVA followed by Duncan's multiple-range tests ($p < 0.05$).

Multiple variance analysis showed that the interaction of O_3 and SCNF had a significant effect on the photosynthetic function of wheat flag leaves and soil chemical properties at the flowering period (Z60) and then on nutrient accumulation and yield at the mature period (Table 1).

Table 1. Multivariate variance analysis of the effects of O_3 combined with SCNF on wheat agronomic parameters and soil properties.

			O_3	SCNF	$O_3 \times$ SCNF
Flowering period (Z60)	Leaf photosynthetic function	LS	**	*	NS
		LMA	**	NS	NS
		Pn	*	*	NS
		SPAD value	NS	NS	**
	Soil enzyme activity	SUE	*	*	NS
		SNR	NS	NS	NS
	Soil chemistry property	pH	*	NS	**
		OM	NS	*	NS
		NO_3^-–N	*	**	**
		NH_4^+–N	NS	NS	NS
		AK	*	*	NS
		AP	NS	NS	**
Mature period (Z92)	Yield structure	Spike	NS	*	NS
		Grain per spike	**	*	*
		TGW	*	*	NS
		Yield	*	**	NS
	Plant nutrient accumulation	TN	**	**	*
		TP	NS	*	**
		TK	**	NS	NS
	Soil chemistry property	pH	**	**	*
		OM	*	NS	NS
		NO_3^-–N	NS	**	NS
		NH_4^+–N	NS	NS	NS
		AK	NS	**	NS
		AP	NS	**	*

Note: O_3, ozone-free-air controlled enrichment; SCNF, sulfur-coated controlled-release nitrogen fertilizer. LS, leaf size; LMA, leaf mass per area; *Pn*, net photosynthetic rate; SUE, soil urease; SNR, soil nitrate reductase; OM, organic matter; NO_3^-–N, nitrate nitrogen; NH_4^+–N, ammonium nitrogen; AK, available potassium; AP, available phosphorus; TGW, 1000-grain weight; TN, total nitrogen; TP, total phosphorus; TK, total potassium. NS, no significance; asterisk mark denotes the significance level: ** $p < 0.01$ and * $p < 0.05$.

2.3. *Effect of O_3 and SCNF on Microbial Diversity and Community Structure*

2.3.1. Quality Evaluation of Metagenomic Sequences

Nucleobase quality and distribution maps of the original sequences (Figure S1) showed that contents of base pair G–C and A–T were equal and remained stable throughout the entire sequencing process with horizontal lines. It is indicated that the construction quality of the metagenomic library was relatively high and the results were reliable. The genetic libraries of bacteria, fungi, and archaea were extracted and constructed for subsequent analysis.

2.3.2. α Diversity Analysis of Soil Microbial Communities

High α diversity is a reflection of the health of microbial communities and the stability of soil ecosystem functions [36]. In this study, the commonly used α diversity metrics (Chao, Shannon, and Simpson indexes) were used to measure the richness and diversity of microbial communities in each treatment. The Chao index measures the richness of the number of species in the community; the Shannon and Simpson index reflects the diversity and stability of the community from the perspective of the status and role of dominant species in the community. In general, the Simpson index is negatively correlated with other α diversity indices.

In this study, it was observed that the response of the soil bacterial community to O_3 was not reflected in α diversity. However, O_3 had obvious stress on soil fungal and archaeal communities, reducing the Chao and Shannon index (Table 2). Replacing urea with SCNF significantly changed the diversity and stability of soil bacterial and archaeal communities, while the effect on α diversity of fungal community was not obvious.

Table 2. α diversity indices of microbial communities at the species level.

	Treatment	Chao Index	Shannon Index	Simpson Index
Bacteria	A_CK	21,435 ± 50 a	5.70 ± 0.01 b	0.016 ± 0.000 ab
	A_S	21,073 ± 84 a	5.79 ± 0.01 a	0.015 ± 0.000 b
	E_CK	20,507 ± 217 a	5.71 ± 0.01 b	0.017 ± 0.001 a
	E_S	20,794 ± 199 a	5.74 ± 0.00 ab	0.016 ± 0.000 ab
Fungi	A_CK	157.00 ± 2.33 a	3.60 ± 0.01 a	0.061 ± 0.003 a
	A_S	148.00 ± 1.86 a	3.54 ± 0.02 ab	0.068 ± 0.001 a
	E_CK	134.00 ± 2.91 b	3.48 ± 0.01 b	0.069 ± 0.002 a
	E_S	133.33 ± 1.07 b	3.53 ± 0.02 ab	0.066 ± 0.002 a
Archaea	A_CK	629.33 ±2.50 a	3.83 ± 0.01 a	0.004 ± 0.001 c
	A_S	602.00 ± 4.06 ab	3.69 ± 0.01 b	0.060 ± 0.002 bc
	E_CK	587.67 ± 7.56 b	3.61 ± 0.02 bc	0.072 ± 0.003 ab
	E_S	594.67 ± 8.70 ab	3.54 ± 0.03 c	0.086 ± 0.004 a

Note: Means are followed by ±s.e.m. Different lowercase letters mean significant difference based on a one-way ANOVA followed by Duncan's multiple-range tests ($p < 0.05$).

2.3.3. Structural Compositions and Differences in Soil Microbial Communities

The top 10 dominant phyla were selected for comparison (Figure 3). Each treatment had the same dominant strains, and the total relative abundance of the top five dominant phyla was more than 90%. However, the relative abundance of the same dominant strain in each treatment was different. The use of SCNF increased the relative abundance of dominant phyla in soil bacterial and fungal communities (Figure 3a,b). While the soil archaeal community showed a very different response (Figure 3c), the relative abundance of the top two archaea (*p_Euryarchaeota* and *p_Thaumarchaeota*) was significantly affected by O_3 and SCNF.

2.3.4. β Diversity Analysis of Soil Microbial Communities

The principal co-ordinates analysis (PCoA) was carried out on the four treated soil microorganisms at the species level. The P_values of permutational multivariate analysis of variance (PERMANOVA) of bacterial and archaeal communities were less than 0.05, indicating that there were significant differences among the sample groups (Figure 4). The results display that the samples in different treatments represent obvious intra-group aggregation and inter-group dispersion.

It is worth noting that the PCoA results of the fungal community showed that the explanation of the PC1-axis (21.50%) and R^2 value (0.33) were significantly lower than those of bacterial and archaeal communities, and the P_value (0.061) was greater than 0.05 (Figure 4b), indicating that the effects of O_3 and SCNF on soil fungal community were much weaker than those on bacterial and archaeal communities. The explanation for the differences in soil fungal communities among the four treatments may not be the different atmospheric environments or nitrogen management models.

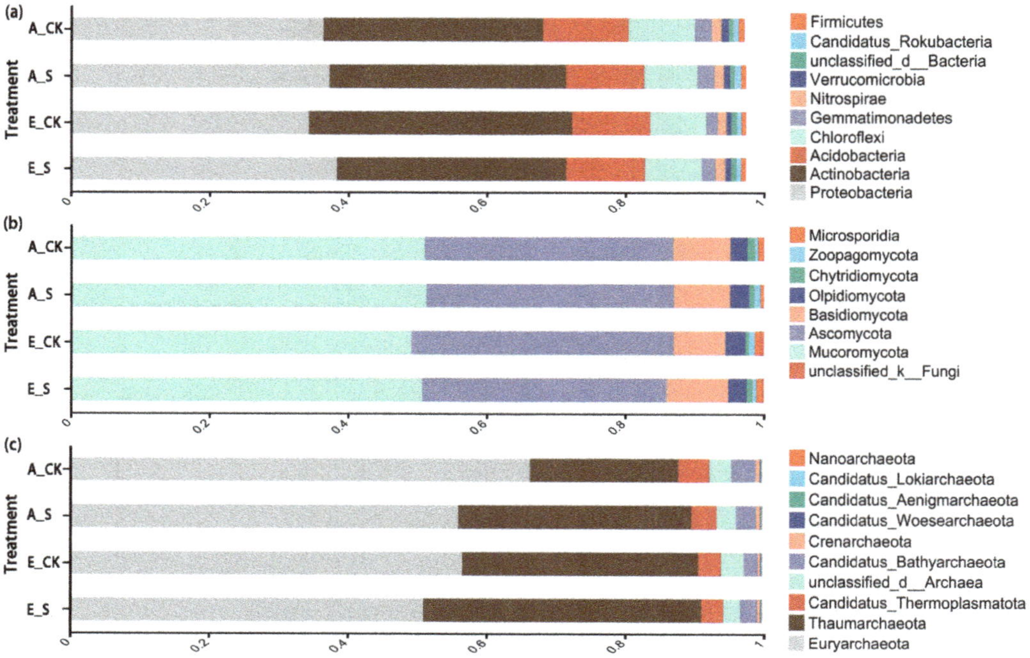

Figure 3. Influence of O_3-FACE combined with SCNF on the relative abundance of soil (**a**) bacterial; (**b**) fungal; and (**c**) archaeal phyla. Only the phyla with RPKM $\geq$ 1% are presented in this figure.

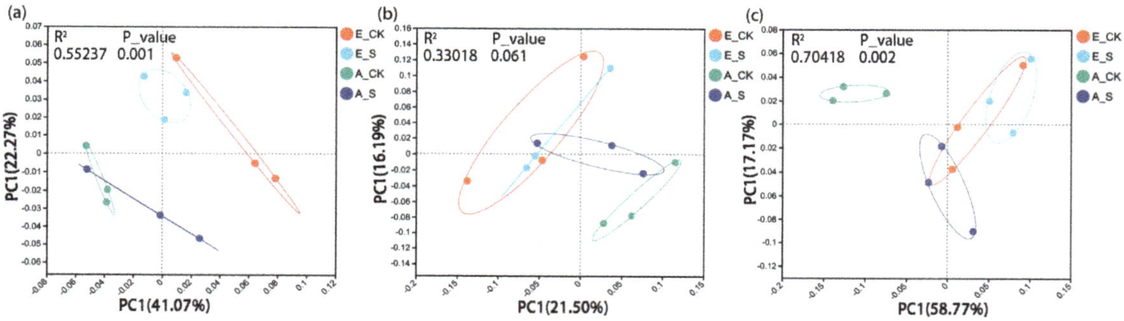

Figure 4. The principal component analysis (PCoA) and PERMANOVA at 99% level based on Bray-Curtis distance of soil (**a**) bacterial; (**b**) fungal; and (**c**) archaeal communities at the species level in the various treatments.

2.4. Interaction among Soil Microbial Communities, Soil Chemical Properties, and Agronomic Parameters

Many of the environmental factors commonly analyzed in the study related to the changes in microbial communities are autocorrelated. Therefore, before the interaction analysis of microorganisms and environmental factors, the variance inflation factor (VIF) is used to screen environmental factors to avoid the influence of autocorrelation environmental factors on the accuracy of subsequent analysis results [37]. The VIF values of the eight environmental factors (pH, OM, NO_3^-–N, NH_4^+–N, AK, AP, SUE, and SNR) selected in this experiment were all less than 10 (Table S1), indicating that the selection of environmental factors was scientific and there was no autocorrelation among them.

Redundancy analysis (RDA) was used to quantify the effect of soil environmental factors on soil microbial communities at the genus level (Figure 5). The two environmental factors (pH and SUE) showed a positive correlation (the arrows were in the same direction) and their effects on the microbial communities showed opposite trends to four environmental factors (AK, OM, NO_3^-–N, and NH_4^+–N) (the arrows were in the opposite direction). Soil pH, SUE, and AK are all long arrows in Figure 5a–c, which can be considered as the three important environmental factors that have great impacts on soil microbial communities. The effect of soil NH_4^+–N on bacterial and fungal communities was stronger than that of NO_3^-–N; however, in the archaeal community, the opposite was true, and it was found that SCNF had a very significant effect on a high-abundance archaea genus (*g_unclassified_f_Nitrososphaeraceae*), which is directly involved in the N-cycling.

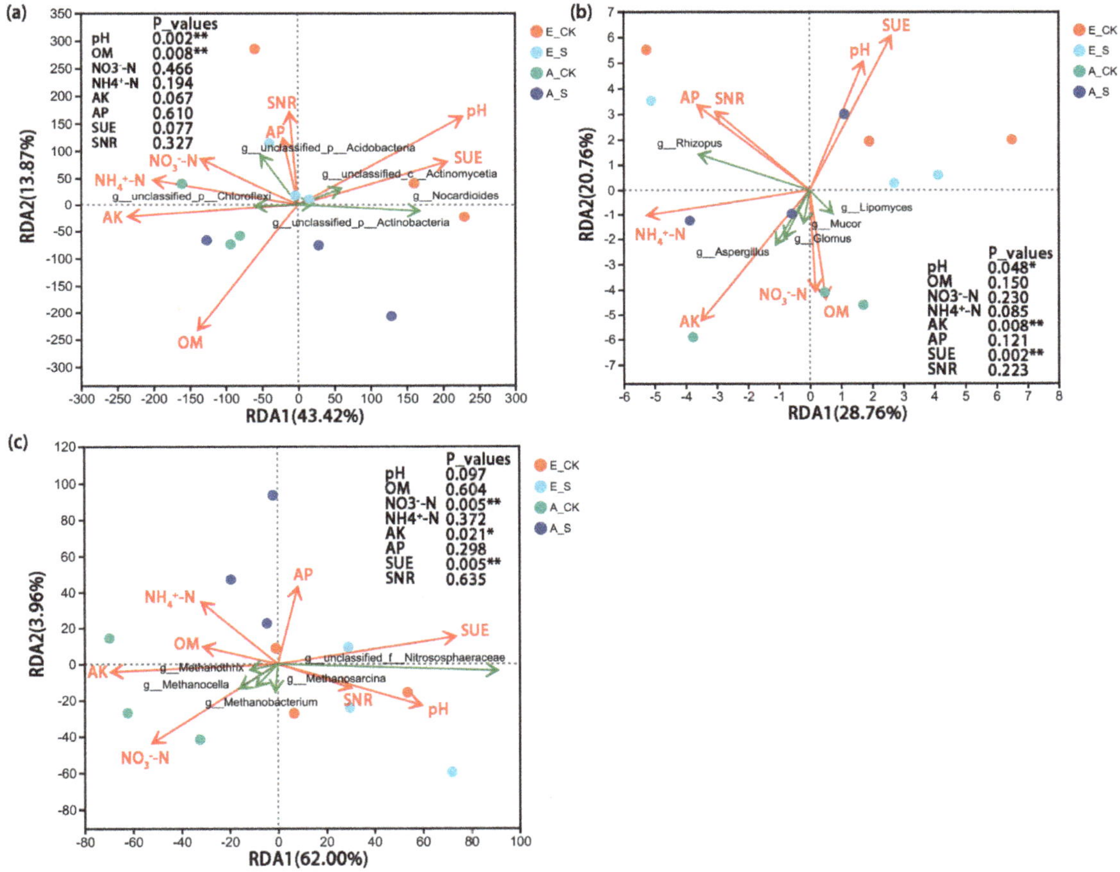

Figure 5. Redundancy analysis (RDA) of soil (**a**) bacterial; (**b**) fungal; and (**c**) archaeal genera with soil chemical properties. The soil biochemical properties were fitted to the ordination plots using a 999–permutation test (P_value). OM, soil organic matter; NO_3^-–N, nitrate nitrogen; NH_4^+–N, ammonium nitrogen; AP, available phosphorus; AK, available potassium; SUE, soil urease; SNR, soil nitrate reductase. Asterisks indicate significant differences at * $p < 0.05$ and ** $p < 0.01$.

Based on the Spearman correlation coefficient between soil microorganisms and soil chemical properties, the correlation networks of soil microbial genera–soil environmental factors were constructed to analyze the possible interactions. In addition to soil pH, SUE and AK, NO_3–N, and OM were also important soil environmental factors affecting bacteria community structure at the genus level (Figure 6a). Many genera of *p_Actinobacteria* were positively correlated with soil pH and SUE but negatively correlated with AK and OM. There was maybe an inhibitory relationship between *p_Proteobacteria* and *p_Actinobacteria*, which showed that many of these genera tend to adapt to an environment of low SUE and pH and a high AK and OM environment. Many bacterial genera in *p_Chloroflexi* only showed significant positive correlations with the soil NO_3^-–N level and were not significantly affected by other soil environmental factors. The correlation networks between soil fungal and archaeal communities and environmental factors were clearer and simpler than those of soil bacteria. In soil fungal communities, *p_Ascomucota* and *p_Mucoromucota* showed strong responses to changes in soil environmental factors (Figure 6b). For the correlation network of archaeal genera, many genera of *p_Thaumarchaeota* and *p_Euryarchaeota* showed an obvious preference for an environment of high NO_3^-−N, AK, and OM, and low AP and SUE (Figure 6c).

O_3 and SCNF jointly affect the wheat yield by affecting the above-underground ecological processes. Spearman correlation analysis and mantel test were used to examine the possible interaction effects between environmental factors and soil microbial communities on the agronomic parameters of wheat in this study (Figure 7). Among these soil environmental factors, pH and OM showed opposite influencing mechanisms on photosynthetic function (LS, LMA, *Pn*, and SPAD value) of flag leaves during the flowering period (Z60) and wheat yield. Unsurprisingly, the photosynthetic function of flag leaves showed extremely strong correlations with yield.

From another aspect, the results of mantel tests verify the reliability of RDA analysis results (Table S2). The community structure of soil microbial phyla and metabolic function at KEGG level 2 were strongly correlated with soil chemical properties and wheat agronomic parameters, respectively. The difference was especially shown in the results of mantel tests on wheat yield. Although the soil microbial community at the flowering period (Z60) did not show a significant association with yield, the metabolic functions of soil microorganisms did show responses.

2.5. Comparison of Abundance Differences in N-Cycling Genes

At the same N supply level in all treatments, we believe that in addition to the effect of O_3, the difference in the release rate of inorganic-N by urea and SCNF would also affect the N-cycling process of soil microorganisms, and the significant difference in SUE was evidence (Figure 2c). By comparing the sequences of metagenomic sequencing with the KEGG library, the metabolic pathways related to the N-cycling (*ko00910*) were screened for significant difference tests (Figure S2). The results showed that each process of soil N-cycling was affected by environmental factors, and the abundances of related functional genes were significantly different among different treatments.

The abundance of 14 key functional genes involved in the important processes of N-cycling was selected for variance analysis, including ammoxidation (*amoA*, *amoB*), denitrification (nitrosation (*napA*, *narG*), NO_2 reduction to NO (*nirK*, *nirS*), and N_2O reduction (*nosZ*)), assimilatory (*nirB*, *nrfA*) and dissimilatory (*nirA*, *nasA*) nitrate reduction ammonia, and nitrogen fixation (*nifD*, *nifH*, *nifK*) (Figure 8) [22]. Under the O_3-FACE condition, except for the *amoA* gene, the abundances of the other 13 functional genes were significantly reduced, indicating that the activities and metabolic functions of soil microorganisms were significantly inhibited. In addition, under a normal atmospheric environment, the responses of N-cycling-related genes to different fertilizer management modes were not significant. However, under the O_3 stress, SCNF could significantly increase the abundance of these genes, which once again proved the mitigation effects of SCNF on O_3 inhibition.

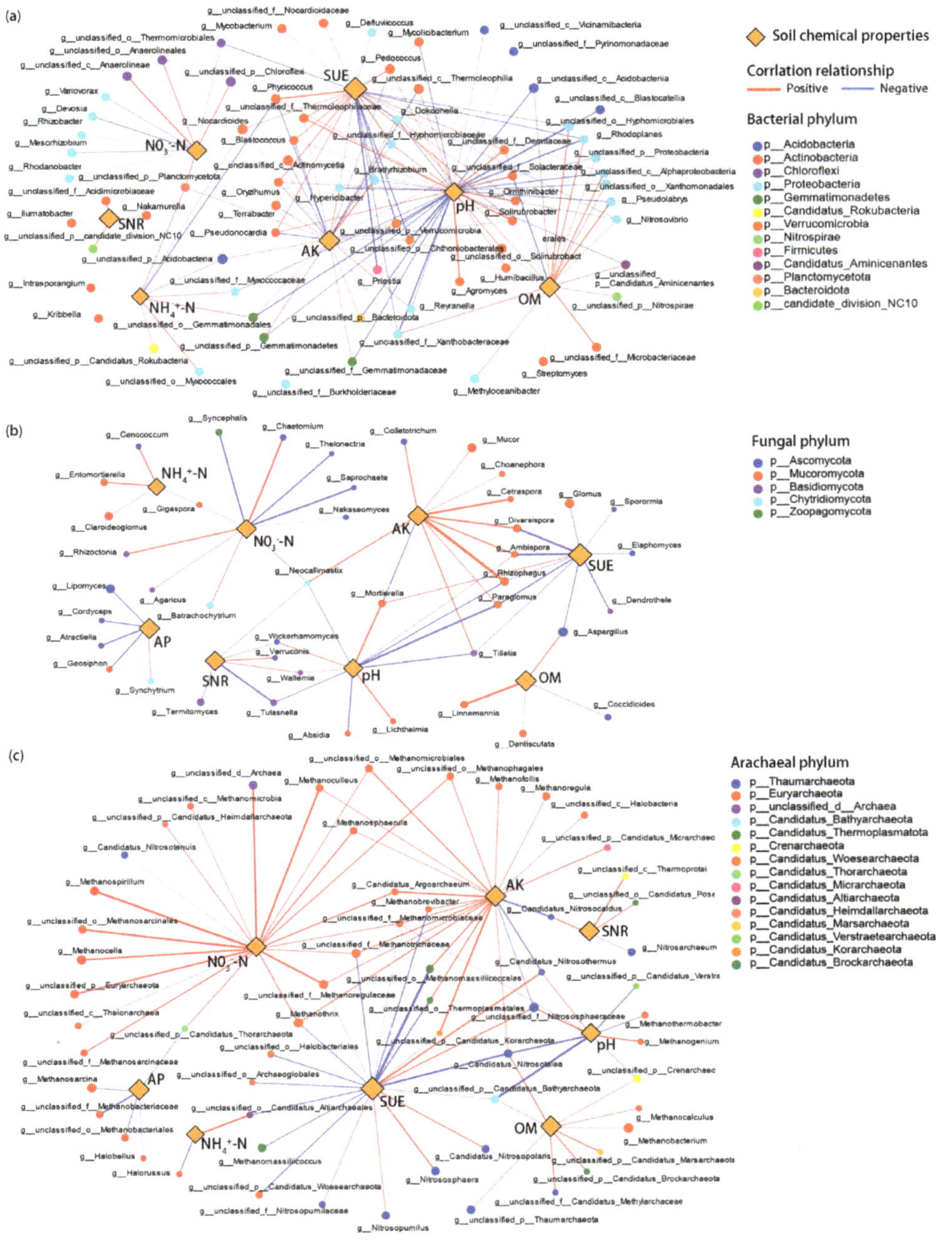

Figure 6. Correlation network between soil (**a**) bacterial; (**b**) fungal; and (**c**) archaeal genera and soil properties based on Gephi 0.9.2 software. Each network node represents a genus; its color and size correspond to the phylum to which it belongs and the relative abundance, respectively. The color and thickness of the network edge are expressed in the Spearman correlation and *r* value between the genus and the environmental factor, respectively.

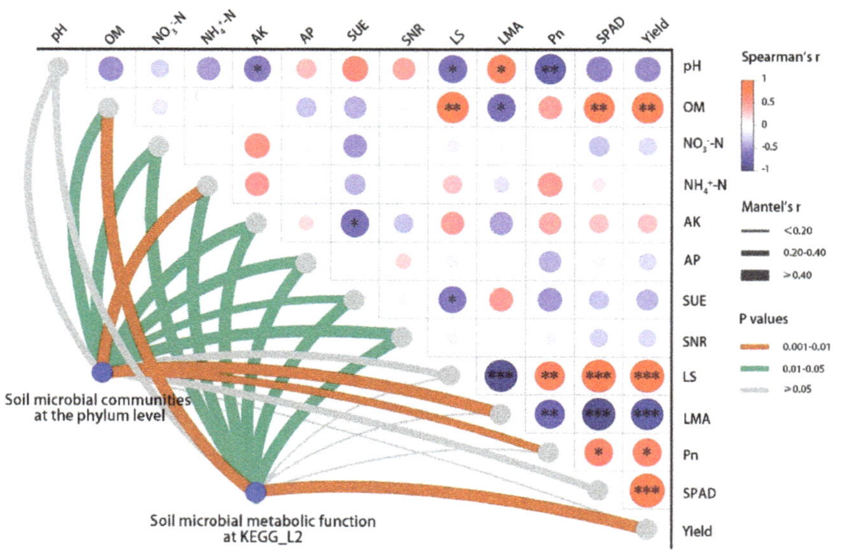

Figure 7. Pairwise comparisons of soil variables, agronomic parameters, and wheat yield. LS, leaf size; LMA, leaf mass per area; *Pn*, net photosynthetic rate. Mantel tests depict the association between soil microbial phyla and metabolic functions at the KEGG level 2 with environmental factors, respectively. The width of each edge matches Mantel's *r* statistic, and the color represents Mantel's P value (Table S2). Asterisks indicate significant differences at * $p < 0.05$, ** $p < 0.01$ and *** $p < 0.001$.

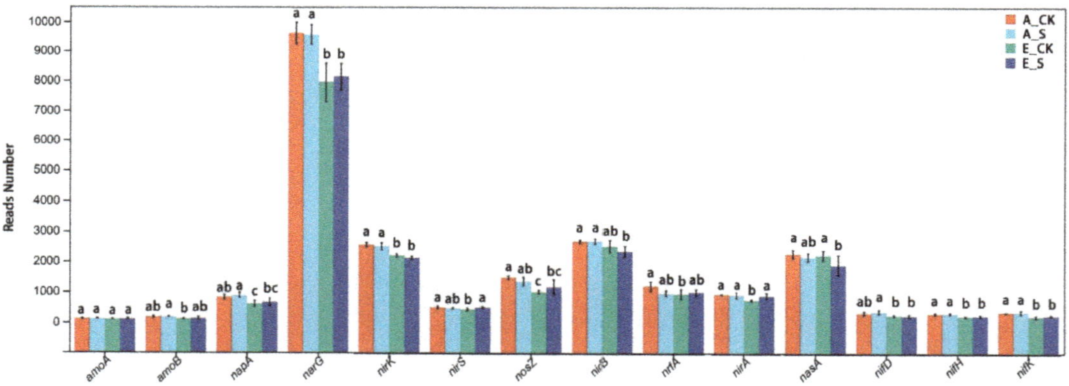

Figure 8. Variance analysis of 14 key functional genes involved in N-cycling. DNRA, dissimilatory nitrate reduction ammonia; ANRA, assimilatory nitrate reduction ammonia. Different lowercase letters mean significant difference based on a one-way ANOVA followed by Duncan's multiple-range tests ($p < 0.05$).

3. Discussion

3.1. Effects of O_3 and SCNF on Photosynthetic Parameters of Wheat Flag Leaves

O_3-induced changes in parameters of leaf photosynthetic function are important indicators for assessing crop adaptive traits and can be widely used in conventional field surveys [38]. In this study, the O_3 stress environment had obvious damage to the photosynthetic function of wheat flag leaves at the flowering period (Z60), especially for LS and LMA (Table 1). LS is an indicator to measure the material transmission capacity between leaves and the environment; LMA is negatively correlated with photosynthetic rate, but positively correlated with leaf longevity [39]. The changes in LS and LMA in this study indicated that wheat sacrificed the photosynthetic rate self-regulation strategy to reduce ozone uptake amount and prolong leaf life under ozone stress. The changes in LS and LMA in this study indicated the self-regulation strategy of wheat sacrifice to the photosynthetic rate in order to reduce ozone absorption and prolong leaf life under O_3 stress. Similar results were reported by Feng et al. [39] and Li et al. [40]: as an important leaf attribute index, LMA indicates the risk of a stressful environment and the improvement in LMA is a "slow return" strategy for leaves to improve resource utilization in stress environments.

The photosynthetic parameters of flag leaves at the flowering stage (Z60) showed very strong correlations with the wheat yield at the maturity stage (Z92; Figure 7). Multivariate variance analysis showed that different nitrogen fertilizer treatments had no significant effect on LMA and SPAD value (Table 1), indicating that the nitrogen fertilizer management mode of SCNF instead of urea could not delay the O_3-induced leaf senescence of winter wheat. However, SCNF could increase nutrient accumulation amount by increasing LS and *Pn*, which was the physiological basis for alleviating the inhibitory effect of O_3 on wheat yield formation.

3.2. Regulation of O_3 and SCNF on Nutrient Accumulation and Wheat Yield

The nitrogen fertilizer management method used in A_CK and E_CK treatments is one of the high-yield practices vigorously promoted in long-term agricultural production in Jiangsu Province. In this experiment, under a normal atmospheric environment, compared with the four times application of urea, the twice-split application of SCNF can reduce the management cost and provide a higher yield (Figure 1h–k). At the same level of nitrogen supply, we believe that the fundamental difference between the urea and SCNF treatments is essentially related to the level and timing of the soil's inorganic-N supply [29].

But surprisingly, SCNF showed effectiveness in alleviating O_3 stress. Compared with that of E_CK treatment, the accumulation amount of nutrients and components of wheat yield (Figure 1) of the E_S treatment were significantly increased. The number of spikes per hectare is one of the key factors of wheat yield components. Related studies have reported that O_3 has a great side effect on the formation of productive ears from tillers during the reproductive growth phase [41]. Under the condition of unifying the planting density of basic seedlings and the same field management, O_3 significantly inhibited the formation of productive ears [42], while SCNF could increase the tiller number and ear-bearing tiller rate of wheat and greatly alleviate the decrease in wheat yield caused by O_3 (Figure 1).

3.3. Effects of O_3 and SCNF on Soil Chemistry Properties

Through conventional statistical analysis, it was found that the soil pH and OM content were relatively stable, while the inorganic-N content fluctuated significantly during the flowering (Z60) and maturity (Z92) stages of winter wheat (Figure 2). Multivariate variance analysis showed that soil pH and OM responded strongly ($p < 0.01$) to O_3 (Table 1). Similar conclusions have reported that O_3, as a strong oxide, can increase soil pH, accelerate the oxidation and decomposition of organic matter, and reduce the soil soluble organic carbon by increasing the soil redox potential [43]. More importantly, O_3-induced soil pH increase is one of the most important factors affecting the denitrification process and N_2O production [44]. However, the effect mechanism of SCNF on N-cycling needs to be further discussed in combination with changes in the abundance of related genes.

Multivariate variance analysis also showed that the changes in soil inorganic-N did not show interaction with O_3, but were significantly affected by SCNF (Table 1). Consistent with previous studies, although the use of SCNF could not rapidly increase the content of soil NO_3^-–N in the short term compared with urea, it provided a long-term N supply and increased the retained soil inorganic-N by delaying the N release rate (Figure 2) [30].

3.4. Responses of the Soil Microbial Community Structure to O_3 and SCNF

Soil microbial communities are very sensitive to the change in soil environment. The soil's physical and chemical properties will directly affect the nutrient supply capacity and living environment of soil microorganisms [45]. Studies have mentioned that O_3 exposure and different fertilizer management models can affect the structure and diversity of soil bacterial and fungal communities [7]. Similar results were also found in this study. Although the compositions and dominant strains of soil microorganisms did not change, different treatments had significant effects on the α (Table 2) and β (Figure 4) diversity. In addition, PCoA results showed that the β diversity of soil microorganisms was more sensitive to the effect of O_3 than SCNF (Figure 4).

p_Proteobacteria, *p_Actinobacteria*, and *p_Acidobacteria* were the three most abundant bacterial phyla in the soil, accounting for more than 80% of the total relative abundance, which was consistent with the results of previous studies [46]. *p_Actinobacteria* is a microbial community with a strong stress tolerance in the soil bacterial community [47,48]. In this study, the relative abundance of *p_Actinobacteria* in E_CK was significantly higher than that in the other three treatments (Figure 3), indicating that *p_Actinobacteria* was obviously more adapted to the stress environment of elevated O_3 than other soil microbial phyla. Moreover, it is also proven that SCNF can adapt microorganisms to a high O_3 environment by regulating the soil environment [45].

Many crop-related studies have discussed the effects of different field management methods on soil fungal communities and explored the interaction between crop production and soil fungi [7,49]. In this study, the PCoA results showed that the PC1-axis interpretation of fungi was much lower than that of bacteria and archaea (Figure 4), indicating that soil bacteria and archaea were more sensitive to changes in environmental factors than fungi. Relevant studies have proposed an explanation that the main reason affecting the soil fungal community structure is soil-borne fungal diseases rather than different fertilization systems or rotation patterns [50].

3.5. Interactions between Environmental Factors and Soil Microorganisms

The results of RDA showed that there were obvious symbiotic or inhibitory relationships between some genera in the bacteria (*p_Actinobacteria*, *p_Actinomycetia*, *p_Chloroflexi*, etc.), fungi ((*p_Mucoromycota*, *p_Ascomycota*, etc.), and archaea phyla (*p_Euryarchaeota*, *p_Thaumarchaeota*, etc.) (Figure 5). The correlation network confirmed that soil pH, AK, SUE, and NO_3^-–N are the key soil environmental factors that changed the soil microbial community structure in this study (Figure 6). Aller and Kemp [51] also concluded that pH and C/N were the key soil factors affecting the archaeal community structure. Unsurprisingly, the results of mantel tests further verified the correlations between key soil environmental factors and microbial genera.

It is worth noting that SCNF could affect the relative abundance of soil N-cycling-related microorganisms by changing the nitrogen release rate. Many microbial genera were identified to be significantly positively (*p_Chloroflexi*, *p_Euryarchaeota*, etc.) or negatively (*p_Proteobacteria*, *p_Ascomycota*, etc.) related to soil NO_3^-–N, respectively (Figure 6), suggesting that these genera may play an important role in soil N-cycling or have a strong response to changes in soil NO_3^-–N content [30]. SCNF had a significant effect on *p_Thaumarchaeota*, a high-abundance ammonia-oxidizing archaea (AOA) directly involved in N-cycling, in which multiple biomarkers (*f_Nitrosopumilaceae*, *f_Nitrososphaeraceae*, etc.) with significantly increased relative abundance were identified, suggesting a high impact of SCNF on the N-cycling process [52].

3.6. Responses of Soil Microbial Nitrogen Metabolism to O_3 and SCNF

The response of soil microorganisms to environmental stress usually reflects the community structure and biological processes and the response of biological processes often occurs earlier than the change in community structure [15]. Relevant studies have suggested that O_3 will destroy the N-cycling process driven by soil microbial activity, which may inhibit many important processes of N transformation such as nitrogen fixation, nitrification, and denitrification and have a negative impact on N_2O emissions from farmland [37]. Similar results were obtained in this study. The genes controlling related enzymes in the nitrogen metabolism pathway (*ko00910*) were significantly affected by O_3 (Figure S2) and the abundances of functional genes related to N-cycling were significantly reduced (Figure 8). Moreover, SCNF significantly alleviated the inhibitory effects of O_3 on these related genes.

It is worth noting that under the same supply level of soil inorganic nitrogen, the nitrogen fertilizer management mode of SCNF instead of urea could significantly increase the inorganic nitrogen retained in the soil at the maturity stage (Z92) of winter wheat (Figure 2). We believed that SCNF could increase the abundance of genes related to nitrogen fixation (*nifD*) and ammoxidation (*amoB*) processes to retain more NO_3^-–N. In addition, it may be possible for N_2O emissions to be indirectly reduced by inhibiting the abundance of genes related to NO_2 reduction to NO (*nirK*, *nirS*) and N_2O reduction (*nosZ*) processes.

3.7. Applicability and Limitations

From the perspective of wheat production and soil biological characteristics, our study confirmed that the use of SCNF instead of urea can promote the photosynthetic function of flag leaves, improve the soil environment, stabilize the microbial community, and increase the yield of winter wheat by providing long-term N supply. This will become an effective mitigation measure to cope with the situation of elevated ambient O_3. However, these conclusions will require years of experimental verification. These results will be presented in a future series of reports.

Additionally, it is undeniable that there are still some shortcomings in this study. Some studies have proposed that the main climatic factors affecting the rice-wheat rotation ecosystem in Jiangsu Province are not only the elevated ambient O_3, but also the continuous increase in atmospheric CO_2 and land surface temperature [53]. Miao et al. [54] suggested that CO_2 and temperature even showed an interactive superposition effect on farmland ecosystems. Therefore, the study of the impact of climate change on farmland ecosystems urgently needs to consider the combined effects of multiple climate factors on the entire structure and function of farmland ecosystems.

4. Materials and Methods

4.1. Experimental Site

The experimental site is located in the research and demonstration base of green agriculture (119°43′ E, 32°25′ N) in Jiangdu District, Yangzhou City, China. The local implementation of a long-term rice-wheat rotation system is one of the typical farmland ecosystems in China [42].

The ozone-free-air controlled enrichment (O_3-FACE) simulation platform was built in 2019 and has been pre-tested for many years to consistently simulate elevated ambient O_3 in field environments. Each built-up FACE zone is an octagonal structure with a diameter of 12 m and an effective area of about 120 m^2 (Figure 9a). In order to avoid the influence of gas diffusion on the control field of the normal atmospheric environment, the distance between each FACE zone is more than 90 m.

The experiment used the WJ-H-Y5 O_3 generator (Wanjie Ozone Electromechanical Equipment Factory, Nanjing, China) to produce O_3 by electrolysis of oxygen in the atmosphere. The O_3 release tubes equipped in each FACE zone were consistent in model and layout and were kept at a height of 30–50 cm from the ground during the wheat growing season. O_3 was ejected from these tubes above the wheat canopy and diffused freely into the FACE zones to achieve a stable concentration increase. The valves and O_3 concentration sensors (accuracy: 1 nL L^{-1}) were set in the tubes to realize the automatic control of O_3 concentration in the O_3-FACE zone. In addition, the experimental site was also equipped with a small meteorological station, using O_3 detectcrs (Model S200, aeroQUAL Co., Auckland, New Zealand) to monitor the O_3 concentration in each FACE zone, to provide data for the control system of simulating different atmospheric environments.

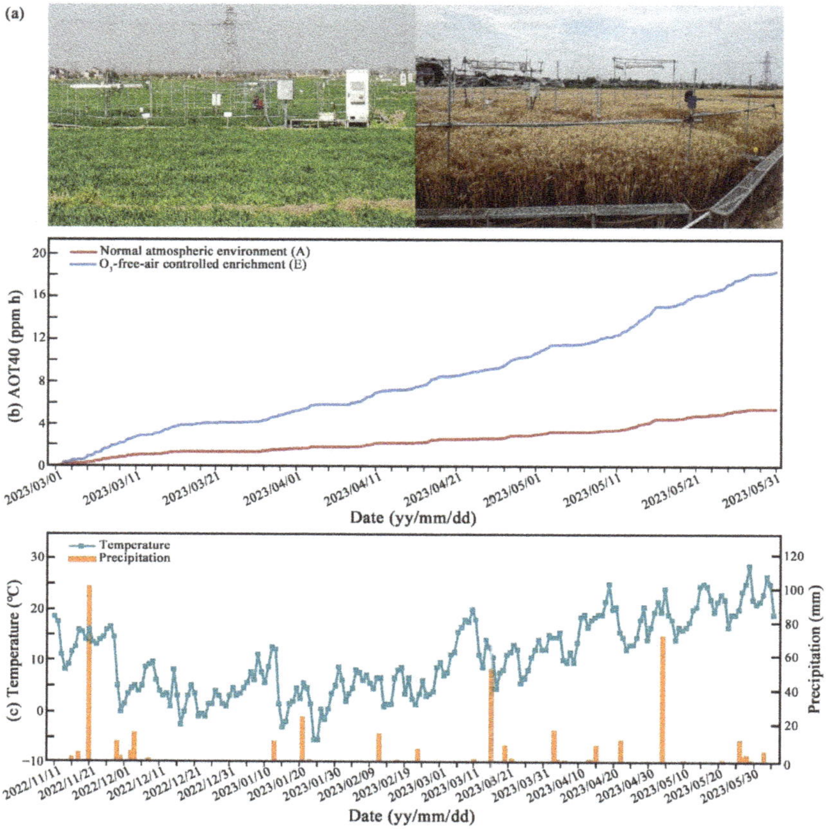

Figure 9. (**a**) The ozone-free-air controlled enrichment (O_3-FACE) simulation platform. (**b**) AOT40 value during the treatment phase of elevated ozone. (**c**) The average temperature and total precipitation at the experimental field during the wheat growing season in 2022–2023.

From 1 March 2023 to the wheat harvest (3 June 2023), O_3 gas was sprayed daily from 8 a.m. to 5 p.m. and the real-time monitoring data of the accumulated hourly ozone concentrations over a cut-off threshold of 40 ppb (AOT40) are shown in Figure 9b. During the wheat-growing season from 2022 to 2023, the average temperature and total precipitation in the experimental site were 10.6 °C and 412.3 mm, respectively (Figure 9c).

4.2. Materials and Treatments

The experimental material was Nongmai-88, which is one of the main and superior cultivated varieties of winter wheat in Jiangsu Province, China. Nongmai-88 ranks first among the strong gluten varieties in China and its grain quality reaches the first-class strong gluten standard of the national standard GB/T 17892-1999 [55]. Nongmai-88 is a medium–

low growth and medium–early maturity wheat variety, which has shown advantages of high and stable yield in regional experiments and production experiments in Jiangsu Province. The whole growth period was about 208 days, and the planting density was 225×10^4 per hectare after removing excess seedlings manually at the three-leaf stage. The growth stage of wheat corresponds to the Zodaks scale.

The experimental soil was silt loam, and the basic soil productivity of the topsoil (0–20 cm) was evaluated before sowing (Table 3).

Table 3. Primary properties of topsoil (0–20 cm) at the test field.

pH	Organic Matter (g kg^{-1})	Nitrate Nitrogen (mg kg^{-1})	Ammonium Nitrogen (mg kg^{-1})	Available Phosphorus (mg kg^{-1})	Available Potassium (mg kg^{-1})
6.7	30.07	43.26	2.26	17.11	60.38

A two-factor randomized block design was used to explore the interaction effects between O_3 and SCNF on the physiology of winter wheat and soil biological characteristics. A treatment structure comprising two main-plot factors (normal atmospheric environment (A) and elevated ambient O_3 concentration (E)) and two sub-plot factors (urea (CK) and SCNF (S)) arranged in a split-plot design was used for this study. (1) A_CK, normal atmospheric environment + urea; (2) A_S, normal atmospheric environment + SCNF; (3) E_CK, elevated ambient O_3 concentration + urea; and (4) E_S, elevated ambient O_3 concentration + SCNF (Table S3). Each treatment was repeated three times.

According to the optimal fertilizer dosage for local wheat cultivation in the wheat industry development report of Jiangsu Province (http://www.jsnjy.net.cn/newsDetail.html?newId=3cf5ac14-869a-4af9-942c-861b6ec8b180, accessed on 10 September 2022), the fertilizer application amount of this experiment was determined to be 225–100–120 kg ha^{-1} (N–P_2O_5–K_2O). The N fertilizers included in this study were SCNF (37% N) and common urea (46% N). The nutrient release longevity of SCNF was 3–4 months, and the sulfur coating begins to weather after about 6 months and is broken down by microorganisms to provide sulfur nutrients to crops after 10 months (purchased from Hanfeng slow-release fertilizer Co., Ltd., Taizhou, China). Urea, superphosphate (12% P_2O_5), and potassium chloride (60% K_2O) were purchased from the local fertilizer distributor.

In treatments A_CK and E_CK, urea was used as the nitrogen fertilizer, and the application proportion of basal fertilizer: tillering fertilizer: jointing fertilizer: panicle fertilizer was 5:1:2:2. This method can achieve better coordination between quality and yield, has great economic benefits, and is one of the high-yield practices vigorously promoted in long-term agricultural production in Jiangsu Province [30]. In treatments A_S and E_S, SCNF was used instead of urea, with 60% applied as basal fertilizer and 40% topdressing at the re-greening stage (Z30). There was no additional irrigation except for rainwater, and field management followed a local high-yield cultivation management program. The phosphate and potassium fertilizers in each treatment were all applied as base fertilizers before sowing (Z00).

4.3. Plant Sampling and Analysis

The net photosynthetic rate (*Pn*) of wheat flag leaf was measured by the Li-6400 portable photosynthesis system (Li COR, Lincoln, NE, USA) under natural light from nine to eleven a.m. every 7 days in sunny weather [56]. The photosynthetic light-response curves were created using ten light intensity gradients at 1800, 1500, 1200, 900, 600, 300, 100, 50, 20, and 0 μmol m^{-2} s^{-1}. Each treatment was repeated three times.

Twenty flag leaves were randomly selected for each treatment to determine the relative content of chlorophyll using a SPAD-502 Plus chlorophyll analyzer made by Minolta, Japan [57].

Plant samples were collected at the flowering (Z60) and maturity (Z92) stages, and the dry weight, nutrient content, and accumulation (total nitrogen, TN; total phosphorus,

TP; total potassium, TK) of wheat organs (stem and sheath, leaf, grain, ear axis and glume) were measured after drying. The contents of TN, TP, and TK were determined by the indophenol blue method, molybdenum antimony colorimetric method, and flame photometry, respectively [58]. Ten representative rows were selected in each field, the wheat yield components were measured at the maturity stage (Z92), and the 1000-grain weight and yield were calculated at 13% moisture content.

4.4. Soil Sampling and Assessment of Chemical Properties

At the flowering (Z60; 19 April 2023) and mature period (Z92; 3 June 2023), topsoil (0–20 cm) samples were collected by a five-point sampling method [59]. After visible plant roots and organic residues were removed, fresh soil was taken by the quartering method, and the enzyme activities (soil urease, SUE; soil nitrate reductase, SNR) of soil samples were determined according to the instructions of the detection kits (Keming Biotechnology Co., Ltd., Suzhou, China). The remaining soil samples were dried and the soil chemical properties were determined. The soil pH was measured by the glass electrode method. A continuous flow injection analyzer (Model AA3–A001–02E, Bran–Luebbe, Norderstedt, Germany) was used to determine the contents of nitrate nitrogen (NO_3^-–N) and ammonium nitrogen (NH_4^+–N) in soil [60]. The potassium dichromate volumetric method, flame photometric method, and molybdenum antimony colorimetric method were adopted to measure the contents of soil organic matter (OM), available potassium (AK), and available phosphorus (AP), respectively [58].

4.5. Soil Metagenomic DNA Extraction, Sequencing and Data Analysis

Part of the mixed fresh soil samples were sub-packed in a sterilized 15 mL centrifuge tube and stored in liquid nitrogen [61]. After about 3 h, they were transported to the laboratory (Major Bio-Pharm Technology, Co., Ltd., Shanghai, China) for soil microbial metagenomic sequencing. After DNA extraction and sequencing, data quality control, and splicing assembly, non-redundant gene sets were constructed and the species annotation information and corresponding functional information were obtained by comparing with the NR and KEGG database by DIAMOND [62–64].

4.6. Statistical Analyses and Bioinformatics

The data of plant agronomic parameters and soil chemical properties were analyzed and processed with the SPSS 20.0 software (IBM Corp, Armonk, NY, USA), and the related images were drawn in Origin 8.0 (Origin Lab Corporation, Northampton, MA, USA). The significantly different means were separated using variance analysis (one-way ANOVA) followed by Duncan's multiple-range tests at a 5% level of probability [65].

The microbial abundance was calculated by the reads per kilobase million (RPKM) method [66]. Bioinformatics data were analyzed using the R statistical software package in RStudio version 0.99.446 (Rstudio, Inc., Boston, MA, USA, 2015).

5. Conclusions

In conclusion, the results of the field experiment showed that SCNF could promote the photosynthetic function of wheat flag leaves and improve the soil environment by providing long-term nitrogen supply and demonstrated the obvious potential of SCNF to alleviate O_3 stress on winter wheat production. O_3 had significant inhibitory effects on the physiology and production capacity of winter wheat, the stability of soil microbial communities, and the abundance of N-related metabolic functional genes; soil microbial metabolic functions responded earlier than the community structure of soil microorganisms. Although SCNF could not delay the O_3-induced leaf senescence of winter wheat, it could effectively alleviate the damage to the photosynthetic function of wheat flag leaves, increase the leaf size to enhance the accumulation of dry matter and nutrients, lay the foundation for yield, and offset some of the effects on the wheat yield. In addition, SCNF significantly alleviated the inhibitory effect of O_3 on functional genes related to N-cycling, which means that SCNF has great potential in regulating the soil N transformation process and increasing soil N retention. The results showed that the replacement of urea by SCNF would be an effective field nitrogen management strategy with both agronomic and ecological benefits under the situation of a continuous increase in ambient O_3 in the future.

Supplementary Materials: The following supporting information can be downloaded at https://www.mdpi.com/article/10.3390/ijms25179381/s1.

Author Contributions: Conceptualization, N.Z., Y.Q. and Q.Y.; Methodology, N.Z., H.S. and X.Z.; Software, N.Z. and L.S.; Formal analysis, Y.Q. and Q.Y.; Resources, Y.Q., L.S. and Q.Y.; Writing—original draft, N.Z., Y.Q. and Q.Y.; Writing—review & editing, N.Z. and Y.L.; Visualization, N.Z.; Supervision, H.S. and Y.L.; Project administration, Y.L. and X.Z. All authors have read and agreed to the published version of the manuscript.

Funding: This work was financially supported by the Special Technology Innovation Fund of Carbon Neutrality in Jiangsu Province (BE2022312).

Institutional Review Board Statement: Not applicable.

Informed Consent Statement: Not applicable.

Data Availability Statement: The data files (reads in FASTQ format) were deposited at the NCBI SRA database under the BioProject, accession No. PRJNA999761 (https://www.ncbi.nlm.nih.gov/sra/PRJNA999761, accessed on 29 July 2023).

Conflicts of Interest: The authors declare no conflicts of interest.

References

1. Ashmore, M.; Emberson, L.; Karlsson, P.E.; Hakan, P. New Directions: A new generation of ozone critical levels for the protection of vegetation in Europe. *Atmos. Environ.* **2004**, *38*, 2213–2214. [CrossRef]
2. Biswas, D.K.; Xu, H.; Li, Y.; Sun, J.; Wang, X.; Han, X.; Jiang, G. Genotypic differences in leaf biochemical, physiological and growth responses to ozone in 20 winter wheat cultivars released over the past 60 years. *Global Change Biol.* **2008**, *14*, 46–59. [CrossRef]
3. Lu, X.; Hong, J.; Zhang, L.; Cooper, O.; Schultz, M.; Xu, X.; Wang, T.; Gao, M.; Zhao, Y. Severe surface ozone pollution in China: A global perspective. *Environ. Sci. Technol.* **2018**, *5*, 487–494. [CrossRef]
4. Wang, N.; Lyu, X.; Deng, X.; Huang, X.; Jiang, F.; Ding, A. Aggravating O_3 pollution due to NO_X emission control in eastern China. *Sci. Total Environ.* **2019**, *677*, 732–744. [CrossRef] [PubMed]
5. Araminienė, V.; Sicard, P.; Anav, A.; Agathokleous, E.; Stakėnas, V.; DeMarco, A.; Varnagirytė-Kabašinskienė, I.; Paoletti, E.; Girgždienė, R. Trends and interrelationships of ground-level ozone metrics and forest health in Lithuania. *Sci. Total Environ.* **2019**, *658*, 1265–1277. [CrossRef]
6. Aguiar-Silva, C.; Brandão, S.E.; Domingos, M.; Bulbovas, P. Antioxidant responses of Atlantic Forest native tree species as indicators of increasing tolerance to oxidative stress when they are exposed to air pollutants and seasonal tropical climate. *Ecol. Indic.* **2016**, *63*, 154–164. [CrossRef]
7. Chen, Z.; Maltz, M.R.; Cao, J.; Yu, H.; Shang, H.; Aronson, E. Elevated O_3 alters soil bacterial and fungal communities and the dynamics of carbon and nitrogen. *Sci. Total Environ.* **2019**, *677*, 272–280. [CrossRef]
8. Zhu, X.; Feng, Z.; Sun, T.; Liu, X.; Tang, H.; Zhu, J.; Guo, W.; Kazuhiko, K. Effects of elevated ozone concentration on yield of four Chinese cultivars of winter wheat under fully open-air field conditions. *Global Change Biol.* **2011**, *17*, 2697–2706. [CrossRef]
9. Mills, G.; Sharps, K.; Simpson, D.; Pleijel, H.; Broberg, M.; Uddling, J.; Jaramillo, F.; Davies, W.J.; Dentener, F.; Van den Berg, M.; et al. Ozone pollution will compromise efforts to increase global wheat production. *Global Change Biol.* **2018**, *24*, 3560–3574. [CrossRef]
10. Ainsworth, E.A. Understanding and improving global crop response to ozone pollution. *Plant J.* **2017**, *90*, 886–897. [CrossRef]
11. Booker, F.; Muntifering, R.; Mcgrath, M.; Burkey, M.; Decoteau, D.; Fiscus, E.; Manning, W.; Krupa, S.; Chappelka, A.; Grantz, D. The ozone component of global change: Potential effects on agricultural and horticultural plant yield, product quality and interactions with invasive species. *J. Integr. Plant Biol.* **2009**, *51*, 337–351. [CrossRef] [PubMed]
12. Mccrady, J.K.; Andersen, C.P. The effect of ozone on below-ground carbon allocation in wheat. *Environ. Pollut.* **2000**, *107*, 465–472. [CrossRef] [PubMed]
13. Li, X.; Deng, Y.; Li, Q.; Lu, C.; Wang, J.; Zhang, H.; Zhu, J.; Zhou, J.; He, Z. Shifts of functional gene representation in wheat rhizosphere microbial communities under elevated ozone. *ISME J.* **2013**, *7*, 660–671. [CrossRef]
14. Miltner, A.; Bombach, P.; Schmidt-Brucken, B.; Kastner, M. SOM genesis: Microbial biomass as a significant source. *Biogeochemistry* **2012**, *111*, 41–55. [CrossRef]

15. Andersen, C.P. Source-sink balance and carbon allocation below ground in plants exposed to ozone. *New phytol.* **2003**, *157*, 213–228. [CrossRef]
16. Liu, P.; Huang, J.J.; Cai, Z.Y.; Chen, H.T.; Huang, X.; Yang, S.N.; Su, Z.X.; Azam, M.; Chen, H.B.; Shen, J.Y. Influence of girdling on growth of litchi (*Litchi chinensis*) roots during cold-dependent floral induction. *Sci. Hortic.* **2022**, *297*, 110928. [CrossRef]
17. Wang, Z.; Zong, H.; Zheng, H.; Liu, G.; Chen, L.; Xing, B. Reduced nitrification and abundance of ammonia-oxidizing bacteria in acidic soil amended with biochar. *Chemosphere* **2015**, *138*, 576–583. [CrossRef]
18. Baker, B.J.; Anda, V.D.; Seitz, K.W.; Dombrowski, N.; Santoro, A.E.; Lloyd, K.G. Diversity, ecology and evolution of Archaea. *Nat. Microbiol.* **2020**, *5*, 887–900. [CrossRef]
19. Chelsea, C.J.; Dove, N.C.; Beman, J.M.; Stephen, C.H.; Emma, L.A. Meta-analysis reveals ammonia-oxidizing bacteria respond more strongly to nitrogen addition than ammonia-oxidizing archaea. *Soil Biol. Biochem.* **2016**, *99*, 158–166. [CrossRef]
20. Berg, I.A.; Kockelkorn, D.; Buckel, W.; Fuchs, G. A 3-hydroxypropionate/4hydroxybutyrate autotrophic carbon dioxide assimilation pathway in Archaea. *Science* **2007**, *318*, 1782–1786. [CrossRef]
21. Pei, Y.; Yu, Z.; Ji, J.; Khan, A.; Li, X. Microbial community structure and function indicate the severity of chromium contamination of the Yellow River. *Front. Microbiol.* **2018**, *9*, 38. [CrossRef] [PubMed]
22. Zou, Y.; Lin, M.; Xiong, W.; Wang, M.; Zhang, J.; Wang, M.; Sun, Y. Metagenomic insights into the effect of oxytetracycline on microbial structures, functions and functional genes in sediment denitrification. *Ecotoxicol. Environ. Saf.* **2018**, *161*, 85–91. [CrossRef]
23. Wang, M.; Li, G.; Feng, Z.; Liu, Y.; Yuan, X.; Uscola, M. A wider spectrum of avoidance and tolerance mechanisms explained ozone sensitivity of two white poplar ploidy levels. *Ann. Bot.* **2023**, *131*, 655–666. [CrossRef]
24. Tahamolkonan, M.; Ghahsareh, A.M.; Ashtari, M.K.; Honarjoo, N. Tomato (*Solanum lycopersicum*) growth and fruit quality affected by organic fertilization and ozonated water. *Protoplasma* **2021**, *259*, 291–299. [CrossRef] [PubMed]
25. Wang, Q.; Wang, D.; Agathokleous, E.; Cheng, C.; Shang, B.; Feng, Z.Z. Soil microbial community involved in nitrogen cycling in rice tields treated with antiozonant under ambient ozone. *Appl. Environ. Microb.* **2023**, *89*, 00180. [CrossRef]
26. Manning, W.J.; Paoletti, E.; Sandermann, H.; Ernst, D. Ethylenediurea (EDU): A research tool for assessment and verification of the effects of ground level ozone on plants under natural conditions. *Environ. Pollut.* **2011**, *159*, 3283–3293. [CrossRef]
27. Peng, J.; Xu, Y.; Shang, B.; Qu, L.; Feng, Z. Impact of ozone pollution on nitrogen fertilization management during maize (*Zea mays* L.) production. *Environ. Pollut.* **2020**, *266*, 115158. [CrossRef]
28. Guo, J.; Wang, Y.; Blaylock, A.D.; Chen, X. Mixture of controlled release and normal urea to optimize nitrogen management for high-yielding (>15 Mg ha^{-1}) maize. *Field Crops Res.* **2017**, *204*, 23–30. [CrossRef]
29. Li, R.; Gao, Y.; Chen, Q.; Li, Z.; Gao, F.; Meng, Q.; Li, T.; Liu, A.; Wang, Q.; Wu, L.; et al. Blended controlled-release nitrogen fertilizer with straw returning improved soil nitrogen availability, soil microbial community, and root morphology of wheat. *Soil Tillage Res.* **2021**, *212*, 105045. [CrossRef]
30. Ma, Q.; Qian, Y.; Yu, Q.; Cao, Y.; Tao, R.; Zhu, M.; Ding, J.; Li, C.; Guo, W.; Zhu, X. Controlled-release nitrogen fertilizer application mitigated N losses and modified microbial community while improving wheat yield and N use efficiency. *Agr. Ecosyst. Environ.* **2023**, *349*, 108445. [CrossRef]
31. Hu, T.; Liu, S.; Xu, Y.; Feng, Z.; Calatayud, V. Assessment of O_3-induced yield and economic losses for wheat in the North China Plain from 2014 to 2017, China. *Environ. Pollut.* **2020**, *258*, 113828. [CrossRef] [PubMed]
32. Ren, X.; Shang, B.; Feng, Z.; Calatayud, V. Yield and economic losses of winter wheat and rice due to ozone in the Yangtze River Delta during 2014–2019. *Sci. Total Environ.* **2020**, *745*, 140847. [CrossRef] [PubMed]
33. Avnery, S.; Mauzerall, D.L.; Liu, J.; Horowitz, L.W. Global crop yield reductions due to surface ozone exposure: 2. Year 2030 potential crop production losses, economic damage, and implications for world hunger under two scenarios of O_3 pollution. *Atmos. Environ.* **2011**, *45*, 2297–2309. [CrossRef]
34. Gelang, J.; Sild, E.; Danielsson, H.; Younis, S.; Sellden, G. Rate and duration of grain filling in relation to flag leaf senescence and grain yield in spring wheat exposed to different concentrations of ozone. *Physiol. Plantarum* **2000**, *110*, 366–375. [CrossRef]
35. Yang, X.; Chen, X.; Ge, Q.; Li, B.; Tong, Y.; Li, Z.; Kuang, T.; Lu, C. Characterization of photosynthesis of flag leaves in a wheat hybrid and its parents grown under field conditions. *Plant Physiol.* **2007**, *164*, 318–326. [CrossRef]
36. Delgado-Baquerizo, M.; Giaramida, L.; Reich, P.B.; Khachane, A.N.; Hamonts, K.; Edwards, C.; Lawton, L.A.; Singh, B.K. Lack of functional redundancy in the relationship between microbial diversity and ecosystem functioning. *J. Ecol.* **2016**, *104*, 936–946. [CrossRef]
37. Feng, P.; Wang, B.; Liu, D.; Xing, H.; Ji, F.; Macadam, I.; Ruan, H.; Yu, Q. Impacts of rainfall extremes on wheat yield in semi-arid cropping systems in eastern Australia. *Climatic Change* **2018**, *147*, 555–569. [CrossRef]
38. Agathokleous, E.; Feng, Z.; Oksanen, E.; Sicard, P.; Wang, Q.; Saitanis, C.J.; Araminiene, V.; Blande, J.D.; Hayes, F.; Calatayud, V.; et al. Ozone affects plant, insect, and soil microbial communities: A threat to terrestrial ecosystems and biodiversity. *Sci. Adv.* **2020**, *6*, eabc1176. [CrossRef]
39. Feng, Z.; Büker, P.; Pleijel, H.; Emberson, L.; Karlsson, P.E.; Uddling, J. A unifying explanation for variation in ozone sensitivity among woody plants. *Glob. Change Biol.* **2018**, *24*, 78–84. [CrossRef]
40. Li, X.; Song, X.; Zhao, J.; Lu, H.; Qian, C.; Zhao, X. Shifts and plasticity of plant leaf mass per area and leaf size among slope aspects in a subalpine meadow. *Ecol. Evol.* **2021**, *11*, 14042–14055. [CrossRef]

41. Zhang, X.; Jia, H.; Li, T.; Wu, J.; Nagarajan, R.; Lei, L.; Powers, C.; Kan, C.C.; Hua, W.; Liu, Z.; et al. TaCol-B5 modifies spike architecture and enhances grain yield in wheat. *Science* **2022**, *376*, 180. [CrossRef] [PubMed]
42. Shang, B.; Fu, R.; Agathokleous, E.; Dai, L.; Zhang, G.; Wu, R.; Feng, Z. Ethylenediurea offers moderate protection against ozone-induced rice yield loss under high ozone pollution. *Sci. Total Environ.* **2022**, *806*, 151341. [CrossRef] [PubMed]
43. Li, P.; Yin, R.; Zhou, H.; Yuan, X.; Feng, Z. Soil pH drives poplar rhizosphere soil microbial community responses to ozone pollution and nitrogen addition. *Eur. J. Soil Sci.* **2021**, *73*, 13186. [CrossRef]
44. Cuhel, J.; Simek, M.; Laughlin, R.J.; Bru, D.; Cheneby, D.; Watson, C.J.; Philippot, L. Insights into the effect of soil pH on N_2O and N_2 emissions and denitrifier community size and activity. *Appl. Environ. Microb.* **2010**, *76*, 1870–1878. [CrossRef]
45. Wang, H.; Boutton, T.; Xu, W.; Hu, G.; Jiang, P.; Bai, E. Quality of fresh organic matter affects priming of soil organic matter and substrate utilization patterns of microbes. *Sci. Rep.* **2015**, *5*, 10102. [CrossRef]
46. Nacke, H.; Thürmer, A.; Wollherr, A.; Will, C.; Hodac, L.; Herold, N.; Schoning, L.; Schrumpf, M.; Daniel, R. Pyrosequencing-based assessment of bacterial community structure along different management types in German forest and grassland soils. *PLoS ONE* **2011**, *6*, e17000. [CrossRef]
47. Huang, Q.; Wang, J.; Wang, C.; Wang, Q. The 19-years inorganic fertilization increased bacterial diversity and altered bacterial community composition and potential functions in a paddy soil. *Appl. Soil Ecol.* **2019**, *144*, 60–67. [CrossRef]
48. Zhou, J.; Guan, D.; Zhou, B.; Zhao, B.; Ma, M.; Qin, J.; Jiang, X.; Chen, S.; Cao, F.; Shen, D.; et al. Influence of 34-years of fertilization on bacterial communities in an intensively cultivated black soil in northeast China. *Soil Biol. Biochem.* **2015**, *90*, 42–51. [CrossRef]
49. Ai, C.; Zhang, S.; Zhang, X.; Guo, D.; Zhou, W.; Huang, S. Distinct responses of soil bacterial and fungal communities to changes in fertilization regime and crop rotation. *Geofis. Int.* **2018**, *319*, 156–166. [CrossRef]
50. He, Z.; Huang, X.; Fan, Y.; Yang, M.; Zhou, E. Metatranscriptomic analysis reveals rich mycoviral diversity in three major fungal pathogens of rice. *Int. J. Mol. Sci.* **2022**, *23*, 91–92. [CrossRef]
51. Aller, J.Y.; Kemp, P.F. Are Archaea inherently less diverse than Bacteria in the same environments? *FEMS Microbiol. Ecol.* **2008**, *65*, 74–87. [CrossRef] [PubMed]
52. Reji, L.; Francis, C.A. Metagenome-assembled genomes reveal unique metabolic adaptations of a basal marine *Thaumarchaeota* lineage. *ISME J.* **2020**, *14*, 2105–2115. [CrossRef]
53. Joshi, B.; Chaudhary, A.; Varma, A.; Tripathi, S.; Bhatia, A. Elevated CO_2, O_3 and their interaction have differential impact on soil microbial diversity and functions in wheat agroecosystems. *Rhizosphere* **2023**, *27*, 100777. [CrossRef]
54. Miao, S.; Qiao, Y.; Jin, J.; Wang, Y.; Tang, C. Greater variation of bacterial community structure in soybean- than maize-grown Mollisol soils in responses to seven-year elevated CO_2 and temperature. *Sci. Total Environ.* **2021**, *764*, 142836. [CrossRef]
55. *GB/T 17892-1999*; High quality wheat—Strong gluten wheat. The State Bureau of Quality and Technical Supervisor: Beijing, China, 1999.
56. Cai, F.; Zhang, Y.S.; Mi, N.; Ming, H.Q.; Zhang, S.J.; Zhang, H.; Zhao, X. Maize (*Zea mays* L.) physiological responses to drought and rewatering, and the associations with water stress degree. *Agr. Water Manag.* **2020**, *241*, 106379. [CrossRef]
57. Liu, J.; Song, M.; Wei, X.; Zhang, H.; Bai, Z.; Zhuang, X. Responses of phyllosphere microbiome to ozone stress: Abundance, community compositions and functions. *Microorganisms* **2022**, *10*, 680. [CrossRef] [PubMed]
58. Bao, S.D. *Soil Agricultural Chemistry Analysis*; China Agriculture Press: Beijing, China, 2000.
59. Zhou, J.; Deng, Y.; Shen, L.; Wen, C.; Yan, Q.; Ning, D.; Qin, Y.; Xue, K.; Wu, L.; He, Z. Temperature mediates continental-scale diversity of microbes in forest soils. *Nat. Commun.* **2016**, *7*, 12083. [CrossRef]
60. Zhang, Z.; Yu, Z.; Zhang, Y.; Shi, Y. Impacts of fertilization optimization on soil nitrogen cycling and wheat nitrogen utilization under water-saving irrigation. *Front. Plant Sci.* **2022**, *13*, 878424. [CrossRef]
61. Zhong, Y.; Yan, W.; Shangguan, Z. Impact of long-term N additions upon coupling between soil microbial community structure and activity, and nutrient-use efficiencies. *Soil Biol. Biochem.* **2015**, *91*, 151–159. [CrossRef]
62. Schmidt, R.; Mitchell, J.; Scow, K. Cover cropping and no-till increase diversity and symbiotroph: Saprotroph ratios of soil fungal communities. *Soil Biol. Biochem.* **2019**, *129*, 99–109. [CrossRef]
63. Mueller, R.; Paula, F.; Mirza, B.; Rodrigues, J.; Nüsslein, K.; Bohannan, B. Links between plant and fungal communities across a deforestation chronosequence in the Amazon rainforest. *ISME J.* **2014**, *8*, 1551. [CrossRef]
64. Buchfink, B.; Reuter, K.; Drost, H. Sensitive protein alignments at tree-of-life scale using DIAMOND. *Nat. Methods* **2021**, *18*, 366–368. [CrossRef] [PubMed]
65. Chen, Y.; Jiang, Y.; Huang, H.; Mou, L.; Ru, J.; Zhao, J.; Xiao, S. Long-term and high-concentration heavy-metal contamination strongly influences the microbiome and functional genes in Yellow River sediments. *Sci. Total Environ.* **2018**, *637*, 1400–1412. [CrossRef]
66. Lawson, C.E.; Wu, S.; Bhattacharjee, A.S.; Hamilton, J.J.; McMahon, K.D.; Goel, R.; Noguera, D.R. Metabolic network analysis reveals microbial community interactions in anammox granules. *Nat. Commun.* **2017**, *8*, 15416. [CrossRef] [PubMed]

International Journal of
Molecular Sciences

MDPI

Article

Cloning and Functional Study of *AmGDSL1* in *Agropyron mongolicum*

Xiuxiu Yan [1,†], Xiaojuan Wu [1,†], Fengcheng Sun [2], Hushuai Nie [1], Xiaohong Du [1], Xiaolei Li [1], Yongyu Fang [2], Yongqing Zhai [1], Yan Zhao [3], Bobo Fan [1] and Yanhong Ma [1,*]

1 Agricultural College, Inner Mongolia Agricultural University, Hohhot 010019, China; yanxx411@126.com (X.Y.); wuxj@emails.imau.edu.cn (X.W.); niehs@imau.edu.cn (H.N.); duxiaohong2024@126.com (X.D.); lixiaoleihlj@163.com (X.L.); zhaiyongqing66@126.com (Y.Z.); fanbobo19@126.com (B.F.)

2 Inner Mongolia Academy of Agricultural & Animal Husbandry Sciences, Hohhot 010031, China; sfcnmnky@sina.com (F.S.); 1984fyy@163.com (Y.F.)

3 College of Grassland, Resources and Environment, Inner Mongolia Agricultural University, Hohhot 010010, China; zhaoyannmg@sina.com

* Correspondence: mayanhong80@imau.edu.cn

† These authors contributed equally to this work.

Abstract: *Agropyron mongolicum* Keng is a diploid perennial grass of triticeae in gramineae. It has strong drought resistance and developed roots that can effectively fix the soil and prevent soil erosion. GDSL lipase or esterases/lipase has a variety of functions, mainly focusing on plant abiotic stress response. In this study, a *GDSL* gene from *A. mongolicum*, designated as *AmGDSL1*, was successfully cloned and isolated. The subcellular localization of the *AmGDSL1* gene (pCAMBIA1302-*AmGDSL1*-EGFP) results showed that the AmGDSL1 protein of *A. mongolicum* was only localized in the cytoplasm. When transferred into tobacco (*Nicotiana benthamiana*), the heterologous expression of *AmGDSL1* led to enhanced drought tolerance. Under drought stress, *AmGDSL1* overexpressing plants showed fewer wilting leaves, longer roots, and larger root surface area. These overexpression lines possessed higher superoxide dismutase (SOD), peroxidase (POD), catalase (CAT), and proline (PRO) activities. At the same time, the malondialdehyde (MDA) content was lower than that in wild-type (WT) tobacco. These findings shed light on the molecular mechanisms involved in the *GDSL* gene's role in drought resistance, contributing to the discovery and utilization of drought-resistant genes in *A. mongolicum* for enhancing crop drought resistance.

Keywords: *Agropyron mongolicum*; *AmGDSL1* gene; drought resistance; antioxidant activity

1. Introduction

During agricultural development, the demand for water is high, and drought seriously affects agricultural production [1]. Generally, massive amounts of ROS (reactive oxygen species) are produced when plants are exposed to drought stresses [2]. Plants have developed intricate defense mechanisms to cope with excess ROS and protect cells from damage, which include superoxide dismutase (SOD), peroxidase (POD), and catalase (CAT) antioxidant enzyme systems [3]. In plants, a large number of drought-related genes have been identified and validated, such as the *MYB*, *NAC*, *WRKY* [4,5], and *bHLH* genes [6,7]. Three *TaMYB31* homeologous genes from hexaploid wheat (*Triticum aestivum* L.), were cloned and characterized. Ectopic expression of the *TaMYB31-B* gene in Arabidopsis (*Arabidopsis thaliana*) affected plants' growth and enhanced drought tolerance [8]. *TaMYBsm1* genes from wheat play an important role in transgenic Arabidopsis drought stress tolerance through upregulation of *DREB2A*, *P5CS1*, and *RD29A* [9]. Wheat under drought stress was analyzed in a genome-wide association study, and it was found that the wheat *NAC* gene (*TaNAC071-A*) is closely associated with drought resistance. Knockdown of the *TaNAC071-A* gene reduces wheat drought resistance, whereas overexpression of *TaNAC071-A* enhances plant drought resistance [10]. Furthermore, overexpression of *TtNAC2A* in Arabidopsis plants showed enhanced drought resistance [11]. *TaWRKY1-2* silencing in wheat increases the MDA content, reduces the contents of proline and chlorophyll and the activities of antioxidant enzymes, and inhibits the expression levels of antioxidants (POD, CAT, and SOD) under drought stress [12]. *TabHLH27* knockout reduced wheat drought tolerance, yield, and water use efficiency [13].

Notably, *GDSL* genes are involved in various key physiological processes in plants, such as cuticle development, seed oil storage, male fertility, and biotic and abiotic stress responses [14,15]. The *GDSL* esterases/lipase (GELP) has been reported to be one of the largest hydrolase superfamilies. Its name is based on the N-terminal, which has a conserved GDSL motif (Gly-Asp-Ser-Leu) [16,17]. Under stress conditions such as drought and high salt, the expression level of *GDSL* lipase will undergo significant changes. Overexpression of *GhirGDSL26* in *Arabidopsis thaliana* resulted in longer root lengths, higher germination rates, and survival rates under drought resistance [18]. The function of the *CcGDSL* gene was analyzed through a cis-elements study. The expression of the *CcGDSL* gene is mainly influenced by fluctuating stressful situations, as well as by light and plant hormones. *CcGDSL21*, *CcGDSL63*, *CcGDSL64*, and *CcGDSL92* may be involved in regulating drought stress in pigeon pea (*Cajanus cajan* (L.) Druce) [19]. After salt stress treatment, weighted gene co-expression network analysis (WGCNA) showed that *CilGDSL41.11*, *CilGDSL39.49*, *CilGDSL34.85*, and *CilGDSL41.01* were significantly associated with salt stress in pecan leaves (*Carya illinoensis* K. Koch) [20]. *CaGLIP1* transgenic plants showed drought tolerance during seed germination and plant growth. Differential expression of *AtRD26A*, *AtADH*, *and AtRab18* genes was responsive in drought [21]. The expression of *CpGLIP1* was increased by treatments of drought, cold, jasmonic acid, and gibberellic acid. After drought and cold treatment, overexpression of *CpGLIP1* in Arabidopsis thaliana and poplar (*Populus tremula* × *Populus alba*) plants resulted in higher rates of growth and survival and better physiological indices than in the wild-type (WT) plants. Overexpression of *CpGLIP1* can improve tolerance of drought and cold [22]. The *GDLS* genes will be useful as targets for genetic modification in improving crop yield, quality, or both.

A. mongolicum Keng diploid (2n = 14) is a perennial herbaceous plant of the triticeae tribe in the poaceae family. As a high-quality forage, it has characteristics such as drought resistance, cold resistance, and wind and sand resistance. *A. mongolicum* Keng is also considered a primitive close relative of some crops, which can provide an abundance of drought stress-responsive genes [23]. An abundance of drought stress-responsive genes in *A. mongolicum* Ken were excavated by sequencing technology. In *A. mongolicum* during drought stress, there were 2171 unigene differences of expression [24]. The gene *AmbHLH148* from *A. mongolicum* was cloned and isolated; the overexpression of *AmbHLH148* in transgenic tobacco acted as a positive regulator under drought stress by enhancing the plant's antioxidant capacity [25]. An *AP2/EREBP* transcription factor gene, cDNA from *Agropyron mongolicum* Keng, was isolated through RT-PCR and RACE techniques and named *MwAP2/EREBP*. *MWAP2/EREBP* has tissue and temporal specificity, which suggested that the *MwAP2/EREBP* gene might be involved in the drought-resistant physiological process of *A. mongolicum* Keng [26]. In addition, some miRNAs have also been identified as being involved in the drought stress response in *A. mongolicum* Keng. Under drought conditions, a total of 114 miRNAs were identified in *A. mongolicum*: amo-miR21, amo-miR5, and amo-miR62 were participants in drought stress response and genetically stable [27]. Fan et al. identified the key miRNA-target gene under drought stress by using multi omics; osa-miR444a-3p.2-*MADS47*, bdi-miR408-5p_1ss19TA-*CCX1*, tae-miR9774_L-2R-11ss11GT-*CARC*, ata-miR169a-3p-*PAO2*, and bdi-miR528-p3_2ss15TG20CA-*HOX24* were discovered to play central roles in the regulation of drought-responsive genes [28]. A large number of drought-resistance genes were excavated in *A. mongolicum* Keng. In previous studies based on transcriptome sequencing data obtained from *A. mongolicum* in drought stress (NCBI, PRJNA742257) [29], 52 *GDSL* genes were identified through bioinformatics screening. Physical and chemical properties analysis, phylogenetic tree analysis, and qRT-PCR analysis were performed on 52 *GDSL* lipase genes. *AmGDSL1* (DN29240_c0_g3) exhibited significant upregulation in response to drought stress. To understand its potential role in drought tolerance, *AmGDSL1* was heterologously inserted into tobacco plants. This research offers valuable insights into the function of *AmGDSL1* in drought tolerance and offers genetic resources that could enhance drought resistance in other crops.

2. Results

2.1. Analysis of AmGDSL1 Gene Expression in A. mongolicum

Under normal growth conditions, the *AmGDSL1* gene is expressed in the roots, stems, and leaves of *A. mongolicum* (Figure 1A). The *AmGDSL1* gene had the highest expression level and showed significant differences in leaves. The expression level of *AmGDSL1* gene in leaves was 2.7 times higher than that in stems and is 1.9 times higher than that in roots. The expression level of the *AmGDSL1* gene in the stem was 1.4 times higher than that in the root.

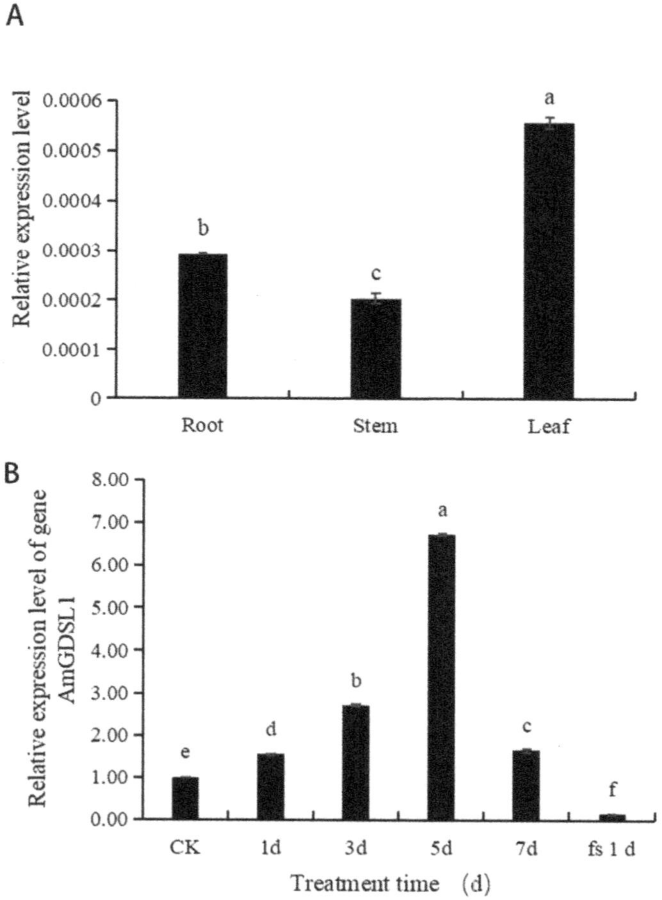

Figure 1. Expression of the *AmGDSL1* gene. (**A**) Expression level of the *AmGDSL1* gene in roots, stems, and leaves of *A. mongolicum*. (**B**) Relative expression of the *AmGDSL1* gene in *A. mongolicum* under drought stress. Different lowercase letters indicate significant differences at the 0.05 level (Duncan test).

At the same time, the seedlings of *A. mongolicum* were subjected to a simulated osmotic stress treatment with 25% PEG-6000, and the expression level of the *AmGDSL1* gene was significantly different in dehydration stress periods. There was a significant difference between the expression level of the *AmGDSL1* gene in the *A. mongolicum* control (CK) and during the different osmotic stress periods (Figure 1B). At different osmotic stress periods, the expression level of the *AmGDSL1* gene was significantly different. Notably, as the osmotic stress duration increased, the expression level of the *AmGDSL1* gene revealed an overall increasing trend. At 5 days, the gene expression reached its highest level, approximately 6.6 times that of the control sample. The results showed that during the simulated osmotic stress treatment of *A. mongolicum* with 25% PEG-6000, the AmGDSL1 gene was upregulated in response to drought.

2.2. Cloning of AmGDSL1 Gene in A. mongolicum and Overexpression Vector Construction

The specific band obtained after PCR amplification using cDNA as a template is shown in Figure 2A. The band size was 634 bp, which is consistent with the size of the *AmGDSL1* gene. The sequencing results of the *AmGDSL1* gene and transcription sequence alignment showed a matching degree of 96.44% and proved that the *AmGDSL1* gene was successfully cloned (Figure 2C). The *AmGDSL1* gene and pCAMBIA1302-EGFP expression vector were cleaved using *NcoI* and *SpeI*. Simultaneously, the T4 ligase was used to ligase cleavage products. The ligase cleavage products amplified a specific band of about 634 bp, which preliminarily confirmed that the plant overexpression vector pCAMBIA1302-*AmGDSL1* was successfully constructed (Figure 2B).

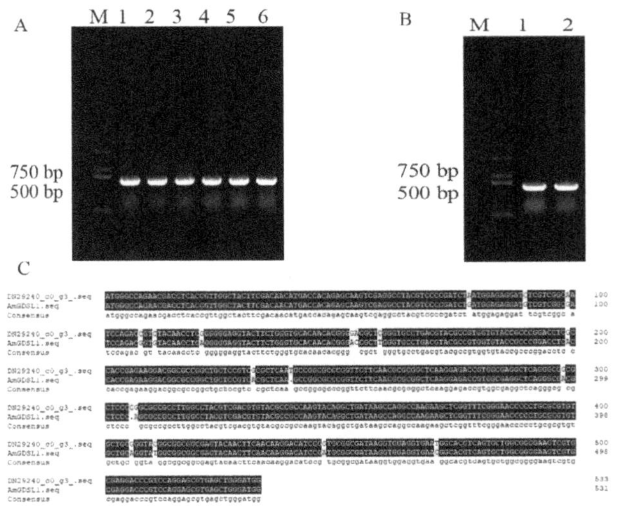

Figure 2. Cloning of *AmGDSL1* gene and vector construction. (**A**) PCR amplification electrophoresis detection of *AmGDSL1*. M:D2000 DNA Maker, 1–6: *AmGDSL1*. (**B**) PCAMBIA1302-*AmGDSL1* transformed Agrobacterium LBA4404 colony PCR electrophoretogram.M:D2000 DNA Maker, 1–6: *AmGDSL1*. (**C**) Alignment of *AmGDSL1* transcribed sequence with sequencing sequence. Black: Consistent sequence alignment. Blue: Base mutation.

2.3. Subcellular Localization of AmGDSL1 Gene

To verify the subcellular localization of the *AmGDSL1* gene, an expression vector, pCAMBIA1302-*AmGDSL1*-EGFP, was constructed, which had a 35S promoter and fused with green fluorescent protein (EGFP). Subsequently, this recombinant vector was injected into onion epidermal cells. Observations made through confocal microscopy revealed that the GFP signals originating from the control pCAMBIA1302-EGFP empty vector were distributed throughout the cell in the cell membrane, nucleus, and cytoplasm of the onion epidermal cells. In contrast, PCAMBIA1302-*AmGDSL1*-EGFP is mainly distributed in the cytoplasm. After treatment with sucrose solution, the onion cytoplasmic wall was separated. Clearly, pCAMBIA1302-*AmGDSL1*-EGFP fusion protein was exclusively localized in the cytoplasm (Figure 3).

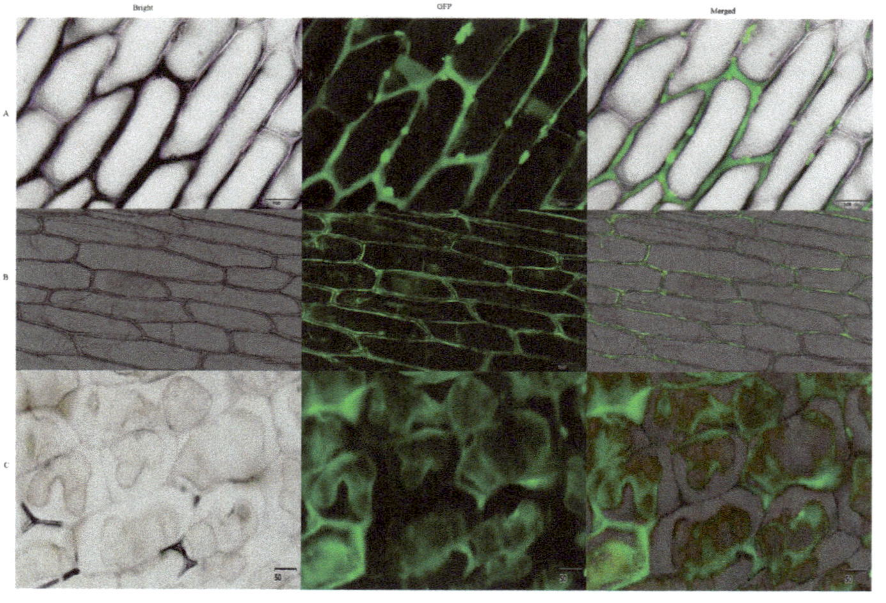

Figure 3. Subcellular localization of *AmGDSL1*. (**A**) EGFP positive control (pCAMBIA1302-EGFP). (**B**) pCAMBIA1302-*AmGDSL1* EGFP is independently expressed in onion epidermal cells. (**C**) Expression of pCAMBIA1302-*AmGDSL1*-EGFP after plasma wall separation. Scale bar = 50 μm.

2.4. Production and Identification of AmGDSL1 Transgenic Tobacco

The Agrobacterium liquid (LBA4404) containing the pCAMBIA1302-*AmGDSL1* recombinant plasmid was used to transform the sterile young tobacco leaves via the leaf disc method. Afterwards, co-culture, differentiation culture, and rooting culture were carried out to obtain regenerated tobacco plants. Using pCAMBIA1302-*AmGDSL1* plasmid as a positive control (P), wild-type tobacco as negative control (WT), and ddH_2O as a blank control (CK), PCR amplification and agarose gel electrophoresis were performed. The results showed that the regenerated plants obtained specific amplified bands around 634 bp, with no bands in WT and CK, indicating that the *AmGDSL1* gene had been integrated into the tobacco genome (Figure 4A). qRT-PCR detection was performed on the leaves of transgenic tobacco-positive plants, and the results showed that among the 15 transgenic positive strains, the OE5 and OE6 strains had the highest expression level of the *AmGDSL1* gene (Figure 4B).

2.5. Phenotype Observation of Transgenic Tobacco under Drought Stress

To verify the drought resistance of transgenic tobacco, *AmGDSL1* overexpression and wild-type tobacco (WT) plants were subjected to natural drought treatment by discontinuing watering. Under normal growth conditions, there is a significant difference between phenotypes in *AmGDSL1*-OE and WT plants. It was found that *AmGDSL1* overexpression tobacco plants have larger leaves and higher stems. Under drought conditions of 7 days, WT plants exhibited a high number of wilted leaves overall. In contrast, *AmGDSL1* overexpression tobacco showed a remarkable reduction in wilted leaves, and they were found at the bottom (Figure 5A). Simultaneously, the transgenic tobacco exhibited longer roots and a greater root surface area compared to the WT plants. This indicates that *AmGDSL1* overexpression can improve tobacco drought tolerance and enable a higher survival rate under drought stress by promoting growth of more roots (Figure 5B,C).

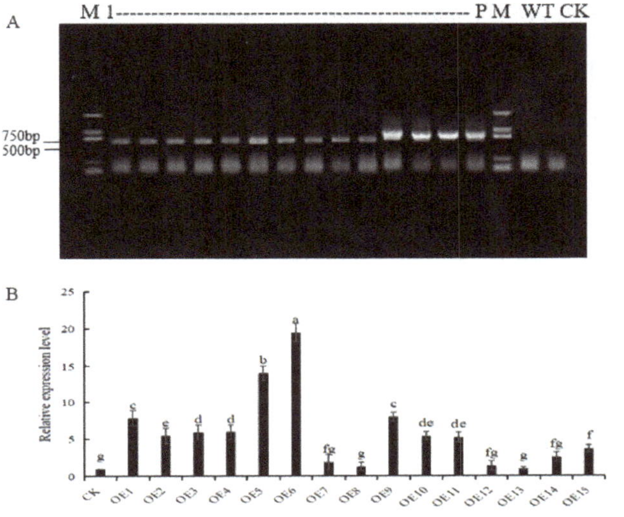

Figure 4. Obtaining and identifying of *AmGDSL1* transgenic tobacco plants. (**A**) Electrophoretogram of PCR assay of positive plants of *AmGDSL1* transgenic tobacco. (**B**) Relative expression levels of the *AmGDSL1* gene in transgenic plants. Different lowercase letters indicate significant differences at the 0.05 level (Duncan test).

Figure 5. *Cont.*

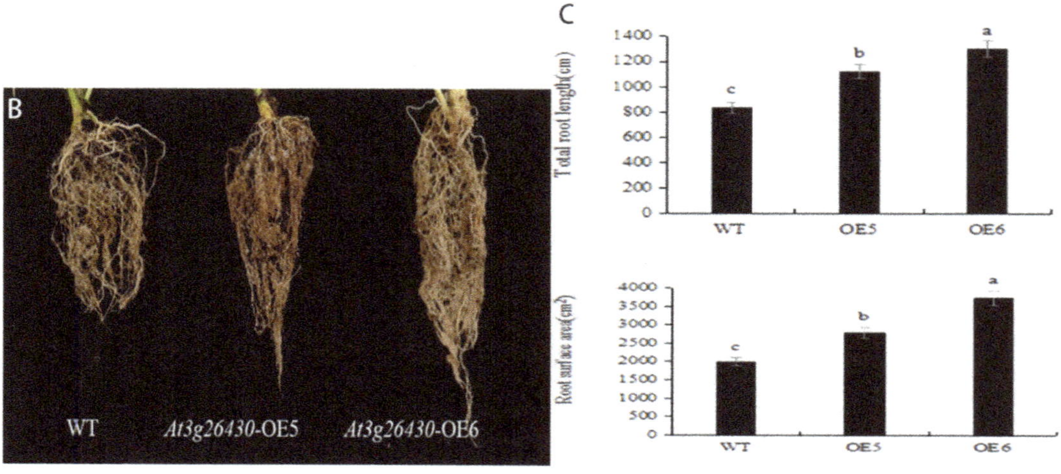

Figure 5. Drought resistance of *AmGDSL1*-transformed tobacco. (**A**) Phenotypic characteristics of transgenic plants under drought stress. (**B**) Root phenotype of transgenic. Root phenotypes of wild-type and transgenic tobacco under drought for 7 days. (**C**) Root system analysis of transgenic plants. Total root length of WT and transgenic tobacco under drought conditions for 7 days. Root surface area of WT and transgenic tobacco under drought conditions for 7 days. Different lowercase letters indicate significant differences at the 0.05 level (Duncan test).

2.6. Overexpression of AmGDSL1 Enhanced Antioxidant Capacity of Tobacco

In order to verify the function of the *AmGDSL1* gene in the antioxidant defense mechanism, SOD, POD, CAT, PRO, and MDA measurements were performed on drought-treated *AmGDSL1* overexpression tobacco. In the WT and *AmGDSL1* transgenic lines, the trend of SOD, POD, and CAT activity was slightly higher than that in WT plants under normal conditions. Apparently, the *AmGDSL1* overexpression significantly increases the SOD, POD, CAT activity in drought stress. As shown in Figure 6, activity of SOD was induced by 1.32-fold in OE5 and 1.41-fold in OE6, activity of POD was induced by 1.13-fold in OE5 and 1.26-fold in OE6, and activity of CAT was induced by 1.42-fold in OE5 and 1.26-fold in OE6. In the WT and *AmGDSL1* transgenic lines, the trend of PRO content was consistent under normal conditions. After 7 days of drought stress, the content of PRO in *AmGDSL1* transgenic lines significantly increased, indicating that transgenic plants have stronger osmotic regulation levels under drought stress conditions. The content of PRO was induced by 1.27-fold in OE5 and 1.17-fold in OE6. In the WT and *AmGDSL1* transgenic lines, the trend of MDA content was slightly lower than that in WT plants under normal conditions. Conversely, the content of MDA in WT tobacco was significantly higher than that in *AmGDSL1* transgenic lines in drought stress. The content of MDA in WT tobacco was 1.51 times that in OE5 and 1.22 times that in OE6. Obviously, under drought stress conditions, the damage to the cell membrane of transgenic plants is relatively mild, while the damage to the membrane system of WT plants is severe.

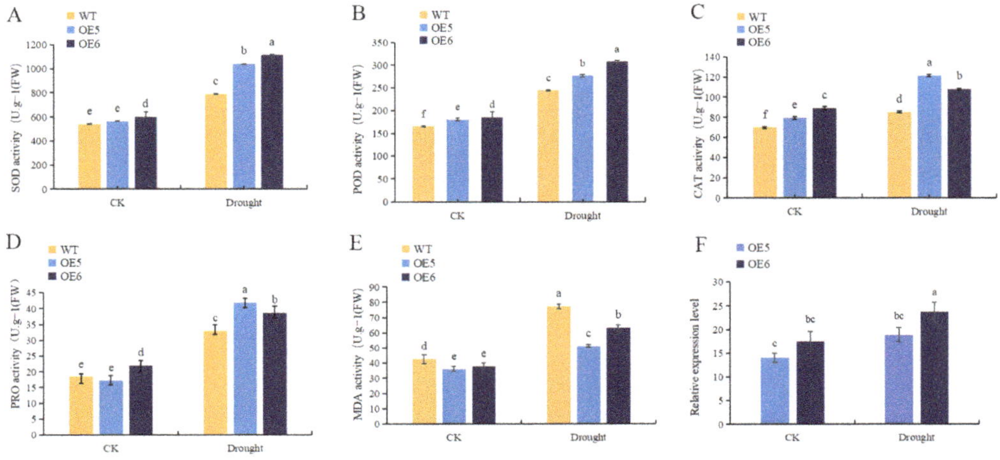

Figure 6. The antioxidant capacity and the AmGDSL1 expression in transgenic plants. (**A**) SOD activity of WT and transgenic plants. (**B**) POD activity of WT and transgenic plants. (**C**) CAT activity of WT and transgenic plants. (**D**) PRO content of WT and transgenic plants. (**E**) MDA content of WT and transgenic plants. Lowercase letters in the above graph represent significant differences at the 0.05 level. All data are expressed as the mean + SD. (**F**) The expression levels of the AmGDSL1 gene and in AmGDSL1-OE plants. Data are mean ± standard error of three replicates. Lowercase letters in the above graph represent significant differences at the 0.05 level (Duncan test).

2.7. Expression Regulation in AmGDSL1 Transgenic Tobacco

To further explore the function of *AmGDSL1* gene, the expression patterns of stress-responsive genes were detected in overexpression lines and WT plants using qRT-PCR (Figure 6F). Under normal conditions, the expression level of overexpressed tobacco's *AmGDSL1* gene is significantly higher than that of wild-type tobacco. In drought stress, the expression levels of *AmGDSL1* genes were significantly induced in overexpression lines compared to that in WT plants. Correlation analysis showed that under drought stress, the SOD, POD, CAT, and MDA activity and proline content were positively correlated with gene expression (Figure 7). Apparently, the *AmGDSL1* in transgenic tobacco has a positive regulatory effect on drought stress.

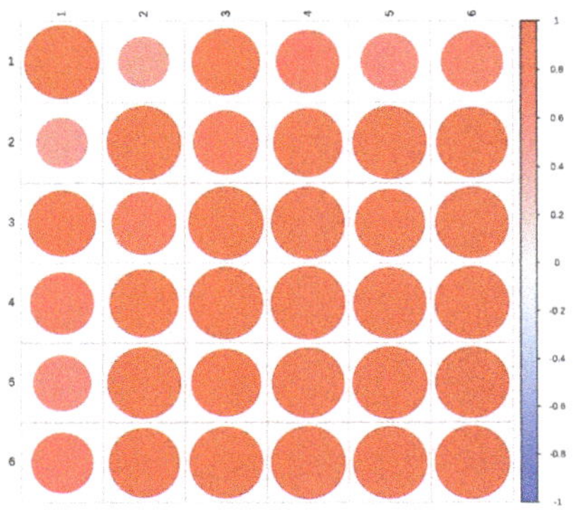

Figure 7. Correlation analysis in individual characteristics. (1) Gene expression levels under drought stress. (2) CAT activity under drought stress. (3) MDA activity under drought stress. (4) POD activity

under drought stress. (5) PRO content under drought stress. (6) PRO activity under drought stress. The horizontal and vertical axes represent different samples, and the color difference in the pie chart represents a comparison of the Pearson correlation coefficient of gene expression in the two samples. The redder the color, the stronger the positive correlation, and the bluer the color, the weaker the correlation, while the larger the circle size, the stronger the correlation.

3. Discussion

With the development of plant genome sequencing and bioinformatics, GDSL lipases have been discovered in various plants. There were 108 GDSL lipase proteins identified in Arabidopsis and 114 GDSL lipase proteins identified in *Oryza sativa* L [28]. There were 53, 65, and 96 *GDSL* genes identified in maize (*Zea mays* L.), watermelon (*Citrus lanatus*), and grape (*Vitis vinifera* L.) [30,31]. In addition, 597 *GELP* genes were identified from six Rosaceae genomes (i.e., *Fragaria vesca*, *Prunus persica*, *Prunus avium*, *Prunus mume*, *Pyrus bretschneideri*, and *Malus domestica*), based on phylogenetic tree analysis, and all *GELP* genes were further divided into ten subfamilies [31]. The diversity of these genes was revealed in terms of gene structure, evolutionary relationships, and expression patterns [32]. In previous studies, differentially expressed *GDSL* genes were identified in the transcriptome data of drought-treated *A. mongolicum* [18]. In this study, the *AmGDSL1* gene was identified in *A. mongolicum* under drought treatment, which may be associated with drought stress.

The expression of the *GDSL* gene shows diversity in different tissues and organs; for instance, *BrEXL6* is only highly expressed in buds and has lower expression levels in other tissues [33]. Lee et al. found that *GLIP1* is expressed at different developmental stages, such as in *Arabidopsis thaliana* seedling, flower, leaf, stem, and root, while *GLIP2* is only highly expressed at the root, stem, and seedling stages [17]. Twelve *GDSL* genes were analyzed by quantitative real-time PCR, and the results showed the expression levels of the 12 genes were different in the russet exocarp and green exocarp of pear (*Pyrus* spp.) at different fruit development stages [34]. Our study shows that the *AmGDSL1* gene has the highest expression level in the *A. mongolicum* leaves, followed by the roots, and the lowest expression level in the stems. To identify whether the *AmGDSL1* gene is induced by drought stress, a 25% PEG-6000-simulated drought stress treatment was performed on *A. mongolicum*. The results showed that except for on the fs1 day, the *AmGDSL1* expression levels were higher than in CK, showing an upregulated expression with prolonged drought time. Obviously, the *AmGDSL1* gene participates in plant drought stress.

The *GDSL* gene is widely distributed in plants and is involved in various abiotic stresses, antiviral activities, in seed germination, and biological metabolism. However, so far, only a few *GDSL* lipase genes have been cloned and validated in plants. Su et al. analyzed the expression patterns of soybean (*Glycine max*) *GmGELP* genes under different tissues and stress conditions through transcriptome sequencing. Seven stress-responsive *GmGELP* genes were screened and validated by qRT-PCR. Among the seven *GmGELP* genes, *GmGELP28* showed a significant response to drought, salt, and ABA treatments, and exhibited drought and salt tolerance characteristics in transgenic Arabidopsis and soybean plants. Therefore, *GmGELP28* was preliminarily identified as a potential candidate gene for improving soybean drought and salt tolerance [35]. Simone et al. found that *LOC-Os01g11760.1*, *LOC-Os07g44780.1*, and *LOC-Os10g25400.1* all belong to the GDSL lipase family and respond to drought stress in *Oryza sativa*. L roots and leaves [36]. Xiao et al. systematically studied the expression patterns of 19 *GDSL* lipases (*GGLs*) in Arabidopsis at different developmental stages and under hormone and abiotic stress treatments. They found that *GGL7*, *GGL14*, and *GGL26* responded to hormone and abiotic stress. At the same time, under severe drought stress, the drought tolerance of *GGL7*, *GGL14*, and *GGL26* mutants was significantly stronger than that of *GGL14-GGL26*, indicating that *GGL7*, *GGL14*, and GGL26 play a unique role in water retention [37]. Tang et al. identified the GDSL lipase gene (*OSP1*) in Arabidopsis thaliana. In drought stress, *OSP1* mutation resulted in significant reduction in leaf wax synthesis and occlusion of stomata, leading to increased epidermal permeability, decreased transpiration rate, and enhanced drought tolerance [38].

Liu et al. found overexpression of the *GhirGDSL26* gene in Arabidopsis thaliana resulted in enhanced drought resistance under drought stress. Silencing of *GhirGDSL26* resulted in a susceptible phenotype, higher MDA and H_2O_2 contents, and lower SOD activity and proline content [18]. The results demonstrated that *GhirGDSL26* plays a critical role in cotton drought stress tolerance. In our study, under drought stress treatment, compared to the WT, *AmGDSL1*-OE plant lines had fewer wilted leaves, which had longer roots and larger root surface area. Furthermore, *AmGDSL1*-OE plants had higher SOD, POD, CAT, and PRO activities than WT plants and lower MDA content than WT. Notably, a typical *GDSL* gene, *AmGDSL1*, was cloned from *A. mongolicum* and functionally verified in overexpressing tobacco. *AmGDSL1* overexpression lines showed significantly increased resistance to dehydration stresses. In our study, the drought tolerance function of the *AmGDSL1* gene was preliminary identified but its metabolic pathways and molecular mechanisms involved in dehydration stress response needed further research.

4. Materials and Methods

4.1. Plant Materials and Sample Preparation

The seeds of *A. mongolicum* were collected in the germplasm resource garden of the Inner Mongolia Agricultural University Grass Experiment Station. *A. mongolicum* seeds were selected and the palea and lemma were removed, rinsed with running water for 3 h and with 75% ethanol for 30 s, then rinsed with sterilized distilled water 5 times, disinfected with 4.3% NaClO for 4 min, and rinsed with sterilized distilled water 5 times. The seeds were placed in a germination box at 24 °C with 16 h/8h light and darkness in an artificial climate box (Ningbo, Zhejiang, China). When seeds germinated and grew about 8–10 cm, they were transplanted to a new germination box containing 1/5 Hoagland's nutrient solution. When the *A. mongolicum* reached the three-leaf one-center stage of growth, 1/5 Hoagland's nutrient solution containing 0.042 mol/L PEG-6000 (25% PEG-6000) was used to simulate osmotic treatment. Leaves were collected after drought for 0 days (CK), 1 day, 2 days, 3 days, 5 days, 7 days, and fs1 days (fs1 d indicates the sample collected after 24 h of restoration in 1/5 Hoagland's nutrient solution). Three leaves from the same plant were used as three biological replicates. All samples were immediately frozen in liquid nitrogen and stored at −80 °C for RNA extraction.

The seeds of *Nicotiana benthamiana* are preserved in the laboratory (Agricultural College, Inner Mongolia Agricultural University). The seeds of *Nicotiana benthamiana* were vernalized at 4 °C for 2–3 days. They were then disinfected with 75% ethanol for 30 s, rinsed with sterile distilled water 5 times, and 4.3% NaClO for 4 min. The seeds were placed in MS solid culture medium, and after germination, they were transferred to a new MS solid culture medium for genetic transformation experiments. The wild-type tobacco *Nicotiana benthamiana* (WT) and AmGDSL1-OE tobacco were cultured to consistent growth at approximately 6 weeks of age, then subjected to drought treatment by cessation of watering for 7 days. Tobacco plants under drought treatment were observed to identify the phenotype characteristics, take indicator measurements, and extract RNA.

4.2. Cloning of AmGDSL1 Gene and Vector Construction

Total RNA was extracted from the *A. mongolicum* using TRNzol (Tiangen, Beijing, China). The total RNA was used to synthesize the cDNA of the genes using the Fast Quant cDNA Kit KR106 (Tiangen, Beijing, China). The ORF of *AmGDSL1* was amplified by RT-PCR, and the primer sequence (*AmGDSL1*-c-F/R) list in Table S1. The purified PCR products were subcloned in the pEASY®-T1 Simple Cloning Vector, and the constructed vector was then transformed into Escherichia Trans1-T1 competent cells (TransGen, Beijing, China). After sequencing, the recombinant plasmid was named pEASY-*AmGDSL1*. pCAMBIA1302-EGFP plasmids (Abiowell, Changsha, China) and pEASY-*AmGDSL1* plasmids were extracted according to the plasmid extraction kit (TIANGEN DP103, Beijing, China) and were digested using *Nco*I and *Spe*I restriction endonucleases (Takara, Beijing, China). The pCAMBIA1302-*AmGDSL1*-EGFP recombinant vector was constructed by ligating the *AmGDSL1* product without a termination codon into the pCAMBIA1302-EGFP vector fragment. The enzyme digestion system included 0.5 μL *NcoI*, 0.5 μL *SpeI*, 2 μL 2 × K buffer, 10 μL pCAMBIA1302-EGFP/pEASY-*AmGDSL1*, and 7 μL ddH_2O. The digestion process was carried out at 37 °C for 16 h. The enzyme digestion products used were DNA Gel/PCR Purification Miniprep Kit (Beiwo, Hangzhou, China). The ligation system consisted of a 2.5 μL of linearized pCAMBIA1302 vector, 1.5 μL AmGDSL1, 2 μL T4 DNA buffer, 1 μL DNA Ligation Kit (TaKaRa, Beijing, China), and 13 μL ddH_2O. The digestion process was carried out overnight at 16 °C. The ligation was carried out using an ice for 25 min, a water bath at 42 °C for 45 s, and ice for 2 min, which transformed the ligation products into E. coli receptor TOP10 cells. Positive transformants were selected and placed on LB agar plates containing 50 μg/mL kanamycin for PCR analysis. After the PCR detection (AmGDSL1-F/R) of the recombinant plasmid, it was named pCAMBIA1302-*AmGDSL1*. The pCAMBIA1302-*AmGDSL1* plasmid was placed in liquid nitrogen for 5 min, then in a water bath at 37 °C for 5 min, and on ice for 5 min, and transformed into agrobacterium LBA4404 competent cells (Weidi, Shanghai, China). Positive transformants were selected on LB agar plates containing 50 μg/mL kanamycin for PCR analysis.

4.3. Subcellular Localization of AmGDSL1

The gene *AmGDSL1* was inserted into the pCAMBIA1302 vector, which was started by the CaMV-35S promoter and contained the EGFP fragment. The Agrobacterium tumefaciens with pCAMBIA1302-*AmGDSL1*-EGFP vector was activated and resurrected by 1/2MS liquid culture medium. The resuspension was adjusted to OD600:0.6 and the suspension was left at room temperature for 2 h, then used to infect onion epidermal cells. Simultaneously, the empty vector pCAMBIA1302-EGFP was used as a control. We spread the onion epidermis on MS medium and pre-treated it in the dark at 25 °C for 24 h. Then, the pre-treated onion epidermis was soaked in the resuspended Agrobacterium tumefaciens liquid for 20 min. Subsequently, the onion epidermal Agrobacterium liquid was dried using filter paper, and the side with the wound was spread flat on MS solid medium. The medium was incubated in a light incubator at a light cycle of 16/8 h and 25 °C for 48 h. The confocal laser microscope was used to observe the fluorescence signal of the fusion protein. The onion epidermal tissue was treated with 300 g/L sucrose solution for 5 min until the separation of the cell wall was observed.

4.4. Obtaining and Identifying of AmGDSL1 Transgenic Tobacco Plants

The pCAMBIA1302 plasmid containing the *AmGDSL1* gene was transformed into tobacco leaves by the Agrobacterium-mediated method (Figure 1). The immature leaves of the *Nicotiana benthamiana* were cut into 0.5 × 0.5 cm^2 samples and placed in resuspended Agrobacterium solution for 8 min. The Agrobacterium liquid on the leaves was dried using filter paper, and leaves were transferred to a co-culture medium (MS+1 mg/L 6-BA) then incubated at 28 °C in the dark for 2–3 days. The co-cultured tobacco leaves was transferred to a differentiation medium (MS + 1 mg/L 6-BA + 300 mg/L TMT + 50 mg/L Kan) for bud differentiation. Subsequently, the differentiated buds were transferred to rooting medium (MS + 0.5 mg/L IAA + 300 mg/L TMT + 50 mg/L Kan) to induce rooting and enable the formation of small plants (Figure 8). DNA was extracted from regenerated plant leaves using a FastPure Plant DNA Isolation Mini Kit-BOX2 (Vazyme, Nanjing, China). The primer *AmGDSL1*-F/R was used for PCR detection to determine if the plants were positive.

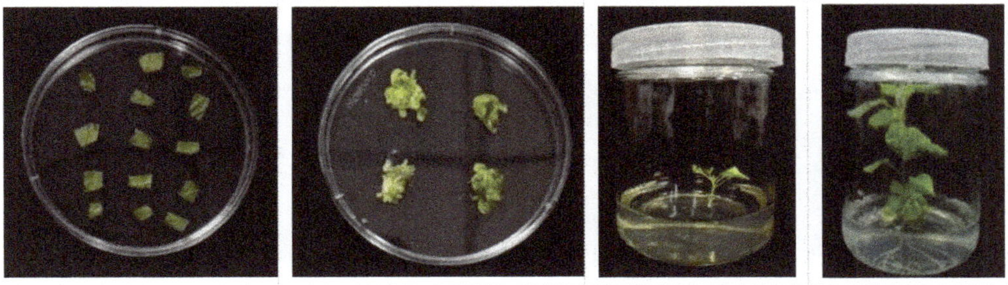

Figure 8. Flow chart of genetic transformation of AmGDSL1 transgenic tobacco. (**A**) Infected leaves. (**B**) Inducing callus. (**C**) Rooting culture. (**D**) Growing into a plant.

4.5. *Drought Stress Tolerance Assays of AmGDSL1 Transgenic Tobacco Plant*

The wild-type tobacco (WT) and *AmGDSL1*-OE tobacco were cultured to consistent growth at approximately 6 weeks of age, then subjected to drought treatment by cessation of watering for 7 days. To investigate the drought resistance function of *AmGDSL1* gene in tobacco, the phenotypes of WT and *AmGDSL1*-OE plant lines were observed after 7 days of drought stress treatment. Three leaves from the same plant were used as three biological replicates. The activities of SOD (A001-3), POD (A084-3), CAT (A007-1-1), MDA (A003-1), and PRO (A107-1-1) in wild-type (WT) and *AmGDSL1*-OE plant lines at 0 d and 7 d of drought were evaluated using specific test kits (Nanjing Jiancheng Bioengineering Research Institute, Nanjing, China).

4.6. *Analysis of AmGDSL1 Gene Expression*

A. mongolicum leaves were collected after stress for 0 days (CK), 1 day, 2 days, 3 days, 5 days, 7 days, and fs1 days (fs1 d indicates the sample collected after 24 h of restoration in 1/5 Hoagland's nutrient solution) for the *AmGDSL1* gene expression under osmotic stress. *A. mongolicum* roots, stems, and leaves were collected to analyze the *AmGDSL1* gene expression in different tissues and organs. WT and *AmGDSL1*-OE plant leaves were collected after drought stress at 0 days and 7 days for the *AmGDSL1* gene expression in WT and transgenic tobacco. All samples from the same plant were used as three biological replicates. The total RNA was used to synthesize the cDNA of the genes using the FastQuant cDNA Kit KR106 (Tiangen, Beijing, China). The qRT-PCR analysis was conducted using an MonAmpTM SYBR® Green qPCR Mix (Monad, Suzhou, China) and the qRT-PCR primer sequence (*AmGDSL1*-q-F/R) list in Table S1. The *U6* gene was used as an internal reference to accurately normalize each reaction for the genes of the *A. mongolicum*, respectively. The relative levels of the *AmGDSL1* gene in *A. mongolicum* were calculated using the $2^{-\Delta\Delta CT}$ [39] method. The *AmGDSL1* gene expression in different organizations of *A. mongolicum* and transgenic tobacco were calculated using the $2^{-\Delta\Delta CT}$ [39] method.

4.7. *Statistical Analysis*

Data were collated and bar charts were drawn using Excel 2010 (Microsoft, Albuquerque, NM, USA). Data were statistically analyzed by one-way variance analysis by Duncan multiple range tests in SPSS 19.0 (IBM Inc, San Francisco, CA, USA) software. Differences were regarded as significant at $p < 0.05$. The Pearson correlation coefficients (r) were computed using R software (psych 2.2.9).

5. Conclusions

In this study, a typical *GDSL* gene *AmGDSL1* was cloned from *A. mongolicum* and functionally verified in overexpressing tobacco. *AmGDSL1* is a cytoplasm-localized protein.

We also examined the increase in *AmGDSL1* gene expression induced by PEG6000 treatment of *A. mongolicum*. *AmGDSL1* overexpression lines showed significantly increased resistance to dehydration stresses. This study explores the *GDSL* gene related to dehydration stresses in *A. mongolicum*, providing reference for crop genetic improvement. Future research may improve crop resistance in gene breeding.

Supplementary Materials: The following supporting information can be downloaded at: https://www.mdpi.com/article/10.3390/ijms25179467/s1.

Author Contributions: Y.M. conceived and designed the research. X.Y. and X.W. performed the experiments. X.Y., X.W., Y.M., F.S., H.N., X.D., X.L., Y.F., Y.Z. (Yongqing Zhai), Y.Z. (Yan Zhao) and B.F. analyzed the data and wrote the manuscript. All authors have read and agreed to the published version of the manuscript.

Funding: This study was funded by the National Natural Science Foundation of China (No. 32360338).

Institutional Review Board Statement: Not applicable.

Informed Consent Statement: Not applicable.

Data Availability Statement: Data are contained within the article.

Conflicts of Interest: The authors declare no conflict of interest.

Abbreviations

2,4-D	2,4-Dichlorophenoxyacetic acid
6-BA	6-Benzylaminopurine
CAT	Catalase
GELP	GDSL-type esterase/lipase
IAA	Indole acetic acid
Kana	Kanamycin
MDA	Malondialdehyde
NAA	1-Naphthaleneacetic acid
PCR	Polymerase chain reaction
POD	Peroxidase
PRO	Proline
qRT-PCR	Real-time quantitative PCR
Rif	Rifampicin
SOD	Superoxide dismutase
TMT	Timentin
WT	Wild-type tobacco

References

1. Yang, P.; Zhai, X.; Huang, H.; Zhang, Y.; Zhu, Y.; Shi, X.; Zhou, L.; Fu, C. Association and driving factors of meteorological drought and agricultural drought in Ningxia, Northwest China. *Atmos. Res.* **2023**, *289*, 106753. [CrossRef]
2. Nadarajah, K.N. ROS Homeostasis in Abiotic stress tolerance in plants. *Int. J. Mol. Sci.* **2020**, *21*, 5208. [CrossRef]
3. Li, X.; Wang, Q.; Sun, J.; Li, Z.; Wang, Y.; Yang, A. NtNAC053, A Novel NAC transcription factor, confers drought and salt tolerances in tobacco. *Front. Plant. Sci.* **2022**, *13*, 817106. [CrossRef] [PubMed]
4. Duan, D.; Yi, R.; Ma, Y.; Dong, Q.; Mao, K.; Ma, F. Apple WRKY transcription factor MdWRKY56 positively modulates drought stress tolerance. *Environ. Exp. Bot.* **2023**, *212*, 105400. [CrossRef]
5. Liu, J.; Wang, Y.; Ye, X.; Zhang, Q.; Li, Y.; Chen, M. Genome-wide identification and expression analysis of the WRKY gene family in response to low-temperature and drought stresses in *Cucurbita pepo* L. *Sci. Hortic.* **2024**, *330*, 113048. [CrossRef]
6. Zhu, C.; Lin, Z.; Liu, Y.; Li, H.; Di, X.; Li, T. A bamboo bHLH transcription factor PeRHL4 has dual functions in enhancing drought and phosphorus starvation tolerance. *Plant Cell Environ.* **2024**, *47*, 3015–3029. [CrossRef]
7. Li, C.; Yan, C.; Sun, Q.; Wang, J.; Yuan, C.; Mou, Y. The bHLH transcription factor AhbHLH112 improves the drought tolerance of peanut. *BMC Plant Biol.* **2021**, *21*, 540. [CrossRef]
8. Zhao, Y.; Cheng, X.Y.; Liu, X.D.; Wu, H.F.; Bi, H.H.; Xu, H.X. The Wheat MYB transcription factor TaMYB31 is involved in drought stress responses in Arabidopsis. *Front. Plant Sci.* **2018**, *9*, 1426. [CrossRef]

9. Li, M.J.; Qiao, Y.; Li, Y.Q.; Shi, Z.L.; Zhang, N.; Bi, C.L.; Guo, J.K. A R2R3-MYB transcription factor gene in common wheat (namely TaMYBsm1) involved in enhancement of drought tolerance in transgenic Arabidopsis. *J. Plant Res.* **2016**, *129*, 1097–1107. [CrossRef]
10. Mao, H.; Li, S.; Chen, B.; Jian, C.; Mei, F.; Zhang, Y. Variation in cis-regulation of a NAC transcription factor contributes to drought tolerance in wheat. *Mol. Plant* **2022**, *15*, 276–292. [CrossRef]
11. Mergby, D.; Hanin, M.; Saidi, M.N. The durum wheat NAC transcription factor TtNAC2A enhances drought stress tolerance in Arabidopsis. *Environ. Exp. Bot.* **2021**, *186*, 104439. [CrossRef]
12. Yang, Y.; Song, T.; Wang, Y.; Zhang, M.; Li, N.; Yu, M. The wheat WRKY transcription factor TaWRKY1-2D confers drought resistance in transgenic Arabidopsis and wheat (*Triticum aestivum* L.). *Int. J. Biol. Macromol.* **2023**, *226*, 1203–1217.
13. Wang, D.; Zhang, X.; Cao, Y.; Batool, A.; Xu, Y.; Qiao, Y. TabHLH27 orchestrates root growth and drought tolerance to enhance water use efficiency in wheat. *J. Integr. Plant Biol.* **2024**, *66*, 1295–1312. [CrossRef]
14. Shen, G.; Sun, W.; Chen, Z.; Shi, L.; Hong, J.; Shi, J. Plant GDSL Esterases/Lipases: Evolutionary, physiological and molecular functions in plant development. *Plants* **2022**, *11*, 468. [CrossRef]
15. Vincent, P.M.; Dae, Y.S. The functions and applications of GDSL esterase/lipase proteins in agriculture. *JSFA Rep.* **2022**, *2*, 304–312.
16. Akoh, C.C.; Lee, G.C.; Liaw, Y.C. GDSL family of serine esterase/lipases. *Prog. Lipid Res.* **2004**, *43*, 534–552. [CrossRef]
17. Lee, D.S.; Kim, B.K.; Kwon, S.J. Arabidopsis GDSL lipase 2 plays a role in pathogen defense via negative regulation of auxin signaling. *Biochem. Biophys. Res. Commun.* **2009**, *379*, 1038–1042. [CrossRef] [PubMed]
18. Liu, J.; Wang, H.; Khan, A.; Xu, Y.; Hou, Y.; Wang, Y. Genome wide identification of GDSL gene family explores a novel GhirGDSL26 gene enhancing drought stress tolerance in cotton. *BMC Plant Biol.* **2023**, *23*, 1–18. [CrossRef]
19. Suman, P.; Kumari, A.; Harsha, S.; Sapna, G.; Kishor, G. Understanding the structure and function of GDSL-type esterase/lipase genes in pigeon pea for their role under moisture stress conditions. *Plant Stress* **2023**, *10*, 100246.
20. Yun, J.; Zhang, J.; Pan, C. Genome-Wide Analysis of the GDSL Genes in Pecan (*Carya illinoensis* K. Koch): Phylogeny, structure, promoter cis-elements, co-Expression networks, and response to salt stresses. *Genes* **2022**, *13*, 1103. [CrossRef]
21. Jeum, K.H.; Hyong, W.C.; Sun, H.; Dae, S.; Nak, H.K.; Du, S.C. Function of a novel GDSL-type pepper lipase gene, CaGLIP1, in disease susceptibility and abiotic stress tolerance. *Planta* **2008**, *227*, 539–558.
22. Liu, D.; Zhao, X.; Liu, Y.; Tian, M.; Zhao, J.; Bai, N. The GDSL lipase CpGLIP1 from Chimonanthus praecox improves drought and cold tolerance in Arabidopsis and poplar. *Ind. Crop. Prod.* **2024**, *215*, 118636. [CrossRef]
23. Du, J.; Li, X.; Li, T. Genome-wide transcriptome profiling provides overwintering mechanism of *Agropyron mongolicum*. *BMC Plant Biol.* **2017**, *17*, 138–151. [CrossRef] [PubMed]
24. Ma, Y.H.; Yu, X.X.; Yu, Z.; Sun, F.C.; Li, X.D.; Li, X.L. RNA_Seq of Agropyron mongolicum Keng in response to drought stress. *Grassl. Sci.* **2018**, *64*, 3–15. [CrossRef]
25. Zhang, X.F.; Ma, Y.H.; Fan, B.B.; Sun, F.C.; Zhai, Y.Q.; Zhao, Y. The Agropyron mongolicum bHLH Gene *AmbHLH148* Positively Involved in Transgenic *Nicotiana benthamiana* Adaptive Response to Drought Stress. *Agronomy* **2023**, *13*, 2918. [CrossRef]
26. Ao, T.G.B.Y.; Gao, L.J.; Wang, L.C.; Li, Y.Q.; Zhao, Y.; Lang, M.L.; Yang, X.J. Cloning and expression analysis of AP2/EREBP transcription factor gene (MwAP2/EREBP) in Agropyron mongolicum Keng. *Nanosci. Nanotechnol. Lett.* **2017**, *9*, 2031–2038. [CrossRef]
27. Zhang, X.T.; Fan, B.B.; Yu, Z.; Nie, L.Z.; Zhao, Y.; Yu, X.X. Functional analysis of three miRNAs in Agropyron mongolicum Keng under drought stress. *Agronomy* **2019**, *9*, 661. [CrossRef]
28. Fan, B.B.; Sun, F.C.; Yu, Z.; Zhang, X.F.; Yu, X.X.; Wu, J. Integrated analysis of small RNAs, transcriptome and degradome sequencing reveal the drought stress network in *Agropyron mongolicum* Keng. *Front. Plant Sci.* **2022**, *13*, 976684. [CrossRef]
29. Chepyshko, H.; Lai, C.P.; Huang, L.M. Multifunctionality and diversity of GDSL esterase/lipase gene family in rice (*Oryza sativa* L. *japonica) genome: New insights from bioinformatics analysis. BMC Genom.* **2012**, *13*, 309.
30. Ren, R.; Yang, X.; Xu, J.H. Genome-wide identification and analysis of promising GDSL-type lipases related to gummy stem blight resistance in watermelon (*Citrullus lanatus*). *Sci. Hortic.* **2021**, *289*, 110461. [CrossRef]
31. Volokita, M.; Rosilio, T.; Rivkin, N. Combining comparative sequence and genomic data to ascertain phylogenetic relationships and explore the evolution of the large GDSL-lipase family in land plants. *Mol. Biol. Evol.* **2011**, *28*, 551–565. [CrossRef]
32. Cao, Y.; Han, Y.; Meng, D. Expansion and evolutionary patterns of GDSL-type esterases/lipases in Rosaceae genomes. *Funct. Integr. Genom.* **2018**, *18*, 673–684. [CrossRef] [PubMed]
33. Ji, R.; Wang, H.; Xin, X. BrEXL6, a GDSL lipase gene of Brassica rapa, functions in pollen development. *Biol. Plant.* **2017**, *61*, 685–692. [CrossRef]
34. Zhang, P.L.; Zhang, H.; Du, J.K. Genome-wide identification and coexpression analysis of GDSI. genes related to suberin formation during fruit russeting in pear. *Hortic. Plant J.* **2022**, *8*, 153–170. [CrossRef]
35. Su, H.G.; Zhang, X.H.; Wang, T.T. Genome-wide identification, evolution, and expression of GDSL-type esterase/lipase gene family in soybean. *Front. Plant Sci.* **2020**, *11*, 726. [CrossRef] [PubMed]
36. Simone, L.; Jean, F.H.; Gea, G. Abiotic stress: Focus on drought and salt stress, recent insights and perspectives. *Front. Plant Sci.* **2017**, *8*, 1214.
37. Xiao, C.L.; Guo, H.M.; Tang, J. Expression pattern and functional analyses of Arabidopsis guard cell-enriched GDSL lipases. *Front. Plant Sci.* **2021**, *12*, 748543. [CrossRef] [PubMed]

38. Tang, J.; Yang, X.; Xiao, C.L. GDSL lipase occluded stomatal pore1 is required for wax biosynthesis and stomatal cuticular ledge formation. *New Phytol.* **2020**, *228*, 1880–1896. [CrossRef] [PubMed]
39. Livak, K.J.; Schmittgen, T.D. Analysis of relative gene expression data using real-time quantitative PCR and the $2^{-\Delta\Delta CT}$ method. *Methods* **2001**, *25*, 402–408. [CrossRef]

International Journal of
Molecular Sciences

MDPI

Article

StEPF2 and StEPFL9 Play Opposing Roles in Regulating Stomatal Development and Drought Tolerance in Potato (*Solanum tuberosum* L.)

Le Kang [1,2,*,†], Junke Liu [3,†], Hongqing Zhu [4], Leqin Liao [1], Muying Ye [3], Yun Wei [1], Nairong Liu [1], Qingbo Ke [3], Ho Soo Kim [2], Sang-Soo Kwak [2] and Quanlu Zhou [4,*]

1 Key Laboratory of Nanchong City of Ecological Environment Protection and Pollution Prevention in Jialing River Basin, College of Environmental Science and Engineering, China West Normal University, Nanchong 637002, China
2 Plant Systems Engineering Research Center, Korea Research Institute of Bioscience and Biotechnology, Daejeon 34141, Republic of Korea
3 Institute of Soil and Water Conservation, Northwest A&F University, Yangling 712100, China
4 Sweetpotato Research Institute, Nanchong Academy of Agricultural Sciences, Nanchong 637000, China
* Correspondence: kangle@cwnu.edu.cn (L.K.); zhouquanlu@163.com (Q.Z.)
† These authors contributed equally to this work.

Abstract: Stomata are essential for photosynthesis and water-use efficiency in plants. When expressed in transgenic *Arabidopsis thaliana* plants, the potato (*Solanum tuberosum*) proteins EPIDERMAL PATTERNING FACTOR 2 (StEPF2) and StEPF-LIKE9 (StEPFL9) play antagonistic roles in regulating stomatal density. Little is known, however, about how these proteins regulate stomatal development, growth, and response to water deficit in potato. Transgenic potato plants overexpressing *StEPF2* (E2 plants) or *StEPFL9* (ST plants) were generated, and RT-PCR and Western blot analyses were used to select two lines overexpressing each gene. E2 plants showed reduced stomatal density, whereas ST plants produced excessive stomata. Under well-watered conditions, ST plants displayed vigorous growth with improved leaf gas exchange and also showed increased biomass/yields compared with non-transgenic and E2 plants. E2 plants maintained lower H_2O_2 content and higher levels of stomatal conductance and photosynthetic capacity than non-transgenic and ST plants, which resulted in higher water-use efficiency and biomass/yields during water restriction. These results suggest that StEPF2 and StEPFL9 functioned in pathways regulating stomatal development. These genes are thus promising candidates for use in future breeding programs aimed at increasing potato water-use efficiency and yield under climate change scenarios.

Keywords: potato; EPIDERMAL PATTERNING FACTOR; stomatal density; water-use efficiency; plant growth

1. Introduction

Potato (*Solanum tuberosum* L.) is the third most consumed staple crop across the world. It is, however, a drought-susceptible plant that favors a cool climate, which makes it difficult to maintain yields given the increased occurrence of extreme weather events, including more and longer periods of low precipitation and high temperature; simultaneously, the growing world population is causing an increased demand for agricultural products [1]. These factors present grave challenges and require plant breeders to develop novel varieties of potatoes with improved stress resistance in response to the detrimental effects of climate change.

Stomata are microscopic pores in the epidermis of the aerial parts of plants that act to maximize gas diffusion, enabling carbon assimilation by photosynthesis, while minimizing water loss through transpiration [2]. Plants must therefore balance gas exchange with transpiration. They do so by altering the apertures of their stomata in response to various environmental cues, while also, over the long term, modifying the stomatal density in developing leaves [1,3]. Natural variation in the anatomical characteristics of stomata provides evidence of a potentially negative relationship between stomatal density and plant water-use efficiency (WUE) [4–6]. Furthermore, some plant species are able to maintain biomass yields at high temperatures due to increased stomatal density, which may aid transpiration and thus improve the process of cooling [7]. Therefore, altering the stomatal density is a potential method for developing crops that are resilient to climate change [8].

Stomatal development involves a series of cell fate transitions and differentiation processes that are regulated by a network of signaling pathways and regulatory factors, including basic helix–loop–helix (bHLH) transcription factors (TFs), the EPIDERMAL PATTERNING FACTOR (EPF)/EPF-LIKE (EPFL) ligand family, TOO MANY MOUTHS (TMMs), the ERECTA receptor complex, and the mitogen-activated protein kinase (MAPK) signaling cascade [9–11]. The ligands–receptors–MAPKs/TFs integrate intrinsic and extrinsic signals, forming a negative feedback loop that precisely controls stomatal development and distribution [12,13]. The EPF/EPFLs are a family of cysteine-rich intracellular signaling peptides that maintain the correct density and spacing of stomatal precursor cells by binding to ERECTA family receptors [14]. The *Arabidopsis thaliana* (Arabidopsis) genome contains, in total, 11 members of the EPFL family; three of these peptides, EPF1, EPF2, and EPFL9/STOMAGEN, play essential roles in stomatal development [15,16]. EPF1 and EPF2 are both secreted from stomatal lineage cells and negatively regulate stomatal density, while EPFL9 is derived from mesophyll cells and promotes stomatal development by competing with EPF2 for binding to the ERECTA receptor [16,17]. *EPFL* genes are conserved across many species of land plants, including both monocots and dicots, and C3 and C4 plants. Genetic manipulation or mutation of the expression patterns of EPF1, EPF2, and EPFL9 alters stomatal development, enabling the desired photosynthetic and transpiration traits to be obtained in Arabidopsis [18,19], rice (*Oryza sativa*) [20], wheat (*Triticum aestivum*) [21], maize (*Zea mays*) [22], and poplar (*Populus* spp.) [23]. Such studies encouraged us to explore whether the orthologous genes in potato have a similar function in the regulation of stomatal development in this crop species.

We reported previously that ectopic overexpression of potato *EPF2* (*StEPF2*) in Arabidopsis resulted in reduced stomatal density and enhanced drought tolerance, whereas plants overexpressing *StEPFL9* displayed the contrary phenotypes [24,25]. In the current study, we wished to determine if the StEPF2 and StEPFL9 function in the regulation of stomatal development was conserved in potato and to determine the impact of manipulating stomatal density on plant growth and drought tolerance. We therefore generated transgenic potato plants overexpressing either *StEPF2* or *StEPFL9* and investigated the effect of changes in the stomatal density on photosynthesis, drought tolerance, potato growth, and tuber yields.

2. Results

2.1. Generation of Transgenic Potato Plants

As shown in Figures S1 and 1A,B, StEPF2 and StEPFL9 proteins were predicted to be composed of three distinct parts: the N-terminal signal peptide, the pro-peptide domain, and the mature peptide, in which six cysteine residues formed a cyclic structure comprising three pairs of disulfide bonds. As shown in Figure S2, seven lines of transgenic plants overexpressing *StEPF2* (E2) and ten lines overexpressing *StEPFL9* (ST) were generated and we used genomic PCR analyses to confirm the presence of the transgene in each line. Subsequently, we selected four lines (E2-1, E2-4, ST-6, and ST-10) for further characterization by qRT-PCR and Western blot analyses (Figure 1C,D).

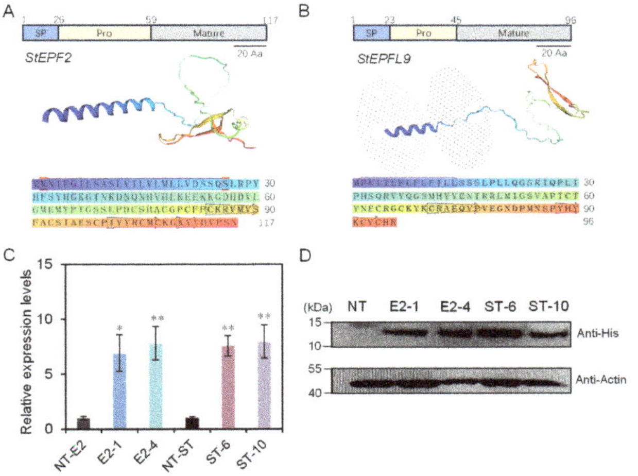

Figure 1. Bioinformatics analysis of the StEPF2 and StEPFL9 proteins. Structural models of (**A**) StEPF2 and (**B**) StEPFL9. SP: signal peptide; Pro: pro-peptide: Mature: mature peptide. (**C**) qRT-PCR analysis of *StEPF2* and *StEPFL9* expression; the potato gene *StEF1α* and *actin* were used as an internal control. (**D**) Western blot analysis of non-transgenic (NT) and transgenic plants. Data show the mean ± SE. Asterisks indicate significant differences between transgenic and NT lines by Duncan's multiple range test; *: $p < 0.05$; **: $p < 0.01$.

2.2. Characterization of Transgenic Potato Plants

As shown in Figure 2A, the stomatal density was reduced by 30.7% in the E2 lines, compared with NT plants, but stomatal density increased by 58.8% in the ST lines (Figure 2B), suggesting that StEPF2 and StEPL9 regulated stomatal density antagonistically in potato. As shown in Figure 3A–C, there was no significant difference in *Pn* under 100 μmol $m^{-2}s^{-1}$ or lower light conditions. The photosynthetic parameters, including *Pmax*, *Isat*, *Ic*, *Cisat*, *Amax*, and Γ, significantly increased in the ST lines, compared with NT plants, whereas they were significantly reduced in the E2 lines. Notably, the values of *CE* and *AQY* in the E2 lines did not differ significantly from those of NT plants, but were higher in the ST lines (Table S2). As shown in Figure 3D, the total water loss in the E2 lines was reduced by 3.5%, compared with NT plants, whereas it increased by 2.15% in the ST lines.

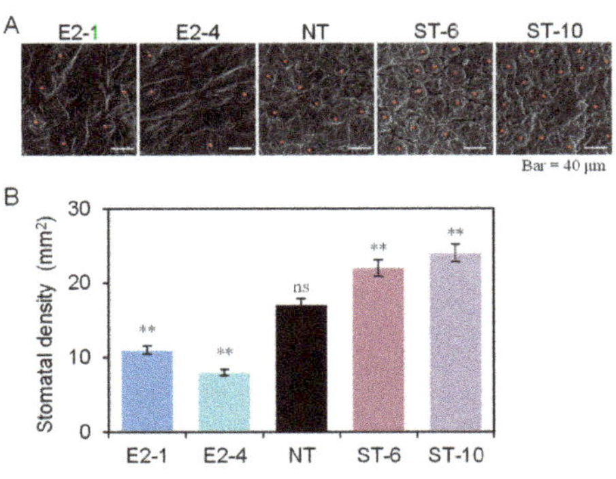

Figure 2. Phenotypic analyses of stomatal density in transgenic potato plants. (**A**) Photographs of the mature abaxial leaf epidermis of 3-week-old plants showing stomata. Scale bars: 40 μm. Red dots denote positions of stomatal complexes. (**B**) Stomatal density of non-transgenic (NT) and transgenic plants. Data show the mean ± SE. Asterisks indicate significant differences between transgenic and NT lines by Duncan's multiple range test; ns: no significant difference; **: $p < 0.01$.

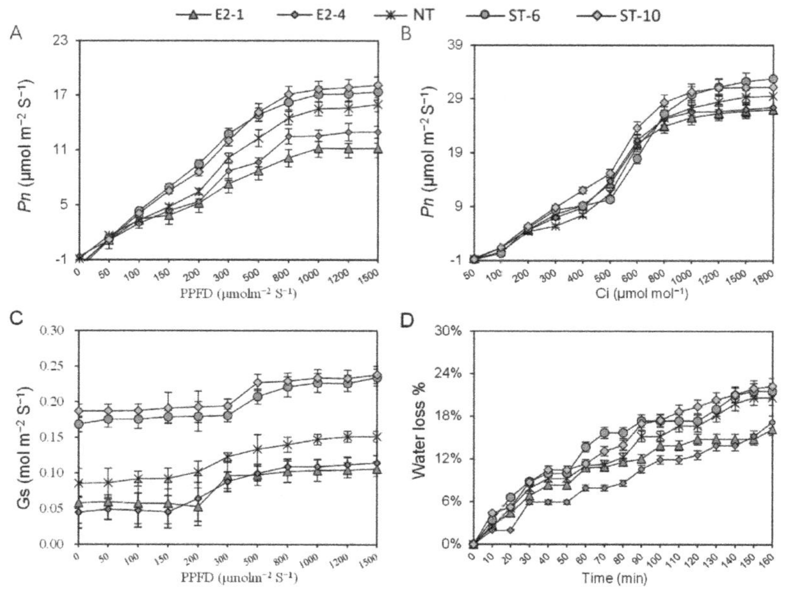

Figure 3. Photosynthetic responses of non-transgenic (NT) and transgenic potato plants. (**A**) *Pn-PPFD* curve. (**B**) *Pn-Ci* curve. (**C**) *Gs-PPFD* curve. (**D**) Water loss from detached leaves of 3-week-old NT and transgenic plants. Data show the mean ± SE. Leaves from the same position on each plant were used in this experiment.

2.3. Relationship between Stomatal Density and Drought Tolerance in Transgenic Potato Plants

As shown in Figure 4A–C, prior to drought treatment, increased levels of *Pn* and Fv/Fm were observed in the ST lines, but not in the E2 lines, compared with NT. Under persistent drought conditions, the leaves of the ST lines exhibited obvious wilting; the values of *Pn* and Fv/Fm in the ST lines decreased significantly compared with NT plants, whereas the values of *Pn* and Fv/Fm increased by 13.4% and 31.9%, respectively, in the E2 lines (Figure 4B,C). The H_2O_2 content in detached leaves of the ST lines was 34.5% higher than in NT plants, whereas it decreased by 22% in the E2 lines, which also accumulated lower levels of H_2O_2, visualized by 3,3′-diaminobenzidine (DAB) staining (Figures 4D and S3).

2.4. Growth of Transgenic Potato Plants under Long-Term Drought Stress

As shown in Figure 5, when compared with NT plants, the values of *Pn*, *Gs*, *Tr*, iWUE, and Fv/Fm in the ST lines in the well-watered condition increased significantly by 26.7%, 22.6%, 11.4%, 7.6%, and 2.1%, respectively; by contrast, the values of these parameters decreased by 7.5%, 36.2%, 18.0%, 4.7%, and 1.3%, respectively, in well-watered plants from the E2 lines. Y(NPQ) increased by 1.2% in the E2 lines but decreased in the ST lines, relative to NT plants (Figure 5E). Under water-restricted conditions, the values of *Pn*, *Gs*, *Tr*, iWUE, and Fv/Fm in the ST lines showed 1.09-, 6.20-, 1.28-, 2.20-, and 1.07-fold decreases, respectively, compared with NT plants, whereas these parameters showed 2.60-, 2.70-, 2.10-, 1.56-, and 1.10-fold increases, respectively, in the E2 lines. Conversely, the ST lines exhibited a significant increase (5%) in Y(NPQ), compared with NT plants, while the E2 lines showed a 10.2% reduction (Figure 5).

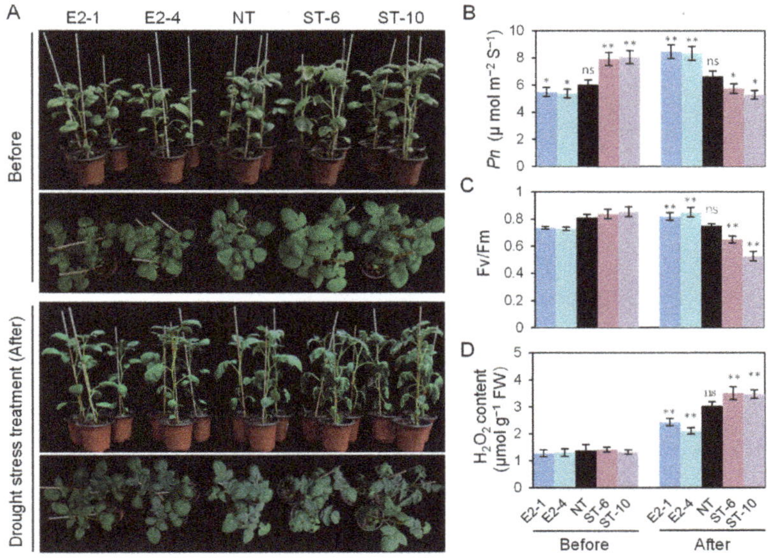

Figure 4. Phenotypic analyses of non-transgenic (NT) and transgenic potato plants under short-term drought stress conditions. (**A**) Appearance of plants before and after drought treatment. (**B**) *Pn*, (**C**) Fv/Fm, and (**D**) H_2O_2 content in non-transgenic (NT) and transgenic plants. Leaves from the same position on each plant were used in this experiment. Data show the mean ± SE. Asterisks indicate significant differences between transgenic and NT lines by Duncan's multiple range test; ns: no significant difference; *: $p < 0.05$; **: $p < 0.01$.

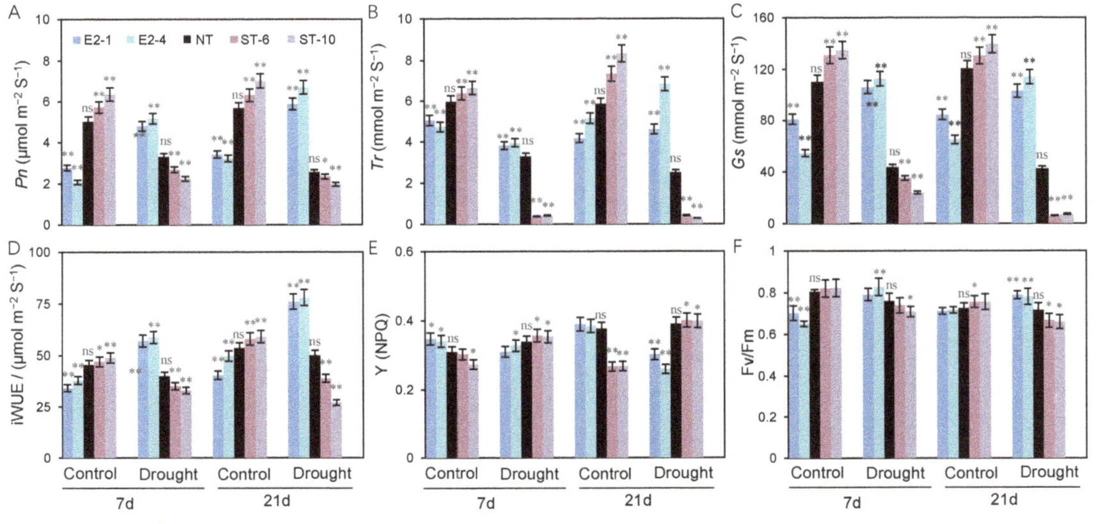

Figure 5. Photosynthetic and chlorophyll fluorescence parameters of transgenic potato plants under long-term drought stress. (**A**) *Pn*, (**B**) *Tr*, (**C**) *Gs*, (**D**) iWUE, (**E**) Y(NPQ), and (**F**) Fv/Fm in non-transgenic (NT) and transgenic plants after 7 or 21 days under well-watered and water-restricted conditions. Leaves from the same position on each plant were used in this experiment. Data show the mean ± SE. Asterisks indicate transgenic and NT lines differed significantly by Duncan's multiple range test; ns: no significant difference; *: $p < 0.05$; **: $p < 0.01$.

Under well-watered conditions, the ST lines grew more vigorously than the NT and E2 plants (Figure 6A), and the daily water consumption, biomass, and yield of the ST lines increased by 23.6%, 9.7%, and 20.6%, compared with these of NT plants, but decreased by 10.2%, 7.1%, and 9.7% in the E2 lines, respectively (Figure 6B–D). Conversely, under water-restricted conditions, compared with NT plants, the biomass and yield of the E2 lines increased by 19.9% and 32.9%, but decreased by 14.2 and 28.3% in the ST lines, respectively (Figure 6B,D). Interestingly, under water-restricted conditions, the daily water consumption of the ST lines increased by 33.8% over that of NT plants, whereas that of the E2 lines decreased by 22.7% (Figure 6C).

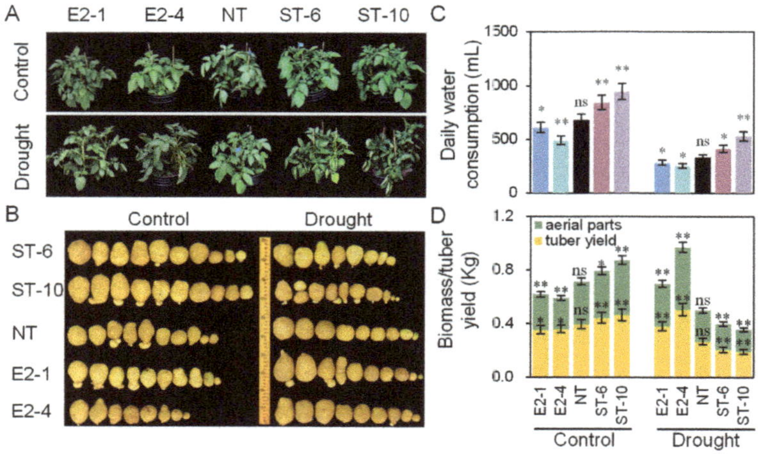

Figure 6. Effect of long-term drought treatment on production of NT and transgenic potato plants. (**A**) Phenotypes of 2-month-old aerial parts of NT, E2, and ST plants under well-watered and water-restricted conditions. (**B**) Phenotypes of tubers produced by NT and transgenic potato plants after harvest. (**C**) Daily water consumption of each line during drought treatment. (**D**) Biomass/tuber yield of NT and transgenic potato plants. Data show the mean ± SE, and each sample (three plants per pot) was replicated five times. Asterisks indicate transgenic and NT lines differed significantly by Duncan's multiple range test; ns: no significant difference; *: $p < 0.05$; **: $p < 0.01$.

3. Discussion

Stomata play a vital role in ensuring plant survival by regulating gas exchange, water loss, and pathogen entry [26]. Our knowledge of the networks of genes implicated in potato stomatal development remains limited, however. In the current study, genetic manipulation of stomatal density was achieved via regulating two EPF orthologues in potato (StEPF2 and StEPFL9) to investigate the impact of stomatal density on potato growth and water-use efficiency.

Several lines of evidence indicate that StEPF2 and StEPFL9 are functional orthologues of EPF2 and EPFL9 from other plants. Mature EPF proteins possess a distinctive knotted structure scaffold, in which three disulfide bonds form between a set of conserved cysteine (C) residues. A variable loop region between C4 and C5 determines the antagonistic activity between EPF2 and EPFL9 [27]. Our study found that the amino acid sequences of StEPF2 and StEPFL9 were highly conserved and closely resembled those from other species. The proteins contained a signal peptide at the N terminus, a pro-peptide domain, and the amino acid residues of the mature protein. Stomatal density was decreased in potato plants overexpressing *StEPF2* and increased in potato plants overexpressing *StEPFL9*, consistent with previous studies of *EPF2* and *EPFL9* orthologues in other plants. Studies manipulating the expression of the *EPF2* and *EPFl9* genes in a range of plant species, including *AtEPF2* (Arabidopsis) [18], *PdEPF2* (white poplar) [28], *OsEPF2* (rice) [20],

SbEPF2 (sorghum) [29], *TaEPF2* (bread wheat) [30], *MdEPF2* (apple) [31], *BnaEPF2* (Brassica napus) [32], *AtEPFL9* [19], *OsEPFL9* [20], and *VvEPFL9* (grapevine) [33], have shown their products, EPF2 and EPFL9, play antagonistic roles in modulating stomatal development. These results provide strong evidence that the *EPF* genes in potato were, at least in part, highly conserved, both structurally and functionally.

It is predicted that, due to climate change, extreme weather events, such as drought and high temperature, will occur at an increasingly high frequency; this will seriously affect the growth and yield of future crops [34]. Stomata are the main facilitators of transpiration as, typically 97% of absorbed water escapes though stomata to the atmosphere, but they are also essential for the uptake of the atmospheric CO_2 used in mesophyll photosynthesis [26]. Plants with lower stomatal density are usually more drought-tolerant and show a higher efficiency in water use under water-deficit conditions [35]. We found that potato plants with reduced stomatal density maintained relative high rates of photosynthesis (*Pn*), transpiration (*Tr*), stomatal conductance (*Gs*), and intrinsic WUE (iWUE) under drought stress treatment. Similarly, genetic manipulation or mutations that reduced stomatal density improve WUE and drought tolerance in several monocot species, including rice [36], sorghum [29], barley [37], and wheat [30], as well as in dicot species such as Arabidopsis [5], poplar [38], grapevine [33], tomato (*Solanum lycopersicum*) [31], and B. napus [32]. Taken together, these findings suggest that reducing the stomatal density of leaves, either by overexpressing *EPF2* or downregulating the expression of *EPFL9*, was an effective strategy for improving drought tolerance and WUE in plants. Conversely, plants with increased stomatal density were likely to exhibit enhanced Gs, carbon dioxide assimilation, and *Pn*. Overexpression of *EPFL9* in Arabidopsis [39], rice [20], poplar [40], and potato (this study), as well as disruption of the Arabidopsis *EPF2* orthologue, increased stomatal density and also enhanced *Pn*, *Gs*, and *Tr* under well-watered conditions. The carboxylation capacity did not differ between *AtEPFL9* overexpression lines and wild-type Arabidopsis plants, which suggested that increasing stomatal density altered photosynthetic capacity by modulating CO_2 diffusion rather than Rubisco activity [39]. The rate of water loss in detached leaves of E2 plants was higher than that in NT and E2 plants, in which lower water potential might have caused stomatal closure. The E2 plants showed consistently lower *Tr* and *Gs* under water-restricted conditions and were much more susceptible to wilting than NT and E2 plants.

Although decreasing stomatal density is an efficient strategy for improving plant tolerance to drought, a reduction in stomatal conductance may affect the rate of CO_2 assimilation, thus reducing biomass production due to the restriction on gas exchange [35]. Investigations in several grass species, however, indicate that a reduction in stomatal density may improve both WUE and drought tolerance without having a negative impact on yields. Overexpression of *HvEPF1* in barley [37], of *OsEPF1* in rice [36], and of *TaEPF2* in bread wheat [30] increases plant resistance to reduced water availability without affecting yield, when the reduction in the number of stomata is moderate. In rice, only lines with a marked reduction (88%) in stomatal densities show a statistically significant decrease in CO_2 assimilation [36]; similarly, wheat lines with >50% reductions in stomatal density were highly susceptible to yield losses [30]. In the current study, the biomass and tuber yield were reduced under well-watered conditions in potato plants with decreased stomatal density, but biomass production and tuber yield were relatively high when the plants were exposed to water-restricted conditions. Several studies that compared stomatal opening and closing responses between grass species with subsidiary cells and those with kidney-shaped stomata suggest that grasses exhibit faster and more efficient stomatal regulation than other plants; this is likely to improve WUE and reduce the risk of disruption to the leaf hydraulic system in grass species [35]. E2 potato plants displayed an enhanced rate of plant growth under optimal growth conditions; this was consistent with a previous study of transgenic poplar that showed plants overexpressing *PagEPFL9* exhibit increased plant height and biomass [40]. It is possible that high stomatal density increases the potential rate of gas exchange, which would have beneficial effects on CO_2 assimilation and also ensure a relatively stable leaf temperature by enabling evaporative cooling to dissipate heat at times of high temperature.

4. Materials and Methods

4.1. Plant Materials and Growth Conditions

All potato plants used in this study were of the Atlantic cultivar (*Solanum tuberosum* L. cv. Atlantic, which is model cultivar for potato transformation and functional analysis of genes and a drought-tolerant cultivar [41]). Plants were sub-cultured monthly by transferring stem cuttings with double nodes to fresh rooting medium (RM; i.e., 1 × Murashige and Skoog (MS) medium). Two groups of experiments were conducted. For short-term measurements of drought stress, following culture on RM for one-month, fifteen rooted plantlets for each line were transplanted into each pot (8 × 9 × 5.5 cm) filled with sterile matrix, vermiculite, and perlite (3:1:1) and placed in a growth chamber at 25 °C, 60% relative humidity, and a 16/8 h (light/dark) photoperiod with a light intensity of 100 μmol m^{-2} s^{-1}.

For long-term drought treatments, fifteen rooted potato plants for each line were transplanted into plastic pots (30 cm diameter × 29 cm height; three seedlings per pot) after a 7-day acclimatization period, cultivated at Yangling (34°17′00″ N, 108°03′42″ E), China. The highest temperature during the period of cultivation was 27 °C, and the lowest was 12.75 °C; shelters were used to keep out the rain. The soil was calcic cambisol [42], pH 8.21, with an organic matter content of 2.26 g kg^{-1}, and total nitrogen (N), phosphorus (P), and potassium (K) contents were 0.22, 0.54, and 0.23 g kg^{-1}, respectively. N, P, and K (0.25, 0.17, and 0.23 g kg^{-1} dried soil) were applied as base fertilizers. Pot weight was measured on a daily basis to determine the percentage of the initial soil water content remaining, as well as water consumption. The soil water content was maintained at 80% in the well-watered control group; fifteen potato plants in tuber initiation stage were exposed to drought treatment; the relative soil water content was allowed to fall progressively for the first 7–8 days and then maintained at approximately 40% until harvest using a daily water supplement (Figure S4).

4.2. Bioinformatics Analysis

BLAST searches were conducted using the GenBank database to identify the nucleotide sequences of EPF2/EPFL9 family members. The online programs ClustalW and ESPript 3.0 (https://espript.ibcp.fr, accessed on 6 June 2024) were used to perform multiple sequence alignments (Figure S1). SignalP 4.1 (http://www.cbs.dtu.dk/services/signalp/, accessed on 6 June 2024) was used to predict and analyze the signal peptides. The three-dimensional structures of the proteins were simulated using SWISS MODEL (https://swissmodel.expasy.org/interactive, accessed on 6 June 2024).

4.3. Potato Transformation

To investigate the function of StEPF2 and StEPFL9 in regulating stomatal density in potato plants, the construct pCAMBIA1380-35S:StEPF2/StEPFL9 was introduced into potato plants using *Agrobacterium tumefaciens* (strain EHA105)-mediated transformation, as described by Ahmad et al. (2008) [43]. The stem internodes from thirty 3-week-old cultured potato seedlings for each gene were transformed; explants were pre-cultured in a culture dish for 2 days and then inoculated with *A. tumefaciens* for 15 min. After incubation for 2 days on co-culture medium, comprising 1 × MS medium containing 2.0 mg/L 2,4-D, the explants were transferred to shoot induction medium containing 2.0 mg/L zeatin, 0.1 mg/L GA3, 0.01 mg/L NAA, and 5 mg/L hygromycin. Hygromycin-resistant shoots were transferred to fresh RM for rooting. Transformation of the rooted plantlets was confirmed using PCR.

4.4. Gene Expression in Potato Plants

The PhirePlant Direct PCR kit (Thermo Scientific, Waltham, MA, USA) was used to analyze genomic DNA. For quantitative real-time (qRT)-PCR analyses, total RNA was extracted from six 3-week-old potato plants using TRIzol reagent and reverse-transcribed using the PrimeScript II 1st strand cDNA synthesis kit (Takara, Beijing, China). qRT-PCR was performed using a QuantStudio™ 5 Real-Time PCR System with SYBR Premix Ex Taq Kit (Takara, Beijing, China) and the gene-specific primers listed in Table S1. The relative level of expression of each gene of interest was quantified using the $2^{-\Delta\Delta CT}$ method with *StEF1a* and *actin* as the internal reference genes [44]. All reactions were performed in triplicate using three independent cDNA samples.

4.5. Protein Immunoblotting

Total protein was extracted from six 3-week-old potato plants grown in pots. Approximately 0.2 g of ground leaf samples was combined with extraction solution containing 50 mM Tris-HCl (pH 7.5), 5 mM EDTA, 5 mM EGTA, 10 mM DTT, and one protease inhibitor tablet (Roche, Mannheim, Germany). Proteins were separated by SDS-PAGE and electro-transferred onto a polyvinylidene fluoride membrane. Membranes were blocked in TBST buffer with 6% skimmed milk at 4 °C overnight. For the immunoblot analyses, we used anti-Actin and anti-His as the primary antibodies and anti-IgG as the secondary antibody (Sangon, Shanghai, China).

4.6. Stomatal Density

To assess the effects of overexpression of *StEPF2* and *StEPFL9* on stomatal density, stomata on the abaxial leaf epidermis of transgenic potato plants were observed using a scanning electron microscope. Briefly, leaf (3rd fully expanded leaves from the top) samples from 3-week-old potato plants were fixed in 4% glutaraldehyde in 0.1 M phosphate buffer (pH 7.2) for 4 h and dehydrated across a gradient of 30% to 100% ethanol; samples were placed in each concentration for 15 min. Finally, samples were moved through three changes of 100% tert-butanol. Samples were oven-dried at 40 °C overnight and coated with a thin layer of gold prior to scanning electron microscopy analysis (TCS SP8, LEICA, Germany). Eight leaf discs (two leaf discs per leaf) per potato line were measured. There were three biological repeats per data point.

4.7. Leaf Gas Exchange and Chlorophyll Fluorescence Measurements

To determine whether the changes in stomatal density affected gas exchange, we measured the Pn-PPFD, Gs-PPFD, and Pn-Ci response curves in fully expanded leaves. Briefly, gas exchange was measured on sunny days in 3rd fully expanded leaves (from the top) from 3-week-old plants using a Li-6800 portable gas exchange system (LI-COR, Lincoln, NE, USA). Leaves located at the same position on each plant were used in this experiment. The environmental conditions within the leaf chamber were light intensity of PPFD = 1000 μmol $m^{-2}s^{-1}$, 50% humidity, 25 °C leaf temperature, 500 μmol s^{-1} airflow, and 400 ppm CO_2. The light intensity gradients were 1500, 1000, 800, 600, 400, 300, 200, 150, 100, 50, and 0 μmol m^{-2} s^{-1}. The net photosynthetic rate (*Pn*), stomatal conductance rate (*Gs*), and transpiration rate (*Tr*) were measured simultaneously. The instantaneous WUE (iWUE) of the leaf was calculated as the rate of *Pn* to *Tr*.

The *Pn-Ci* response curve was generated by measuring the photosynthetic rate in leaves exposed to 400, 200, 100, 50, 300, 500, 600, 800, 1000, 1200, 1500, and 1800 ppm CO_2. The modified rectangular hyperbola model was used to match the Pn-Ci response curve and to calculate the initial carboxylation efficiency (*CE*), photosynthetic capacity (*Amax*), saturated intercellular CO_2 concentration (*Cisat*), CO_2 compensation point (Γ), rate of photorespiration (*Rp*), apparent quantum efficiency (*AQY*), light saturation point (*Isat*), dark respiration rate (*Rd*), and light compensation point (*Ic*). The quantum efficiency of photosystem II (PSII) was determined in expanded leaves after dark adaptation for 30 min by calculating the ratio of variable fluorescence to maximum fluorescence (Fv/Fm), and non-photochemical quenching (NPQ) of leaves was measured using the MINI-PAM-II chlorophyll fluorescence system (Walz, Effeltrich, Germany). All experiments were three biological repeats with five leaves in each replicate.

4.8. Quantification and Detection of Hydrogen Peroxide (H_2O_2)

The 3rd leaves from the top were collected from 3-week-old potato plants cultured in pots and ground in liquid nitrogen. H_2O_2 levels were measured using the Hydrogen Peroxide Assay Kit (Comin, Nanjing, China), according to the manufacturer's protocol. There were five biological repeats with three leaves in each replicate.

4.9. Measurement of Water Loss from Detached Leaves

To assess the effect of the alterations in stomatal density on water loss, water loss was measured in detached leaves (3rd leaves from the top) from 3-week-old potted seedlings. Nail polish was applied to the stem cross-section to prevent transpiration. The weight of the detached leaves was measured at 10 min intervals; water loss between 0 and 3 h was calculated using (initial weight–test weight)/initial weight × 100%; this process was repeated five times for each group of leaves with three leaves in each replicate.

4.10. Measurement of Daily Water Consumption, Biomass, and Tuber Yield

To assess the changes in yield of transgenic potato plants under long-term drought stress conditions, we evaluated the biomass and tuber yields and calculated the mean daily water consumption following a period of drought treatment. Briefly, the daily water consumption of NT and transgenic lines was equivalent to the average amount of water supplement per pot during the drought treatment. The growth cycle of potato plants lasted 84 days; the aerial parts and tubers were harvested, and their fresh weights were immediately measured. Each sample (three plants per pot) was replicated five times.

4.11. Statistical Analyses

Statistical analyses were performed using the SPSS version 26.0 (SPSS, Chicago, IL, USA). One-way ANOVA with Duncan's multiple comparison was used to test for differences between lines of plants. All values are presented as means ± standard error (SE). Statistically significant differences between means of non-transgenic (NT) and transgenic plants are indicated by asterisks: *: $p < 0.05$; **: $p < 0.01$.

5. Conclusions

To conclude, our results show that StEPF2 and StEPFL9 play opposing roles in regulating stomatal development and potato plants with reduced stomatal density improved drought tolerance under deficit conditions, while potato plants with increased stomatal density enhanced plant growth and tuber yields under well-watered conditions. This study increases our understanding of the complex regulatory mechanism underlying stomatal development and demonstrates the correlations among stomatal density, plant growth, and drought tolerance in potato. The potato plants developed in this study must be further evaluated in the field under naturally fluctuating environmental conditions. Furthermore, despite the restrictions on the cultivation of GM plants imposed by many countries, stomatal development can be modified via exogenous application of synthetic EPF peptides. In future, the manipulation of stomatal density may be a useful way for plant breeders to optimize potato yield in face of the challenge posed by climate change.

Supplementary Materials: The following supporting information can be downloaded at https://www.mdpi.com/article/10.3390/ijms251910738/s1.

Author Contributions: Conceptualization, L.K. and Q.Z.; methodology, L.K. and J.L.; software, J.L. and H.Z.; formal analysis, J.L., H.Z., L.L., M.Y., Y.W. and N.L.; resources, Q.K., H.S.K. and S.-S.K.; writing—original draft preparation, L.K., J.L. and Q.Z.; writing—review and editing, L.K., Q.K. and Q.Z. All authors have read and agreed to the published version of the manuscript.

Funding: This research was funded by China Scholarship Council, National Natural Science Foundation of China (31901588), Natural Science Foundation of Sichuan Province (2024NSFSC1216), Natural Science Foundation of Shaanxi Province (2024JC-YBMS-178), Breeding Research Project of Sichuan Provincial Department of Science and Technology (2021YFYZ0020), Sichuan Innovation Team for

Tuber Crops (SCCXTD-2020-09), and the Korea Research Institute of Bioscience and Biotechnology (KRIBB) Research Initiative Program (KGM1002412 and KGM5372423).

Institutional Review Board Statement: Not applicable.

Informed Consent Statement: Not applicable.

Data Availability Statement: No large datasets were created in this study.

Conflicts of Interest: The authors declare no conflicts of interest.

References

1. Dahal, K.; Li, X.Q.; Tai, H.; Creelman, A.; Bizimungu, B. Improving potato stress tolerance and tuber yield under a climate change scenario—A current overview. *Front. Plant Sci.* **2019**, *10*, 563. [CrossRef]
2. Gupta, A.; Rico-Medina, A.; Caño-Delgado, A.I. The physiology of plant responses to drought. *Science* **2020**, *368*, 266–269. [CrossRef]
3. Casson, S.; Gray, J.E. Influence of environmental factors on stomatal development. *New Phytol.* **2008**, *178*, 9–23. [CrossRef]
4. Liu, C.; He, N.; Zhang, J.; Li, Y.; Wang, Q.; Sack, L.; Yu, G. Variation of stomatal traits from cold temperate to tropical forests and association with water use efficiency. *Funct. Ecol.* **2018**, *32*, 20–28. [CrossRef]
5. Franks, P.J.; Doheny-Adams, T.W.; Britton-Harper, Z.J.; Gray, J.E. Increasing water-use efficiency directly through genetic manipulation of stomatal density. *New Phytol.* **2015**, *207*, 188–195. [CrossRef]
6. Faralli, M.; Matthews, J.; Lawson, T. Exploiting natural variation and genetic manipulation of stomatal conductance for crop improvement. *Curr. Opin. Plant Biol.* **2019**, *49*, 1–7. [CrossRef]
7. Chaves, M.; Costa, J.M.; Zarrouk, O.; Pinheiro, C.; Lopes, C.; Pereira, J.S. Controlling stomatal aperture in semi-arid regions–The dilemma of saving water or being cool? *Plant Sci.* **2016**, *251*, 54–64. [CrossRef]
8. Endo, H.; Torii, K.U. Stomatal development and perspectives toward agricultural improvement. *CSH Perspect. Biol.* **2019**, *11*, a034660. [CrossRef]
9. Simmons, A.R.; Bergmann, D.C. Transcriptional control of cell fate in the stomatal lineage. *Curr. Opin. Plant Biol.* **2016**, *29*, 1–8. [CrossRef]
10. Torii, K.U. Stomatal differentiation: The beginning and the end. *Curr. Opin. Plant Biol.* **2015**, *28*, 16–22. [CrossRef]
11. Smit, M.E.; Bergmann, D.C. The stomatal fates: Understanding initiation and enforcement of stomatal cell fate transitions. *Curr. Opin. Plant Biol.* **2023**, *76*, 102449. [CrossRef]
12. Qi, X.; Torii, K.U. Hormonal and environmental signals guiding stomatal development. *BMC Biol.* **2018**, *16*, 21. [CrossRef]
13. Torii, K.U. Stomatal development in the context of epidermal tissues. *Ann. Bot.* **2021**, *128*, 137–148. [CrossRef]
14. Shimada, T.; Sugano, S.S.; Hara-Nishimura, I. Positive and negative peptide signals control stomatal density. *Cell. Mol. Life Sci.* **2011**, *68*, 2081–2088. [CrossRef]
15. Rowe, M.H.; Bergmann, D.C. Complex signals for simple cells: The expanding ranks of signals and receptors guiding stomatal development. *Curr. Opin. Plant Biol.* **2010**, *13*, 548–555. [CrossRef]
16. Torii, K.U. Mix-and-match: Ligand–receptor pairs in stomatal development and beyond. *Trends Plant Sci.* **2012**, *17*, 711–719. [CrossRef]
17. Lin, G.; Zhang, L.; Han, Z.; Yang, X.; Liu, W.; Li, E.; Chang, J.; Qi, Y.; Shpak, E.D.; Chai, J. A receptor-like protein acts as a specificity switch for the regulation of stomatal development. *Genes Dev.* **2017**, *31*, 927–938. [CrossRef]
18. Hara, K.; Yokoo, T.; Kajita, R.; Onishi, T.; Yahata, S.; Peterson, K.M.; Torii, K.U.; Kakimoto, T. Epidermal cell density is autoregulated via a secretory peptide, EPIDERMAL PATTERNING FACTOR 2 in Arabidopsis leaves. *Plant Cell Physiol.* **2009**, *50*, 1019–1031. [CrossRef]
19. Sugano, S.S.; Shimada, T.; Imai, Y.; Okawa, K.; Tamai, A.; Mori, M.; Hara-Nishimura, I. Stomagen positively regulates stomatal density in Arabidopsis. *Nature* **2010**, *463*, 241–244. [CrossRef]
20. Lu, J.; He, J.; Zhou, X.; Zhong, J.; Li, J.; Liang, Y.K. Homologous genes of epidermal patterning factor regulate stomatal development in rice. *J. Plant Physiol.* **2019**, *234*, 18–27. [CrossRef]
21. Sun, Q.; Qu, J.; Yu, Y.; Yang, Z.; Wei, S.; Wu, Y.; Yang, J.; Peng, Z. TaEPFL1, an EPIDERMAL PATTERNING FACTOR-LIKE (EPFL) secreted peptide gene, is required for stamen development in wheat. *Genetica* **2019**, *147*, 121–130. [CrossRef]
22. Xia, H.; Wang, Q.; Chen, Z.; Sun, X.; Zhao, F.; Zhang, D.; Fei, J.; Zhao, R.; Yin, Y. Identification and functional analysis of the EPF/EPFL gene family in maize (*Zea mays* L.): Implications for drought stress response. *Agronomy* **2024**, *14*, 1734. [CrossRef]
23. Xia, Y.; Jiang, S.; Wu, W.; Du, K.; Kang, X. MYC2 regulates stomatal density and water use efficiency via targeting EPF2/EPFL4/EPFL9 in poplar. *New Phytol.* **2024**, *241*, 2506–2522. [CrossRef]
24. Wang, Y.; Xie, T.; Zhang, C.; Li, J.; Wang, Z.; Li, H.; Liu, X.; Yin, L.; Wang, S.; Zhang, S. Overexpression of the potato *StEPF2* gene confers enhanced drought tolerance in Arabidopsis. *Plant Biotechnol. Rep.* **2020**, *14*, 479–490. [CrossRef]
25. Wang, Y.l.; Xie, T.; Zhang, C.l.; Li, J.J.; Liu, Y.J.; Li, H.B.; Yin, L.N.; Wang, S.W.; Deng, X.; Ke, Q.B. Cloning and functional analysis of potato stomatal density epidermal patterning factor StSTOMAGEN. *Agric. Res. Arid. Areas* **2020**, *38*, 235–244.
26. Hasanuzzaman, M.; Zhou, M.; Shabala, S. How does stomatal density and residual transpiration contribute to osmotic stress tolerance? *Plants* **2023**, *12*, 494. [CrossRef]

27. Herrmann, A.; Torii, K.U. Shouting out loud: Signaling modules in the regulation of stomatal development. *Plant Physiol.* **2021**, *185*, 765–780. [CrossRef]
28. Liu, S.; Wang, C.; Jia, F.; An, Y.; Liu, C.; Xia, X.; Yin, W. Secretory peptide PdEPF2 enhances drought tolerance by modulating stomatal density and regulates ABA response in transgenic *Arabidopsis thaliana*. *Plant Cell Tissue Organ Cult.* **2016**, *125*, 419–431. [CrossRef]
29. Ferguson, J.N.; Schmuker, P.; Dmitrieva, A.; Quach, T.; Zhang, T.; Ge, Z.; Nersesian, N.; Sato, S.; Clemente, T.; Leakey, A. Reducing stomatal density by expression of a synthetic EPF increases leaf intrinsic water use efficiency and reduces plant water use in a C4 crop. *bioRxiv* **2024**, *02*, 01.578512.
30. Dunn, J.; Hunt, L.; Afsharinafar, M.; Meselmani, M.A.; Mitchell, A.; Howells, R.; Wallington, E.; Fleming, A.J.; Gray, J.E. Reduced stomatal density in bread wheat leads to increased water-use efficiency. *J. Exp. Bot.* **2019**, *70*, 4737–4748. [CrossRef]
31. Jiang, Q.; Yang, J.; Wang, Q.; Zhou, K.; Mao, K.; Ma, F. Overexpression of *MdEPF2* improves water use efficiency and reduces oxidative stress in tomato. *Environ. Exp. Bot.* **2019**, *162*, 321–332. [CrossRef]
32. Jiao, P.; Liang, Y.; Chen, S.; Yuan, Y.; Chen, Y.; Hu, H. Bna.EPF2 enhances drought tolerance by regulating stomatal development and stomatal size in Brassica napus. *Int. J. Mol. Sci.* **2023**, *24*, 8007. [CrossRef]
33. Clemens, M.; Faralli, M.; Lagreze, J.; Bontempo, L.; Piazza, S.; Varotto, C.; Malnoy, M.; Oechel, W.; Rizzoli, A.; Dalla Costa, L. *VvEPFL9-1* knock-out via CRISPR/Cas9 reduces stomatal density in grapevine. *Front. Plant Sci.* **2022**, *13*, 878001. [CrossRef]
34. Rivero, R.M.; Mittler, R.; Blumwald, E.; Zandalinas, S.I. Developing climate-resilient crops: Improving plant tolerance to stress combination. *Plant J.* **2022**, *109*, 373–389. [CrossRef]
35. Bertolino, L.T.; Caine, R.S.; Gray, J.E. Impact of stomatal density and morphology on water-use efficiency in a changing world. Front. *Plant Sci.* **2019**, *10*, 225.
36. Caine, R.S.; Yin, X.; Sloan, J.; Harrison, E.L.; Mohammed, U.; Fulton, T.; Biswal, A.K.; Dionora, J.; Chater, C.C.; Coe, R.A. Rice with reduced stomatal density conserves water and has improved drought tolerance under future climate conditions. *New Phytol.* **2019**, *221*, 371–384. [CrossRef]
37. Hughes, J.; Hepworth, C.; Dutton, C.; Dunn, J.A.; Hunt, L.; Stephens, J.; Waugh, R.; Cameron, D.D.; Gray, J.E. Reducing stomatal density in barley improves drought tolerance without impacting on yield. *Plant Physiol.* **2017**, *174*, 776–787. [CrossRef] [PubMed]
38. Wang, C.; Liu, S.; Dong, Y.; Zhao, Y.; Geng, A.; Xia, X.; Yin, W. PdEPF1 regulates water-use efficiency and drought tolerance by modulating stomatal density in poplar. *Plant Biotechnol. J.* **2016**, *14*, 849–860. [CrossRef] [PubMed]
39. Tanaka, Y.; Sugano, S.S.; Shimada, T.; Hara-Nishimura, I. Enhancement of leaf photosynthetic capacity through increased stomatal density in Arabidopsis. *New Phytol.* **2013**, *198*, 757–764. [CrossRef]
40. Xia, Y.; Du, K.; Ling, A.; Wu, W.; Li, J.; Kang, X. Overexpression of *PagSTOMAGEN*, a positive regulator of stomatal density, promotes vegetative growth in poplar. *Int. J. Mol. Sci.* **2022**, *23*, 10165. [CrossRef]
41. Zaki, H.E.M.; Radwan, K.S.A. Response of potato (*Solanum tuberosum* L.) cultivars to drought stress under in vitro and field conditions. *Chem. Biol. Technol. Agric.* **2022**, *9*, 1. [CrossRef]
42. IUSS Working Group WRB. World Reference Base for Soil Resources. In *International Soil Classification System for Naming Soils and Creating Legends for Soil Maps*, 4th ed.; International Union of Soil Sciences (IUSS): Vienna, Austria, 2022.
43. Ahmad, R.; Kim, M.D.; Back, K.H.; Kim, H.S.; Lee, H.S.; Kwon, S.Y.; Murata, N.; Chung, W.I.; Kwak, S.S. Stress-induced expression of choline oxidase in potato plant chloroplasts confers enhanced tolerance to oxidative, salt, and drought stresses. *Plant Cell Rep.* **2008**, *27*, 687–698. [CrossRef] [PubMed]
44. Nicot, N.; Hausman, J.F.; Hoffmann, L.; Evers, D. Housekeeping gene selection for real-time RT-PCR normalization in potato during biotic and abiotic stress. *J. Exp. Bot.* **2005**, *56*, 2907–2914. [CrossRef] [PubMed]

International Journal of
Molecular Sciences

MDPI

Article

Genomic Identification and Expression Analysis of Regulator of Chromosome Condensation 1-Domain Protein Family in Maize

Rui Liu [1], Tian Ma [1], Yu Li [1], Xiongbiao Lei [1], Hongjing Ji [1], Hewei Du [1], Jianhua Zhang [2,3,*] and Shi-Kai Cao [1,3,*]

1 School of Life Science, Yangtze University, Jingzhou 434025, China; liuxiaoshuang_6@163.com (R.L.); 19164047603@163.com (T.M.); tinashecool@163.com (Y.L.); leixionbiao627@163.com (X.L.); jjing0105@163.com (H.J.); duhewei666@163.com (H.D.)

2 Department of Biology, Hong Kong Baptist University, Hong Kong, China

3 State Key Laboratory of Agrobiotechnology, The Chinese University of Hong Kong, Hong Kong, China

* Correspondence: jzhang@hkbu.edu.hk (J.Z.); shikaicao@yangtzeu.edu.cn (S.-K.C.)

Abstract: Abiotic stress affects the growth and development of maize (*Zea mays*). The regulator of chromosome condensation 1 (RCC1)-containing proteins (RCPs) plays crucial roles in plant growth and development and response to abiotic stresses. However, a comprehensive analysis of the maize RCP family has not been reported in detail. This study presents a systematic bioinformatics analysis of the ZmRCP family, identifying a total of 30 members distributed across nine chromosomes. The physicochemical properties and *cis*-acting elements in the promoters of *ZmRCP* members are predicted. The results of subcellular localization showed that ZmRCP3 and ZmRCP10 are targeted to mitochondria and ZmRCP2 is localized in the nucleus. A heatmap of expression levels among family members under abiotic stress conditions revealed varying degrees of induced expression, and the expression levels of 10 *ZmRCP* members were quantified using RT-qPCR under abiotic stress and plant hormone treatments. The results showed that *ZmRCP* members exhibit induced or inhibited responses to these abiotic stresses and plant hormones. These results contribute to a better understanding of the evolutionary history and potential role of the ZmRCP family in mediating responses to abiotic stress in maize.

Keywords: RCC1-containing protein; bioinformatic analysis; abiotic stress; maize (*Zea mays*)

1. Introduction

Maize is an important cereal crop that often suffers from abiotic stresses such as drought, high temperature, low temperature, and salinity during its growth and development, leading to severe yield reduction [1]. During long periods of evolution, organisms have evolved a range of mechanisms to cope with a variety of stresses, such as UV protection and osmo-protection mechanisms [2,3]. Regulator of chromosome condensation 1 (RCC1)-containing proteins (RCPs) have been reported to mediate diverse biological processes, including these abiotic stress responses [4–6].

Proteins that contain one or more RCC1/RCC1-like domains/repeats are classified as RCC1 proteins/RCPs [7]. Each repeat contains 51-68 amino acid residues and adopts a seven-bladed β-propeller fold [4,8]. RCP members are commonly found in eukaryotic cells. In humans, RCC1 is a key cell cycle regulatory factor involved in chromosome binding, nuclear cytoplasmic transport, and cell cycle control [8]. In animals, RCC1 is involved in many biological processes coupled with Ran in the nucleus, where RanGDP can produce RanGTP under the action of RCC1 [9,10].

In plants, RCPs have been identified in several species, including rice (*Oryza sativa*), maize (*Zea mays*), cotton (*Gossypium Hirsutum*), *Populus euphratica*, and *Arabidopsis thaliana* [5,6,11–14]. In Arabidopsis, 24 putative RCP family members have been identified. Six AtRCP proteins have been characterized, including UVR8 (ultraviolet receptor 8), tolerant to chilling and freezing (TCF1), RCC1/UVR8/GEF like 1 (RUG1), RUG2, RUG3, and sensitive to ABA 1 (SAB1) [5,11,12,15–18]. Under UV-B irradiation, UVR8 can dissociate from homologous dimers into monomers, which can interact with COP1 and continue to accumulate in the nucleus, triggering the UV-B cascade reaction, so that the plant activates the UV protection mechanism [3,19]. RUG1 encodes bilirubin deaminase (PBGD), which is an enzyme in the tetrapyrrole biosynthesis pathway that produces chlorophyll, heme, and so on. Deficiency of *RUG1* has been shown to lead to leaf damage, abnormal plant development, and down-regulated expression of auxin-related genes [20]. RUG2 encodes a dual localization protein, which localizes in both mitochondria and chloroplasts, and its mutation exhibits pale green leaves, growth retardation, and a reduced number of mesophyll cells [16]. During seed germination and post-germination growth, mutation of *RUG3* enhances the responses to ABA, suggesting that RUG3 as a negative regulatory factor mediates the ABA signaling pathway in Arabidopsis [17]. Arabidopsis SAB1 is also an RCP member as a negative regulator of *ABI5* (*abscisic acid-insensitive 5*); it functions upstream of *ABI5* and regulates promoter binding activity, its stability, and histone-methylation-mediated gene silencing of *ABI5* post-germination [11]. Moreover, Arabidopsis TCF1 encodes a glycosylphosphatidylinositol anchored protein that mediates a plant-specific freezing-resistant transcription program by regulating lignin biosynthesis independently of the CBF pathway [5]. These cases indicate that RCP members are involved in the response to abiotic stresses in Arabidopsis.

In addition, rice *Root Length Regulator 4* (*OsRLR4*) encodes an RCP protein that affects root elongation by negatively regulating RAM (root apical meristem) activity [13]. Overexpression of *Populus euphratica UVR8* (*PeUVR8*) in the Arabidopsis *uvr8* mutant rescued its phenotype and overexpression of *PeUVR8* in wild-type Arabidopsis enhanced anthocyanin accumulation and inhibited hypocotyl elongation, indicating that PeUVR8 functions in controlling plant photomorphogenesis [14]. In maize, several RCP family members have been reported. For example, DEK47 is responsible for splicing four introns of mitochondrial *nad2* in maize; therefore, it is critical for the assembly and activity of mitochondrial complex I and seed development [21]. The down-regulated expression of *ZmRCC1-1* causes cytological changes, such as altering grain orientation and affecting cell cycle progression. *ZmRCC1-2* and *ZmRCC1-3* neutralize the changes caused by *ZmRCC1-1* down-regulation, returning chromosome function to a normal state [22]. Although RCP members have been studied for over thirty years, there is still a lack of systematic identification and an analysis of the RCP family in maize.

In this study, we identified a total of 30 RCP members in maize through a systematic bioinformatics analysis. A comprehensive analysis of the ZmRCP family, including physicochemical properties, a phylogenetic tree, and subcellular localization, is performed. The expression analysis of *ZmRCP* genes were determined by RT-qPCR under abiotic stresses and plant hormones treatments. These data provide theoretical support and optional gene resources for future research on the biological functions of the ZmRCP family and the cultivation of stress-resistant maize varieties.

2. Results

2.1. Identification and Prediction of Protein Physicochemical Properties

To identify the RCP family members in maize, we operated the homologous alignment using the Arabidopsis RCP proteins using bidirectional BLASTp from NCBI. After domain validation, a total of 30 maize RCP family members were identified (Table 1, Figure 1). The number of amino acid residues (aa) in the ZmRCP family ranges from 181 to 1092. The molecular weight of ZmRCP family proteins ranges from 19,246.71 Da to 117,959.71 Da. The isoelectric point is between 4.95 and 9.84; most are alkaline proteins. The instability index of ZmRCP family proteins ranges from 6.81 to 49.16, and 19 members have an instability index lower than 40, indicating their stability. The hydrophilicity index ranges from −0.609 to −0.01.

Table 1. Prediction of physicochemical properties of the maize RCP family members.

Gene ID	Protein Name	Subcellular Localization	Protein Length/aa	Molecular Mass/Da	Theoretical pI	Instability Index	Grand Average of Hydropathicity
Zm00001eb016900	ZmRCP1	Chloroplast	371	41,019.26	4.95	35.26	−0.308
Zm00001eb041760	ZmRCP2	Nucleus	427	45,901.78	8.38	28.27	−0.349
Zm00001eb049950	ZmRCP3	Nucleus	480	51,477.95	5.94	30.7	−0.206
Zm00001eb072860	ZmRCP4	Nucleus	1073	118,737.65	9.00	37.37	−0.468
Zm00001eb075990	ZmRCP5/ZmRLD1	Nucleus	1060	115,768.8	8.47	41.97	−0.458
Zm00001eb082500	ZmRCP6/ZmRCC1-2	Nucleus	443	47,158.43	5.38	37.24	−0.346
Zm00001eb096520	ZmRCP7	Cell wall, nucleus	483	51,229.18	6.62	27.16	−0.010
Zm00001eb140790	ZmRCP8/ZmRLD4	Chloroplast, nucleus	1091	118,047.68	8.89	43.98	−0.448
Zm00001eb148070	ZmRCP9/ZmRCC1-1	Nucleus	503	51,936.8	5.43	45.05	−0.263
Zm00001eb151590	ZmRCP10	Nucleus	404	43,284.09	5.59	29.71	−0.230
Zm00001eb153510	ZmRCP11	Cell wall, nucleus	200	21,328.17	9.84	6.81	−0.177
Zm00001eb154440	ZmRCP12	Nucleus	416	45,496.38	5.21	43.89	−0.181
Zm00001eb179510	ZmRCP13/ZmRCC1-3	Nucleus	389	41,874.04	6.01	28.34	−0.351
Zm00001eb183880	ZmRCP14	Cell wall, mitochon-drion	535	57,269.85	5.65	32.71	−0.093
Zm00001eb187440	ZmRCP15	Nucleus	1067	117,523.14	8.91	43.04	−0.475
Zm00001eb188870	ZmRCP16	Nucleus	181	19,246.71	8.54	19.79	−0.118
Zm00001eb196450	ZmRCP17	Nucleus	448	49,266.91	9.51	24.39	−0.031
Zm00001eb197640	ZmRCP18	Nucleus	462	49,028.32	6.40	43.18	−0.102
Zm00001eb205920	ZmRCP19	Nucleus	442	46,469.45	5.70	36.51	−0.034
Zm00001eb250930	ZmRCP20	Nucleus	784	86,289.9	8.85	39.96	−0.506
Zm00001eb267290	ZmRCP21/ZmRLD2	Nucleus	917	99,449.7	8.51	35.57	−0.400
Zm00001eb294600	ZmRCP22	Nucleus	1009	109,869.18	8.83	42.92	−0.407
Zm00001eb300770	ZmRCP23	Nucleus	1078	116,042.22	9.12	42.32	−0.447
Zm00001eb306630	ZmRCP24	Cell wall, nucleus	483	51,361.1	6.52	29.82	−0.048
Zm00001eb320310	ZmRCP25/DEK47	mitochondria	432	45,433.14	6.49	36.49	−0.153
Zm00001eb360990	ZmRCP26	Nucleus	1054	115,204.69	8.71	39.45	−0.482
Zm00001eb362420	ZmRCP27/ZmRLD3	Nucleus	912	98,988.94	8.70	46.22	−0.470
Zm00001eb362840	ZmRCP28/ZmRLD5	Nucleus	1092	117,959.71	8.95	43.31	−0.445
Zm00001eb367810	ZmRCP29	Nucleus	588	61,993.36	6.22	49.16	−0.350
Zm00001eb432950	ZmRCP30	Cell wall, nucleus	549	58,115.71	8.42	38.59	−0.609

2.2. Phylogenetic Relationship and Collinearity Analysis

In order to explore the specific phylogenetic relationship of RCP members in maize and other species, we constructed a phylogenetic tree of RCP proteins in three species, including *Zea mays*, *Arabidopsis thaliana*, and *Oryza sativa*. After domain validation of the rice protein sequence downloaded from NCBI, it was found that there are 22 members of the same domain in rice. There are 30 members in maize, 24 members in Arabidopsis, and 23 members in rice, indicating that the number of RCP family members is relatively conservative with no significant expansion (Figure 2). After constructing the tree using the adjacency method, all RCP family members in these three species were clustered into eight classes based on their genetic relationships, from Class I to VIII. In maize, there are two members in Class V, three members in Class I, VI, and VII, four members in Class II and III, five members in Class VIII, and seven members in Class IV (Figure 2).

To further understand the evolutionary relationship of the *RCP* gene family in different species, we selected three plants for a species collinearity analysis, *Zea mays*, *Oryza sativa*, and *Arabidopsis thaliana*. The results showed that there are two *RCP* genes in *Zea mays* that are collinear with *Arabidopsis thaliana* (Figure S1). However, there are 22 *RCP* genes in maize that are collinear with *Oryza sativa*. Among them, *ZmRCP9* and *ZmRCP29* are collinear with both *Arabidopsis thaliana* and *Oryza sativa*. Compared with dicotyledonous plants like *Arabidopsis thaliana*, more *RCP* genes in *Zea mays* show collinearity with monocotyledonous plants like *Oryza sativa*, indicating that *Zea mays* and *Oryza sativa*, both of which are monocotyledonous plants, have a higher homology of RCP members. Most *RCP* genes originated from the ancestors of monocotyledonous plants, while a small number of genes may originate from dicotyledonous plants.

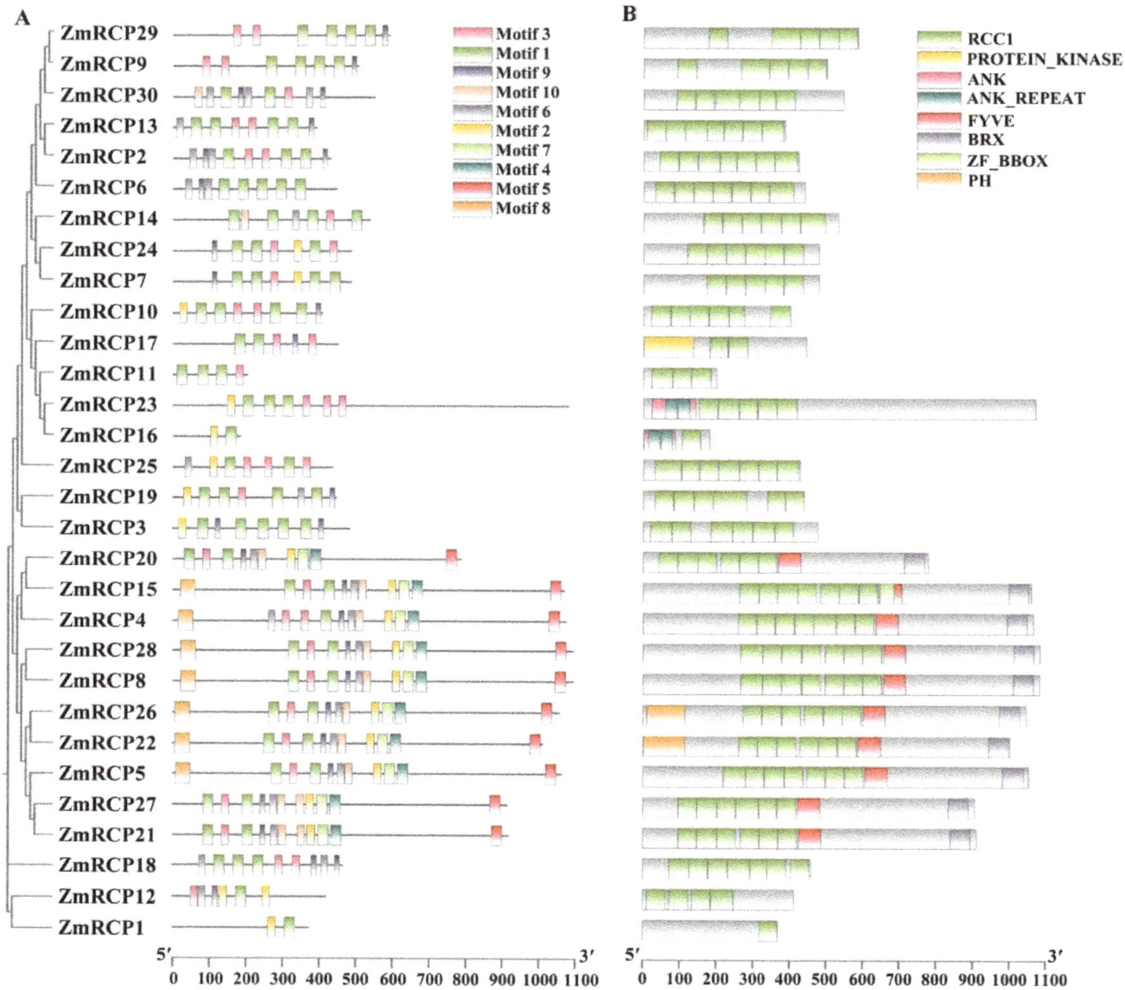

Figure 1. Prediction of conserved motifs and domains of ZmRCP family members in maize. (**A**) The phylogenetic relationship (left panel) and conserved motifs (right panel) of ZmRCP members are listed. (**B**) Prediction of conserved domains of ZmRCP members.

Collinearity analysis is a technique used to identify homologous protein sequences in the genome that are arranged in almost the same order at different positions in order to further reveal the genetic relationships between or within species of these proteins. The results of an intra-species collinearity analysis in maize showed that there were eight fragment duplication events on chromosomes 1, 2, 3, 4, 5, 7, and 8, involving a total of 12 genes (Figure S2). Based on the results of the evolutionary tree, we found that two genes with collinearity are classified in the same class (Figure 2). For example, *ZmRCP2* and *ZmRCP13*/*ZmRCC1-3* both belong to the Class IV subfamily, indicating that their gene structures and conserved domains are similar, which further supports the rationality of evolutionary tree grouping. The members of the *RCP* gene family have a certain degree of conservation and many gene duplications, which not only play an important role in the survival process of maize but also provide possibilities for the diversification of gene functions and gene evolution.

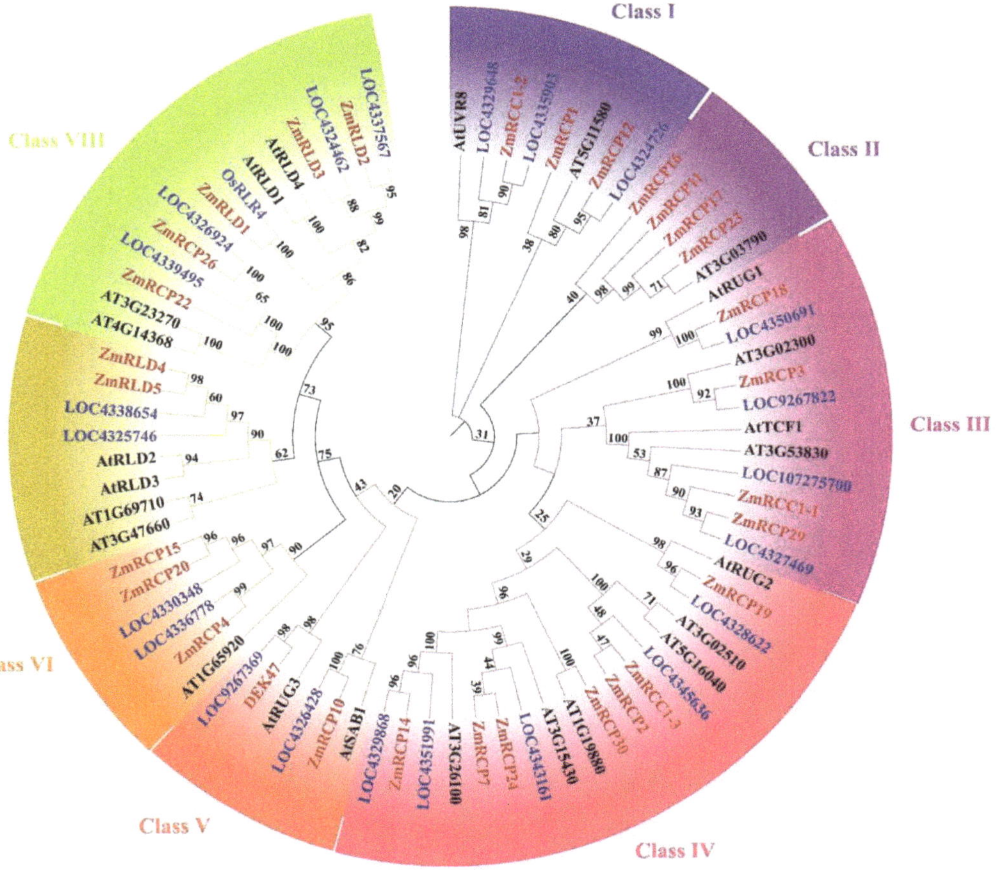

Figure 2. Phylogenetic tree of RCP family members among maize, rice, and Arabidopsis. Bootstrap values are marked at the branches. Different groups are marked with different colors.

2.3. Conserved Motifs, Domains, and Gene Structure Analysis

In order to study the genetic structural changes of the RCP family genes in maize, their gene structures were analyzed. The diagram of gene structure shows that all members of the maize RCP family contain different quantities of introns, and most members contain untranslated regions at the 5′ and/or 3′ ends except for three members (Figure S3). The numbers of introns of RCP members contained are from 2 to 15.

A motif is commonly referred to as a super secondary structure. This structure is formed by the aggregation of secondary structural units of proteins, and the motif plays an important role in the function and structure of proteins. The prediction of motifs of RCP family members showed that the *ZmRCP* gene family in maize contains a total of 10 motifs, with ZmRCP1 and ZmRCP16 having only two motifs (Figure 1A). The vast majority of members contain motif 1 and motif 3, except for ZmRCP1, ZmRCP3, ZmRCP6, and ZmRCP16 (Figure 1A).

The conserved domains of the maize *RCP* gene family were analyzed using the MEME tool of Tbtools. The prediction results of conserved domains showed that all the

ZmRCP members have the RCC1 domain, and the amounts of the RCC1 domain are from 1 to 7 (Figure 1B). In addition, there are the PROTEIN_KINASE, ANK (Ankyrin), ANK-REAPEAT, FYVE, BRX (BREVIS RADIX domain), ZF_BBOX (B-box type zinc finger), and PH (Pleckstrin Homology) domains. Among them, there are members that are relatively close in evolutionary relationships, such as ZmRCP29 and ZmRCP9 and ZmRCP13 and ZmRCP2 (Figure 1B).

2.4. Distribution of ZmRCP Family Members on Chromosomes

After obtaining the chromosomal location information of *ZmRCP* family members from the NCBI and MaizeGDB databases, TBtools software II was used to visualize it. The diagram showed that the 30 members of the *ZmRCP* family are distributed on nine chromosomes (Chr), with 1 member on Chr5 and Chr10; 2 members on Chr6; 3 members on each of Chr1 and Chr7; 4 members on each of Chr2 and Chr8; 5 members on Chr3; and 7 members on Chr4 (Figure 3). Most of the genes are distributed at the ends of the chromosome.

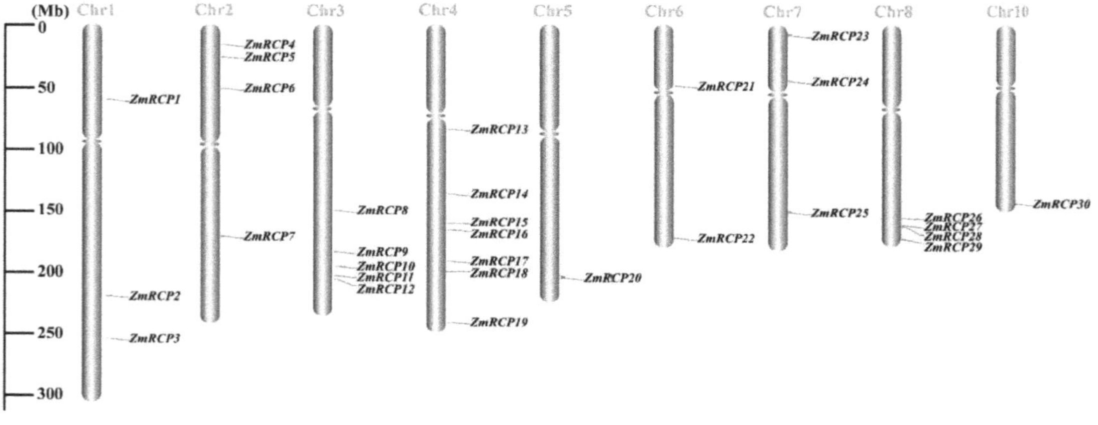

Figure 3. Distribution of *ZmRCP* family members on chromosomes (Chr). Gray lines indicate the location of the genes on the chromosomes, and the length of the chromosomes is shown on the left scale.

2.5. Subcellular Localization of ZmRCP Family Members

In order to understand the subcellular compartments where RCP family members are located, we conducted a subcellular localization prediction (Table 1). The results indicate that there are 22 ZmRCPs localized in the nucleus and four members localized in both the cell wall and nucleus. To verify the accuracy of subcellular localization prediction, we selected three family members for subcellular localization observation, ZmRCP2, 3, and 10. The resulting constructs ZmRCP2-GFP, ZmRCP3-GFP, and ZmRCP10-GFP were transfected to *Nicotiana benthamiana* leaves via *Agrobacterium*-mediated methods. Fluorescence signals were imaged using fluorescence confocal microscopy. The results showed that the green fluorescence signals of ZmRCP2-GFP were observed in the nucleus, whereas ZmRCP3-GFP and ZmRCP10-GFP were detected in dots that were not merged with the chloroplast autofluorescence signals (Figure 4), indicating that ZmRCP3-GFP and ZmRCP10 are targeted to mitochondria.

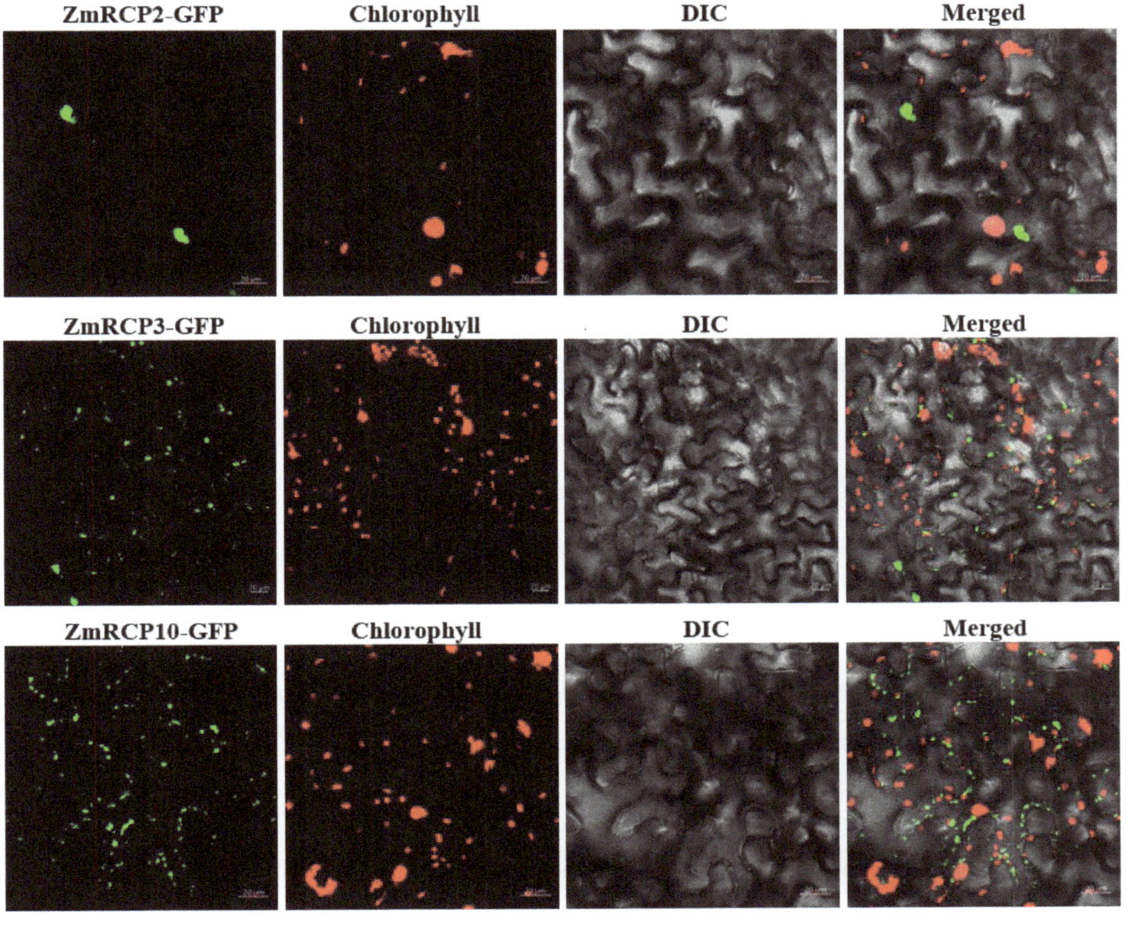

Figure 4. The subcellular localization of ZmRCP2, ZmRCP3, and ZmRCP10 in tobacco leaf epidermal cells. Fluorescence signals from GFP (green) and chlorophyll (red) were imaged under a confocal microscope. Scale bars = 20 μm.

2.6. Prediction of Cis-Elements in the Promoters

The 2000 bp sequences upstream of the start codons of 30 *ZmRCP* genes were extracted as the promoter regions used for *cis*-element prediction (Figure 5). The predicted *cis*-acting elements can be roughly divided into three categories: abiotic stress, plant hormones, and the growth and development process of maize. Abiotic-stress-responsive elements include drought, low-temperature, anoxic, and light-responsive elements. Plant-hormone-responsive elements include abscisic acid, MeJA, gibberellin, salicylic acid, and auxin-inductive or -responsive elements. The *cis*-acting elements during the growth and development of maize include the zein metabolism regulation element and endosperm expression element. All *ZmRCP* family members' promoter regions contain light-responsive elements, and the majority of *ZmRCP* family members' promoters contain multiple light-responsive elements. A total of 14 *ZmRCP* member promoter regions contain low-temperature responsive elements, and 22 *ZmRCP* gene promoter regions contain a MYB binding site involved in drought-inducibility elements. These prediction results indicate that the *ZmRCP* family may play an important regulatory role in the growth and development of and response to abiotic stress and plant hormones. The prediction of *cis*-acting elements can provide ideas for future research on related genes.

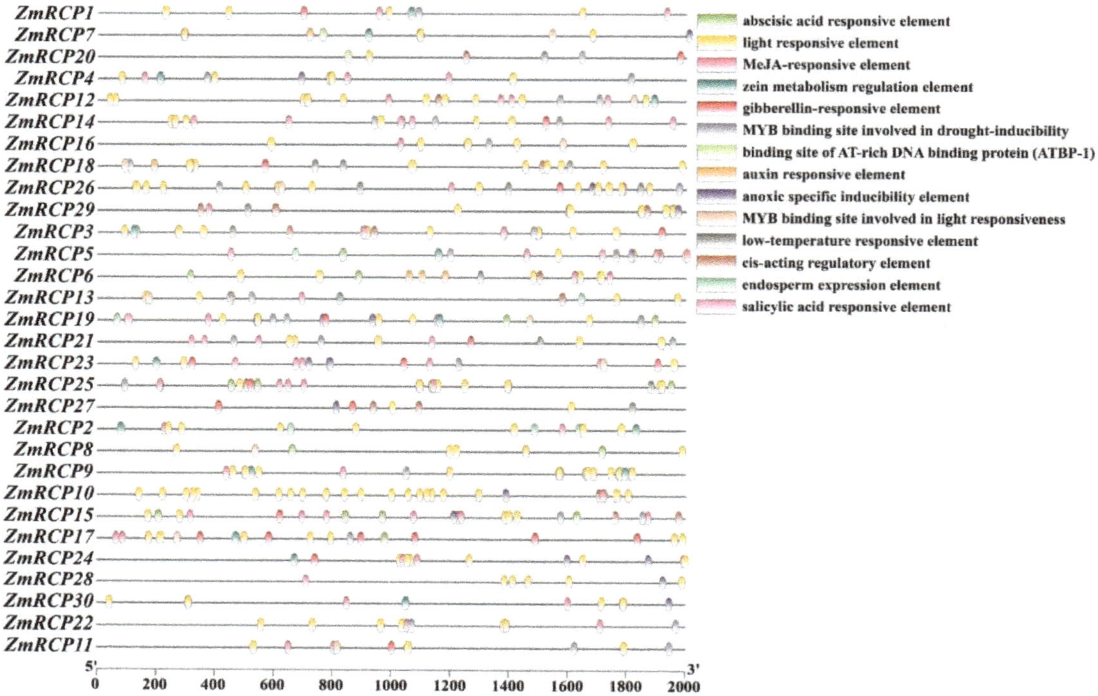

Figure 5. Prediction of *cis*-elements in promoters of 30 *ZmRCP* family members. Various colored boxes indicate different *cis*-elements that correspond to each colored box shown on the right side of the picture, and promoter lengths (base pairs) are shown on the scale below.

2.7. Analysis of Expression Patterns under Different Tissues and Stresses

In plants, gene expression in different tissues is often closely related to their functions. By analyzing transcriptome data from different tissues or developmental stages of plants, the expression levels of gene family members in a certain tissue or developmental stage can be determined, leading to speculation that these genes may play important roles in plant growth and development. The results of an expression pattern heatmap of *ZmRCP* members showed that the relative expression levels of genes in the red area are higher, while those in the blue area indicate lower levels (Figure 6). A total of 24 *ZmRCP* genes have relatively higher expression levels in embryos, endosperms, seeds, or the pericarp aleurone than other genes, indicating that these genes may function in the development of embryos, endosperms, seeds, or seed germination. As shown in Figure 6, *ZmRCP30*, *29*, and *15* have higher expression levels in SAM. *ZmRCP6*, *24*, *7*, *10*, *19*, and *3* have higher expression levels in mature leaf 8 than others. The expression of *ZmRCP24*, *7*, and *25* is higher than other genes in primary roots at day 5 (different days after pollination), indicating that these three genes may play a role in root development.

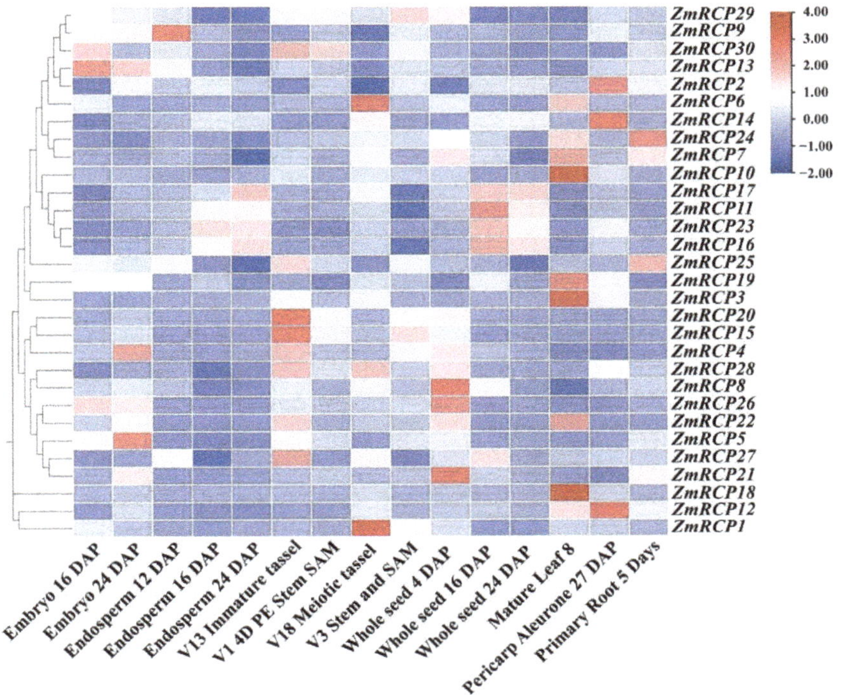

Figure 6. Expression pattern of 30 *ZmRCP* family members in different developmental stages and tissues. The different colors indicate the expression levels of the *ZmRCP* genes in maize. The expression data were downloaded from qTeller in MaizeGDB (https://qteller.maizegdb.org/genes_by_name_B73v5.php (accessed on 27 June 2024)) and the heatmap was generated using Tbtools software.

To investigate the stress response of *ZmRCP* family members under abiotic stresses, we analyzed the RNA-seq data online of *ZmRCP* members under different abiotic stresses, including heat, cold, salt, UV, and drought stress. The results showed that the expression levels of *ZmRCP29, 9,* and *3* were induced under heat stress; the expression of *ZmRCP30, 7, 10, 17, 11, 23, 16,* and *20* was induced by cold stress; the expression of *ZmRCP29, 9, 13, 25, 3, 5, 27, 18, 12,* and *1* was induced by salt stress; and the expression of *ZmRCP30, 13, 17, 23, 16, 25, 20, 15, 4, 28, 26, 5, 27,* and *21* was induced by UV stress (Figure S4A). In addition, the expression of *ZmRCP29, 20, 14, 17, 11, 23, 16, 19, 3, 20, 15, 4, 8, 27,* and *18* was induced by a mild water deficit (24 h). Under a severe water deficit, the expression of *ZmRCP29, 9, 14, 7, 10, 17, 11, 23, 16, 3, 20, 4, 28, 26, 22, 5, 27, 21, 18,* and *12* was induced (24 h) (Figure S4B). These data suggest that ZmRCP family members possibly play important roles in the response to abiotic stresses.

To further study the potential role of *ZmRCPs*, we performed RT-qPCR to detect the expression levels of 10 family members (*ZmRCP3, 4, 10, 14, 16, 17, 18, 20, 23,* and *29*) under different abiotic stresses and plant hormone treatments, which were induced by abiotic stresses based on RNA-Seq data online (Figure S4). The results showed that the transcript of these 10 *ZmRCP* members was decreased under drought stress at 6 h (Figure 7A), while the transcript of *ZmRCP17* was increased under salt stress (Figure 7B). *ZmRCP4, 14,* and *20* were remarkably induced under heat stress (Figure 7C), whereas *ZmRCP17* and *20* were significantly induced under clod stress (Figure 7D). These results suggest that *ZmRCP* members differentially respond to drought, salt, heat, and cold treatments.

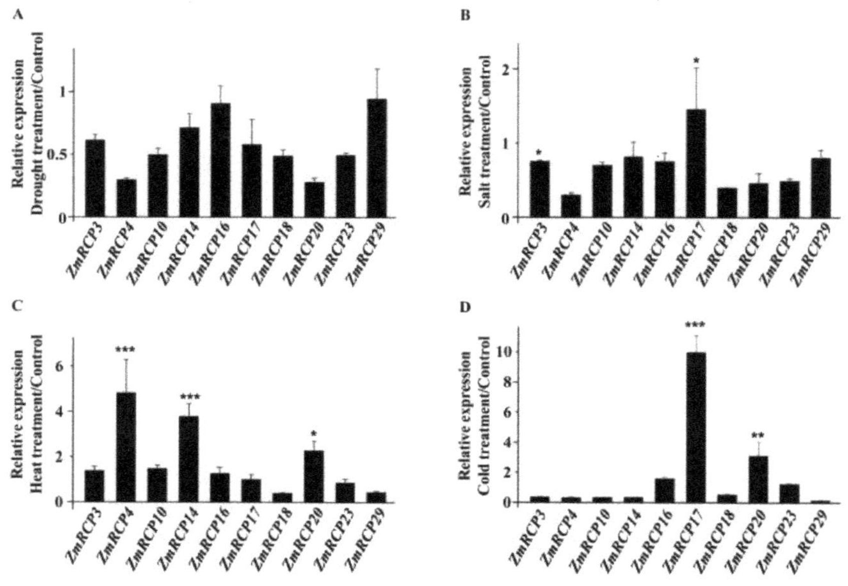

Figure 7. Expression levels of 10 *ZmRCP* members under abiotic stress treatments. Expression levels of 10 *ZmRCP* members were detected by RT-qPCR under drought (**A**), salt (**B**), heat (**C**), and cold (**D**) treatments. *ZmActin* (GRMZM2G126010) was used as the reference gene. Data are means (±SE) of three biological replicates. The comparison groups of Student's *t*-test are treatment and control. Asterisks indicate significant differences between means calculated with Student's *t*-test. * $p < 0.05$; ** $p < 0.01$; *** $p < 0.001$.

In addition, *ZmRCP17* showed induced expression under IAA, GA, and MeJA treatment (Figure 8A,C,E). *ZmRCP14, 17,* and *20* showed induced expression under ABA treatment (Figure 8B), while *ZmRCP17* and *20* showed induced expression under GA treatment (Figure 8E). These results suggest that *ZmRCP* members respond to different hormone treatments.

2.8. Prediction of Protein Interaction Network

In order to understand the potential interacting proteins of RCP gene family members, we used a website to predict the possible interacting proteins of RCP family members and then drew the diagram of the protein interaction network. There are a total of 37 proteins that possibly interact with the RCP family members, with the majority of RCP family members interacting with ATG7 (Figure S5). The ATG7 protein is an autophagy-related protein that allows cells to degrade damaged tissues through autophagy while also providing energy to help plants resist external stress and adverse factors [23,24]. In addition, other possible interacting proteins predicted include PSAH (photosystem I reaction center subunit VI) and TIDP3014 (GTP binding nucleoprotein), etc., most of which are related to stress and plant growth and development [25,26]. These results suggest that the ZmRCP family members may participate in plant growth and development processes and respond to abiotic stress through protein–protein interactions.

A previous study reported that ZmRCP25 (ZmDEK47) is required for the splicing of four introns of mitochondrial *nad2* transcript and seed development. The small PPR protein 2 (SPR2) is required for the splicing of 15 mitochondrial group II introns in maize, including the 4 introns of *nad2* [27], indicating that the functions of these two splicing factors are overlapped with each other. To test whether these two proteins interact with each other, we performed a yeast two-hybrid assay. The results showed that ZmRCP25AD/SPR2BD can grow on QDO or QDO/X/A medium (Figure 9), indicating that ZmRCP25 interacts with SPR2 in the yeast system.

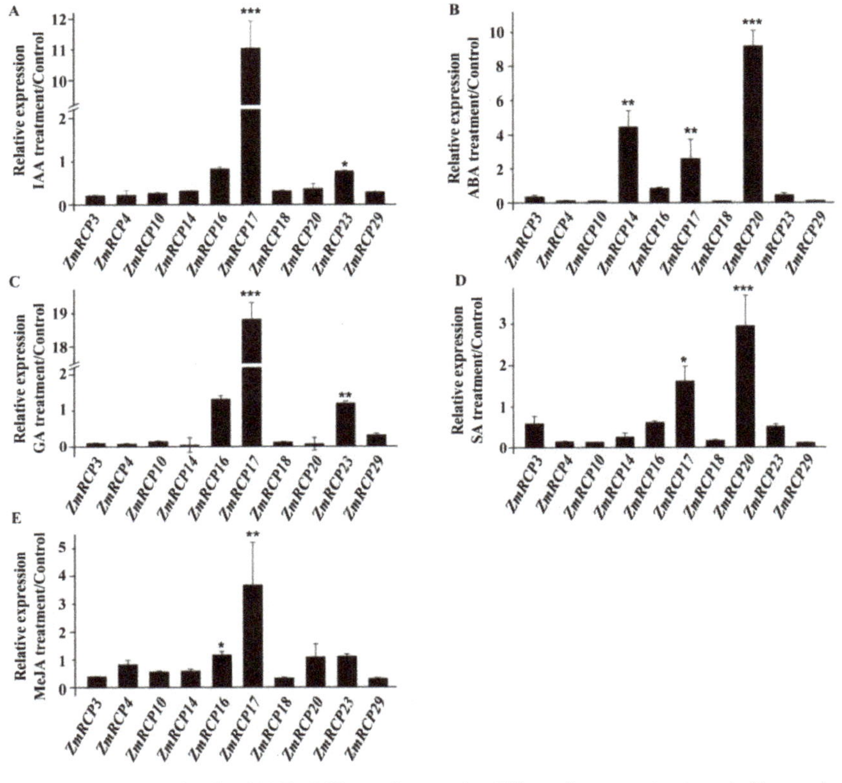

Figure 8. Expression levels of 10 *ZmRCP* members under different hormone treatments. Expression levels of 10 *ZmRCP* members were detected by RT-qPCR under IAA (**A**), ABA (**B**), GA (**C**), SA (**D**), and MeJA (**E**) treatments. *ZmActin* (GRMZM2G126010) was used as the reference gene. Data are means (±SE) of three biological replicates. The comparison groups of Student's *t*-test are treatment and control. Asterisks indicate significant differences between means calculated with Student's *t*-test. * $p < 0.05$; ** $p < 0.01$; *** $p < 0.001$.

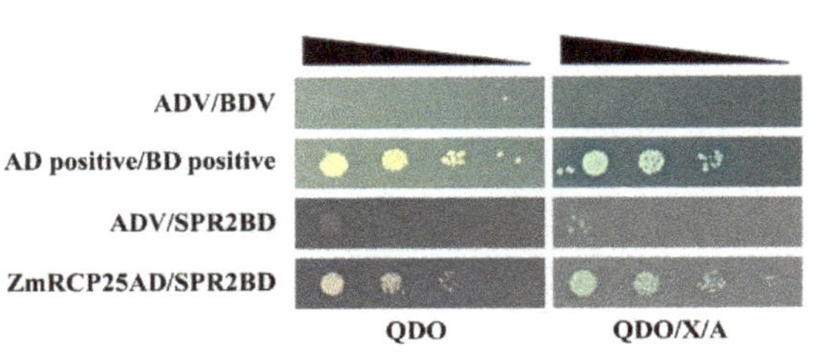

Figure 9. ZmRCP25 interacts with SPR2 in yeast two-hybrid system. Yeast two-hybrid assay was performed to determine interactions between ZmRCP25 and SPR2. These constructs were co-transfected into Y2H Gold strain and QDO (SD/−Ade/-His/−Leu/−Trp) medium with added X-a-gal and AbA (QDO/X/A) medium to reveal protein–protein interactions. AD: GAL4 activation domain; BD: GAL4 DNA binding domain.

3. Discussion

In this study, a phylogenetic tree of RCP family members was constructed among maize, rice, and Arabidopsis (Figure 2). The results showed that there are slightly more members of the RCP family in maize than in rice and Arabidopsis, indicating that the number of members of the RCP family has slightly increased during species evolution and that its functions may be more diverse in maize. Regarding the RCC1-conserved domain, members of the RCP family typically consist of six or seven RCC1 domains [12]. Most members in maize still retained this characteristic, but some members had a reduced number of RCC1 domains, or even only one RCC1 domain remained (Figure 1B). This may be because some conservative structural domains are lost when dealing with changes in the environment or other situations. Gene duplication is important for biological evolution, as it can promote the production of new adaptive functions in plants and expand gene families [28]. In the maize *RCP* gene family, a total of eight fragment duplication events were identified (Figure S2). These repeated fragments may enable maize to adapt to complex environments.

Previous studies reported that RCP members function in various aspects of plant growth and development. For example, the RCC1 domain protein ZmRCP25/DEK47 is required for the splicing of mitochondrial *nad2* introns and kernel development in maize [21]. The small PPR protein SPR2 is also required for the splicing of *nad2* introns and kernel development in maize [27]. Interestingly, ZmRCP25 interacted with SPR2 in the yeast two-hybrid assay (Figure 9), implying that ZmRCP25/SPR2 may form a protein complex in intron splicing. The RCC1-containing protein RUG3 is involved in the splicing of mitochondrial *nad2* intron 2 and 3 in Arabidopsis [12]. AtRUG3 and DEK47 cluster into a branch in Class V (Figure 2), which are both required for the intron splicing of mitochondrial *nad2*, indicating they have a functional homology. Similarly, ZmRCP18 is in Class III with AtRUG1, which is related to chlorophyll synthesis [20], implying that ZmRCP18 may be related to chlorophyll synthesis. Moreover, ZmRCP19, 2, 30, 24, 7, and 14 are in the Class IV, as are AtRUG2 and ZmRCC1-3 (Figure 2). AtRUG2 maintains normal transcriptional function in chloroplasts and mitochondria and ZmRCC1-3 and ZmRCC1-2 attenuate the defects of the cell cycle caused by ZmRCC1-1 dysfunction [16,22], implying that these ZmRCPs may be involved in certain functions in chloroplasts and mitochondria or play a role in cell cycle regulation. Because of the importance of chloroplast photosynthesis and mitochondrial oxidative respiration, loss of these ZmRCPs may cause severe defects in plant growth and development and even embryo lethality in maize.

Several RCP members were reported to be involved in response to abiotic stresses, such as responding to UV stress, cold stress, drought stress, heat stress, and salt stress [5,29,30]. For instance, AtUVR8 is involved in chromatin condensation and plays a role in the UV-B signaling pathway, inducing flavonoid biosynthesis [31]. AtUVR8 can activate a protection mechanism to cope with UV stress. Under drought stress, salt stress, or cold stress, the expression level of the *AtUVR8* gene increased compared to the corresponding sensitive and wild-type plants [3,19,29]. The RCC1-domain protein TCF1 can improve plant cold tolerance by reducing lignin accumulation under low-temperature conditions in Arabidopsis and soybeans [5,30]. In our data, the phylogenetic tree analysis showed that ZmRCC1-1 and ZmRCP29 are clustered in the same branch as AtTCF1, and ZmRCC1-2 is clustered in the same branch as AtUVR8 (Figure 2). Meanwhile, the prediction of *cis*-acting elements shows that promoter regions of multiple *ZmRCP* members contain low-temperature-responsive elements and a MYB binding site involved in drought-inducibility elements (Figure 5). The heatmap of gene expression under abiotic stress shows that the expression levels of most ZmRCP members increase under UV stress (Figure S4A). Moreover, the expression of several *ZmRCP* members was induced under different abiotic stresses (Figures 6 and S4A,B). These results suggest that the *ZmRCP* family may play important regulatory roles in the response to abiotic stresses in maize.

In previous reports, RCP family members have also been shown to respond to plant hormones such as ABA and auxin [5,11]. ABA and auxin are important plant hormones that regulate plant growth, development, and stress response. ABA plays a crucial role in various physiological processes of plants, such as stomatal closure, the accumulation of epidermal wax, leaf senescence, bud dormancy, seed germination, osmotic regulation, and growth inhibition [31]. The RCC1 protein SAB1 is involved in ABA signaling and affects the stability of ABI5 by initiating complex mechanisms, thereby negatively regulating the expression of *ABI5* genes [11]. RCC1 proteins AtRLD1 and AtRLD4 change the gravitropism of roots by altering the distribution of auxin [32]. The prediction of *cis*-acting elements shows that promoters of many ZmRCP members contain elements responding to plant hormones, including auxin, abscisic acid, MeJA, gibberellin, etc. (Figure 5). In addition, RT-qPCR results showed that the expression of *ZmRCP14*, *17*, and *20* was remarkably induced under different plant hormone treatments (Figure 7). These data suggest that *ZmRCP* members are involved in the response to plant hormones in maize.

Based on the above results, it is speculated that the *ZmRCP* gene may play a role in the response to abiotic stress and plant hormones, particularly in drought stress and cold stress. Although the potential functions of ZmRCP family members may be speculated based on the above analysis, whether these members truly participate in these functions still needs to be experimentally proven.

4. Materials and Methods

4.1. Materials and Treatments

Maize inbred line B73 seedlings grown to three leaves were used for abiotic stress and hormone treatments, and RNA was extracted for subsequent RT-qPCR. Seedlings were subjected to drought, salt, or control treatments with 20% (m/v) PEG-6000, 200 mM NaCl, or ddH_2O for 6h, respectively. Cold, high-temperature, and control treatments were carried out in a light incubator at a low temperature of 4 °C, a high temperature of 42 °C, and a control temperature of 25 °C for 24 h, respectively. Hormone treatments were carried out with sterile water (ddH_2O) or abscisic acid (ABA), auxin (IAA), salicylic acid (SA), methyl jasmonate (MeJA), and gibberellin (GA) at concentrations of 100 $\mu mol \cdot L^{-1}$, and the hormone solutions were used to soak roots for 4 h. Each of the above groups was planted with 3 cups of sand-cultured maize seedlings, and 5 seeds were sown in each cup.

4.2. Identification of RCP Family Members

The genome and annotation files of maize were downloaded from Ensembl Plants (http://plants.ensembl.org/info/data/ftp/index.html (accessed on 16 February 2024)). The 24 *Arabidopsis thaliana* RCP family members were searched, and their protein sequences were downloaded from the Arabidopsis TAIR (https://www.arabidopsis.org/ (accessed on 16 February 2024)) website. The extraction of the protein sequences of maize was carried out using TBtools software II [33,34], and the protein sequences of 24 *Arabidopsis thaliana* RCP family members were used as candidate family members using BLASTp bidirectional BLAST from NCBI (https://blast.ncbi.nlm.nih.gov/ (accessed on 16 February 2024)). The InterPro (https://www.ebi.ac.uk/interpro/ (accessed on 18 February 2024)) website was used to detect the RCP structural domain, and the ones that did not contain this conserved domain were removed, which finally determined the maize *RCP* gene family member list.

4.3. Construction of Phylogenetic Tree

The rice RCP members were obtained through a homologous blast. The RCP protein sequences of *Arabidopsis thaliana*, *Oryza sativa*, and *Zea mays* were imported, and a sequence alignment was performed by applying the ClustalW algorithm of MEGA7.0 software, choosing the neighbor-joining method, and setting the step value (Bootstrap) to 1000 to construct the multi-sequence phylogenetic tree [35]. The phylogenetic tree was uploaded to iTOL (https://itol.embl.de/itol.cgi (accessed on 26 February 2024)) to beautify and classify it.

4.4. Collinearity Analysis

The intra-species and inter-species collinearity analyses of maize RCP members were performed using McScanX, and the collinearity diagrams were drawn using the built-in Circos function in Tbtools software [33,34].

4.5. Analysis of Cis-Acting Elements in the Promoters

The 2000 bp promoter sequences upstream the start codon of each of the maize RCP family members were obtained from the official maize database MaizeGDB (https://maizegdb.org/ (accessed on 19 May 2024)), the *cis*-acting elements of each promoter of the maize RCP family members were predicted using the online tool PlantCARE (https://bioinformatics.psb.ugent.be/webtools/plantcare/html/ (accessed on 20 May 2024)), and the drawing/diagram was completed using TBtools software [33,34].

4.6. Prediction of Protein Physicochemical Properties

The protein molecular weight, instability index, average hydropathicity, isoelectric point, and number of amino acid residues of RCP family members were analyzed by uploading each identified protein sequence to the ExPASy ProtParam official website (https://web.expasy.org/protparam/ (accessed on 11 March 2024)). The subcellular localization of each family member was predicted using CELLO (http://cello.life.nctu.edu.tw/ (accessed on 12 March 2024)), wolf psort (https://wolfpsort.hgc.jp/ (accessed on 12 March 2024)), and Cell-Ploc (http://www.csbio.sjtu.edu.cn/bioinf/Cell-PLoc-2/ (accessed on 12 March 2024)).

4.7. Gene Structure, Conserved Motif, and Structural Domain Analysis

The information about the length of open reading frames and the number of exons of maize RCP family members was obtained from the MaizeGDB (https://maizegdb.org/ (accessed on 25 March 2024)). The conserved motifs were predicted by the MEME online tool (https://memesuite.org/meme/tools/meme (accessed on 26 March 2024)), and the number of conserved motifs was set as 10. The conserved domains of the maize RCP family members were further obtained by using the CD-search tool on the official website of NCBI (https://www.ncbi.nlm.nih.gov/ (accessed on 27 March 2024)). Diagrams of gene structure, protein conserved motifs, and conserved domains were visualized using the Gene Structure View function in TBtools software [33,34].

4.8. Location on Chromosome

The diagram of chromosome localization was obtained by uploading GFF3 files and gene IDs in the Gene Location Visualize from GTF/GFF tool of TBtools [33,34].

4.9. Expression Pattern Analysis

The expression data of ZmRCP family members under abiotic stresses and in different tissues were downloaded from qTeller in MaizeGDB. The clustering analysis was performed and the expression heatmaps of maize RCP family members were generated by using Tbtools software [33,34].

4.10. Subcellular Localization

The full-length cDNAs of *ZmRCP-2, 3,* and *10* were cloned into the pSuper1300-GFP vector and fused to the N-terminal of GFP (Green Fluorescence Protein). The constructs were transformed into tobacco leaves through *Agrobacterium EHA105*-mediated transformation as described previously [27]. The subcellular localization of the target proteins was preliminarily determined where the was green fluorescence signal observed.

4.11. Analysis of Protein Interaction Network Prediction

The interaction relationships corresponding to maize RCP proteins were extracted directly from the STRING (https://string-db.org/ (accessed on 29 July 2024)) database and the PlantSPEAD database [36]. The protein interaction network was constructed by inputting the protein interaction information in Cystoscope v3.10.1 software.

4.12. RT-qPCR

Total RNA was extracted from 0.2 g of leaves from seedlings after treatment with different hormones or abiotic stresses using the RNA easy Isolation Reagent kit (Vazyme Biotech Co., Ltd., Nanjing, China). Reverse transcription was conducted after DNA digestion using a HiFi Script cDNA Synthesis Kit (Vazyme Biotech Co., Ltd., Nanjing, China) using 1 μg total RNA as the template. RT-qPCR was performed with a CFX96 Real Time PCR instrument (Bio-Rad, Hercules, CA, USA) and an SYBR Green I Master PCR kit (Roche, Basel, Switzerland) using *ZmActin* (GRMZM2G126010) as the internal reference gene. Three independent biological and technical replicates were performed, and the relative expression level of the gene was calculated using the formula $2^{-\Delta\Delta Ct}$ [37]. The primers used in this study are in Supplementary Table S1.

4.13. Yeast Two-Hybrid Assay

The full-length cDNA of ZmRCP25 and SPR2 was cloned into pGADT7 and pGBKT7, respectively. ZmRCP25-pGADT7 and SPR2-pGBKT7 were co-transformed into the yeast strain *Y2H Gold*. The growth conditions of the co-transformants were imaged after being inoculated onto Minimal Media Quadruple Dropout (QDO, SD-Ade/-His/-Leu/-Trp) plates for 4 days as described previously [27].

5. Conclusions

In this study, 30 *ZmRCP* members were identified and analyzed in maize through a systematic bioinformatics analysis. We performed a physicochemical property analysis, a phylogenetic tree analysis, *cis*-acting elements prediction, and subcellular localization of ZmRCP. Several *ZmRCP* members exhibited induced or inhibited responses to abiotic stresses and plant hormones. These results contribute to a better understanding of the evolutionary history and potential role of the ZmRCP family in mediating responses to abiotic stresses in maize.

Supplementary Materials: The following supporting information can be downloaded at https://www.mdpi.com/article/10.3390/ijms252111437/s1.

Author Contributions: R.L. and S.-K.C. designed the experiments. R.L., T.M., Y.L., X.L., H.J. and S.-K.C. performed the experiments and bioinformatics analysis. R.L., H.D., J.Z. and S.-K.C. analyzed the data and wrote the manuscript. All authors have read and agreed to the published version of the manuscript.

Funding: This research was funded by the Hubei Natural Science Foundation (2022CFB183, 2024AFB301, and 2022CFA030).

Institutional Review Board Statement: Not applicable.

Informed Consent Statement: Not applicable.

Data Availability Statement: Data are contained within the article and Supplementary Materials.

Conflicts of Interest: The authors declare no conflicts of interest.

References

1. Zhang, H.M.; Zhu, J.H.; Gong, Z.Z.; Zhu, J.K. Abiotic stress responses in plants. *Nat. Rev. Genet.* **2022**, *23*, 104–119. [CrossRef] [PubMed]
2. Park, H.J.; Kim, W.Y.; Yun, D.J. A new insight of salt stress signaling in plant. *Mol. Cells* **2016**, *39*, 447–459. [CrossRef] [PubMed]
3. Brown, B.A.; Cloix, C.; Jiang, G.H.; Kaiserli, E.; Herzyk, P.; Kliebenstein¶, D.J.; Jenkins, G.I. A UV-B-specific signaling component orchestrates plant UV protection. *Proc. Natl. Acad. Sci. USA* **2005**, *102*, 18225–18230. [CrossRef] [PubMed]
4. Hadjebi, O.; Casas-Terradellas, E.; Garcia-Gonzalo, F.R.; Rosa, J.L. The RCC1 superfamily: From genes, to function, to disease. *Biochim. Biophys. Acta* **2008**, *1783*, 1467–1479. [CrossRef] [PubMed]

5. Ji, H.; Wang, Y.; Cloix, C.; Li, K.; Jenkins, G.I.; Wang, S.; Shang, Z.; Shi, Y.; Yang, S.; Li, X. The Arabidopsis RCC1 family protein TCF1 regulates freezing tolerance and cold acclimation through modulating lignin biosynthesis. *PLoS Genet.* **2015**, *11*, e1005471. [CrossRef]
6. Liu, X.; Wu, X.; Sun, C.; Rong, J. Identification and expression profiling of the regulator of chromosome condensation 1 (RCC1) gene family in Gossypium Hirsutum L. under abiotic stress and hormone treatments. *Int. J. Mol. Sci.* **2019**, *20*, 1727. [CrossRef]
7. Dasso, M. RCC1 in the cell cycle: The regulator of chromosome condensation takes on new roles. *Trends Biochem. Sci.* **1993**, *18*, 96–101. [CrossRef]
8. Renault, L.; Nassar, N.; Vetter, I.; Becker, J.; Klebe, C.; Roth, M.; Wittinghofer, A. The 1.7 A crystal structure of the regulator of chromosome condensation (RCC1) reveals a seven-bladed propeller. *Nature* **1998**, *392*, 97–101. [CrossRef]
9. Bischoff, F.R.; Ponstingl, H. Catalysis of guanine nucleotide exchange on Ran by the mitotic regulator RCC1. *Nature* **1991**, *354*, 80–82. [CrossRef]
10. Cekan, P.; Hasegawa, K.; Pan, Y.; Tubman, E.; Odde, D.; Chen, J.Q.; Herrmann, M.A.; Kumar, S.; Kalab, P. RCC1-dependent activation of Ran accelerates cell cycle and DNA repair, inhibiting DNA damage-induced cell senescence. *Mol. Biol. Cell* **2016**, *27*, 1346–1357. [CrossRef]
11. Ji, H.; Wang, S.; Cheng, C.; Li, R.; Wang, Z.; Jenkins, G.I.; Kong, F.; Li, X. The RCC1 family protein SAB1 negatively regulates ABI5 through multidimensional mechanisms during postgermination in Arabidopsis. *New Phytol.* **2019**, *222*, 907–922. [CrossRef] [PubMed]
12. Kuhn, K.; Carrie, C.; Giraud, E.; Wang, Y.; Meyer, E.H.; Narsai, R.; des Francs-Small, C.C.; Zhang, B.; Murcha, M.W.; Whelan, J. The RCC1 family protein RUG3 is required for splicing of nad2 and complex I biogenesis in mitochondria of *Arabidopsis thaliana*. *Plant J.* **2011**, *67*, 1067–1080. [CrossRef] [PubMed]
13. Sun, C.; Li, D.; Gao, Z.; Gao, L.; Shang, L.; Wang, M.; Qiao, J.; Ding, S.; Li, C.; Geisler, M.; et al. OsRLR4 binds to the OsAUX1 promoter to negatively regulate primary root development in rice. *J. Integr. Plant Biol.* **2022**, *64*, 118–134. [CrossRef] [PubMed]
14. Mao, K.; Wang, L.; Li, Y.Y.; Wu, R. Molecular cloning and functional analysis of UV RESISTANCE LOCUS 8 (PeUVR8) from Populus euphratica. *PLoS ONE* **2015**, *10*, e0132390. [CrossRef]
15. Rizzini, L.; Favory, J.J.; Cloix, C.; Faggionato, D.; O'Hara, A.; Kaiserli, E.; Baumeister, R.; Schafer, E.; Nagy, F.; Ulm, R.; et al. Perception of UV-B by the Arabidopsis UVR8 protein. *Science* **2011**, *332*, 103–106. [CrossRef]
16. Quesada, V.; Sarmiento-Manus, R.; Gonzalez-Bayon, R.; Hricova, A.; Perez-Marcos, R.; Gracia-Martinez, E.; Medina-Ruiz, L.; Leyva-Diaz, E.; Ponce, M.R.; Micol, J.L. Arabidopsis RUGOSA2 encodes an mTERF family member required for mitochondrion, chloroplast and leaf development. *Plant J.* **2011**, *68*, 738–753. [CrossRef]
17. Su, C.; Yuan, J.; Zhao, H.; Zhao, Y.; Ji, H.; Wang, Y.; Li, X. RUG3 is a negative regulator of plant responses to ABA in *Arabidopsis thaliana*. *Plant Signal Behav.* **2017**, *12*, e1333217. [CrossRef]
18. Su, C.; Zhao, H.; Zhao, Y.; Ji, H.; Wang, Y.; Zhi, L.; Li, X. RUG3 and ATM synergistically regulate the alternative splicing of mitochondrial nad2 and the DNA damage response in *Arabidopsis thaliana*. *Sci. Rep.* **2017**, *7*, 43897. [CrossRef]
19. Yin, R.; Skvortsova, M.Y.; Loubery, S.; Ulm, R. COP1 is required for UV-B-induced nuclear accumulation of the UVR8 photoreceptor. *Proc. Natl. Acad. Sci. USA* **2016**, *113*, E4415–E4422. [CrossRef]
20. Quesada, V.; Sarmiento-Manus, R.; Gonzalez-Bayon, R.; Hricova, A.; Ponce, M.R.; Micol, J.L. PORPHOBILINOGEN DEAMINASE deficiency alters vegetative and reproductive development and causes lesions in Arabidopsis. *PLoS ONE* **2013**, *8*, e53378. [CrossRef]
21. Cao, S.-K.; Liu, R.; Sayyed, A.; Sun, F.; Song, R.; Wang, X.; Xiu, Z.; Li, X.; Tan, B.-C. Regulator of chromosome condensation 1-domain protein DEK47 functions on the intron splicing of mitochondrial nad2 and seed development in maize. *Front. Plant Sci.* **2021**, *12*, 695249. [CrossRef] [PubMed]
22. Zheng, H.; Wu, H.; Pan, X.; Jin, W.; Li, X. Aberrant meiotic modulation partially contributes to the lower germination rate of pollen grains in maize (*Zea mays* L.) under low nitrogen supply. *Plant Cell Physiol.* **2017**, *58*, 342–353. [CrossRef] [PubMed]
23. Minina, E.A.; Moschou, P.N.; Vetukuri, R.R.; Sanchez-Vera, V.; Cardoso, C.; Liu, Q.; Elander, P.H.; Dalman, K.; Beganovic, M.; Lindberg Yilmaz, J.; et al. Transcriptional stimulation of rate-limiting components of the autophagic pathway improves plant fitness. *J. Exp. Bot.* **2018**, *69*, 1415–1432. [CrossRef] [PubMed]
24. Zhu, F.; Alseekh, S.; Wen, W.; Cheng, Y.; Fernie, A.R. Genome-wide association studies of Arabidopsis dark-induced senescence reveals signatures of autophagy in metabolic reprogramming. *Autophagy* **2022**, *18*, 457–458. [CrossRef] [PubMed]
25. Jensen, P.E.; Bassi, R.; Boekema, E.J.; Dekker, J.P.; Jansson, S.; Leister, D.; Robinson, C.; Scheller, H.V. Structure, function and regulation of plant photosystem I. *Biochim. Biophys. Acta* **2007**, *1767*, 335–352. [CrossRef]
26. Haizel, T.; Merkle, T.; Pay, A.; Fejes, E.; Nagy, F. Characterization of proteins that interact with the GTP-bound form of the regulatory GTPase Ran in Arabidopsis. *Plant J.* **1997**, *11*, 93–103. [CrossRef]
27. Cao, S.K.; Liu, R.; Wang, M.; Sun, F.; Sayyed, A.; Shi, H.; Wang, X.; Tan, B.C. The small PPR protein SPR2 interacts with PPR-SMR1 to facilitate the splicing of introns in maize mitochondria. *Plant Physiol.* **2022**, *190*, 1763–1776. [CrossRef]
28. Panchy, N.; Lehti-Shiu, M.; Shiu, S.H. Evolution of Gene Duplication in Plants. *Plant Physiol.* **2016**, *171*, 2294–2316. [CrossRef]
29. Tossi, V.E.; Regalado, J.J.; Iannicelli, J.; Laino, L.E.; Burrieza, H.P.; Escandon, A.S.; Pitta-Alvarez, S.I. Beyond Arabidopsis: Differential UV-B response mediated by UVR8 in diverse species. *Front. Plant Sci.* **2019**, *10*, 780. [CrossRef]
30. Dong, Z.; Wang, H.; Li, X.; Ji, H. Enhancement of plant cold tolerance by soybean RCC1 family gene GmTCF1a. *BMC Plant Biol.* **2021**, *21*, 369. [CrossRef]

31. Chen, K.; Li, G.J.; Bressan, R.A.; Song, C.P.; Zhu, J.K.; Zhao, Y. Abscisic acid dynamics, signaling, and functions in plants. *J. Integr. Plant Biol.* **2020**, *62*, 25–54. [CrossRef] [PubMed]
32. Furutani, M.; Hirano, Y.; Nishimura, T.; Nakamura, M.; Taniguchi, M.; Suzuki, K.; Oshida, R.; Kondo, C.; Sun, S.; Kato, K.; et al. Polar recruitment of RLD by LAZY1-like protein during gravity signaling in root branch angle control. *Nat. Commun.* **2020**, *11*, 76. [CrossRef] [PubMed]
33. Chen, C.; Chen, H.; Zhang, Y.; Thomas, H.R.; Frank, M.H.; He, Y.; Xia, R. TBtools: An integrative toolkit developed for interactive analyses of big biological data. *Mol. Plant* **2020**, *13*, 1194–1202. [CrossRef]
34. Chen, C.; Wu, Y.; Li, J.; Wang, X.; Zeng, Z.; Xu, J.; Liu, Y.; Feng, J.; Chen, H.; He, Y.; et al. TBtools-II: A "one for all, all for one" bioinformatics platform for biological big-data mining. *Mol. Plant* **2023**, *16*, 1733–1742. [CrossRef]
35. Kumar, S.; Stecher, G.; Tamura, K. MEGA7: Molecular evolutionary genetics analysis version 7.0 for bigger datasets. *Mol. Biol. Evol.* **2016**, *33*, 1870–1874. [CrossRef]
36. Chen, M.X.; Mei, L.C.; Wang, F.; Boyagane Dewayalage, I.K.W.; Yang, J.F.; Dai, L.; Yang, G.F.; Gao, B.; Cheng, C.L.; Liu, Y.G.; et al. PlantSPEAD: A web resource towards comparatively analysing stress-responsive expression of splicing-related proteins in plant. *Plant Biotechnol. J.* **2021**, *19*, 227–229. [CrossRef]
37. Song, Y.C.; Chen, M.X.; Zhang, K.L.; Reddy, A.S.N.; Cao, F.L.; Zhu, F.Y. QuantAS: A comprehensive pipeline to study alternative splicing by absolute quantification of splice isoforms. *New Phytol.* **2023**, *240*, 928–939. [CrossRef]

International Journal of
Molecular Sciences

MDPI

Article

Genome-Wide Identification of the DnaJ Gene Family in Citrus and Functional Characterization of *ClDJC24* in Response to Citrus Huanglongbing

Yuzhen Tian [1,†], Xizi Wang [1,†], Huoqing Huang [2], Xin Deng [3,4], Baihong Zhang [5], Yixuan Meng [1], Libo Wu [1], Hang Chen [1], Yun Zhong [2,*] and Wenli Chen [1,*]

1 MOE Key Laboratory of Laser Life Science & Institute of Laser Life Science, Guangdong Provincial Key Laboratory of Laser Life Science, Guangzhou Key Laboratory of Spectral Analysis and Functional Probes, College of Biophotonics, School of Optoelectronic Science and Engineering, South China Normal University, Guangzhou 510631, China; tianyzh@126.com (Y.T.); wangxizi1103@163.com (X.W.); 2021023392@m.scnu.com (Y.M.); 2023023373@m.scnu.edu.cn (L.W.); ccc2020xxsc@126.com (H.C.)

2 Institute of Fruit Tree Research, Guangdong Academy of Agricultural Sciences, Key Laboratory of South Sub-Tropical Fruit Biology and Genetic Research Utilization, Ministry of Agriculture and Rural Affairs, Guangdong Provincial Key Laboratory of Science and Technology Research on Fruit Tree, Guangzhou 510640, China; hqhuang07@163.com

3 Department of Biomedical Sciences, City University of Hong Kong, Kowloon Tong, Hong Kong SAR, China; xindeng@cityu.edu.hk

4 Shenzhen Research Institute, City University of Hong Kong, Shenzhen 518057, China

5 Institute of Nanfan & Seed Industry, Guangdong Academy of Science, Guangzhou 510640, China; zhangbyhome@126.com

* Correspondence: zhongyun@gdaas.cn (Y.Z.); chenwl@scnu.edu.cn (W.C.)

† These authors contributed equally to this work.

Abstract: Citrus Huanglongbing (HLB) is the most destructive citrus disease worldwide. The etiological agent responsible for this disease is "*Candidatus* Liberibacter asiaticus" (CLas), a phloem-restricted bacterium transmitted by psyllid vectors. To date, effective practical strategies for curing Citrus HLB remain elusive. Additionally, no susceptibility genes associated with HLB have been identified in Citrus species, thereby complicating the application of gene-editing techniques such as CRISPR-Cas9 to enhance resistance to HLB. The co-chaperone *DnaJ* plays a crucial role in protein folding and the regulation of various physiological activities, and it is also associated with multiple pathological processes. *DnaJ* has been extensively studied in many species, including Arabidopsis, rice, and wheat. However, there is limited information available regarding the DnaJ gene family in citrus. In this study, we conducted a comprehensive genome-wide analysis of the DnaJ family genes in various Citrus species. The Citrus genome was identified to contain 86 *DnaJ* genes, which were unevenly distributed across nine chromosomes. Phylogenetic analysis indicated that these genes could be classified into six distinct groups. Furthermore, transcriptomic analysis revealed that nine *DnaJ* genes exhibited significantly higher induction in HLB-infected samples relative to non-HLB-infected Citrus. *Cis*-acting elements within the promoters of *DnaJ* genes were also examined, revealing the presence of hormone and defense/stress responsiveness elements (TC-rich) distributed on the *ClDJC24* gene. The results were validated using quantitative real-time PCR (qRT-PCR). Additionally, the silencing of *ClDJC24* suggested that this gene negatively regulates disease resistance in Citrus. Our study provided useful clues for further functional characterization and constructed a theoretical foundation for disease-resistant breeding in Citrus.

Keywords: *DnaJ*; citrus; huanglongbing (HLB); disease resistance

1. Introduction

Heat shock proteins (HSPs), the most extensive class of molecular chaperones, safeguard the interacting surfaces of protein substrates from conformational damage. They achieve this by binding to the substrates and promoting the proper folding of unfolded or nascent polypeptides [1]. HSPs can be categorized into six distinct families based on their molecular weights: HSP100, HSP90, HSP70, HSP60, HSP40, and small heat shock proteins [2,3]. DnaJ is a prototypical member of the HSP40 family, characterized by an N-terminal conserved domain known as the J domain, a glycine- and phenylalanine-rich (G/F-rich) domain, a central cysteine-rich domain (CR-type zinc finger) that includes four repeats of a CxxCxGxG motif (zinc finger-like region), and a substrate-binding C-terminal domain [4–6]. The J domain binds to HSP70 and stimulates the ATPase activity of co-chaperone proteins to facilitate protein folding, unfolding, translocation, and degradation [7]. For example, DnaJ proteins have been documented to facilitate the regulation of growth, development, and environmental adaptation in both prokaryotic and eukaryotic organisms, including fungi and plants [8–10]. The HSP40/DnaJ protein family is one of the most diverse families of auxiliary molecular chaperones [6]. *DnaJ*, also known as HSP40 or J-proteins, recognizes unfolded substrates and delivers them to DnaK. It facilitates ATP hydrolysis in HSP70s and induces conformational changes in stable chaperone proteins [11]. The DnaJ family represents the most diverse group of co-chaperones, distinguished by conserved J-domains initially identified in *Escherichia coli*. Members of the DnaJ family also contain DnaJ protein and a highly conserved HPD tripeptide signature motif [12]. HSP70 plays a crucial role in multiple stages of the viral lifecycle, making it one of the multifunctional proteins involved in regulating viral infection by the host [13]. Research indicates that HSP70 typically does not function in isolation; its activity relies on the assistance of HSP40/DnaJ proteins for regulation [6]. There may be multiple ways of interacting between DnaJ family proteins and HSP70 proteins, and regulating the function of HSP70 through different interaction ways may be the key to the wide range of cellular functions of HSP70 proteins [14].

Previous studies have identified 117 DnaJ homologs in Arabidopsis [15], 115 in rice [16], and 88 in wheat [12]. However, there is limited research about Citrus DnaJ. DnaJ proteins play diverse physiological roles, including hormone regulation and disease resistance [17–19]. The Arabidopsis AtJ1 protein is crucial for the abscisic acid (ABA) response [20], while the expression of the J-protein LeCDJ2 is induced by salicylic acid (SA) [21]. Moreover, the movement protein of the tobacco mosaic virus has been shown to interact with a DnaJ-like protein [22–24], and the interaction between DnaJ and soybean mosaic virus coat protein leads to susceptibility [25]. Additionally, various members of the DnaJ protein family play critical roles in regulating the replication and pathogenesis of diverse viruses [26].

Citrus Huanglongbing (HLB) is caused by the phloem-inhabiting, Gram-negative α-proteobacterium "*Candidatus* Liberibacter asiaticus" (CLas). This pathogen significantly reduces yield and leads to considerable economic losses in citrus production, ultimately jeopardizing the health of the entire tree [27,28]. Despite extensive scientific investigations, no cure for HLB has yet been discovered. To date, the DnaJ gene family has been extensively studied in rice, wheat, and Arabidopsis. DnaJ proteins have been identified in various species and are generally associated with plant growth, development, and stress resistance, among other functions [29]. In addition, research has shown that the DnaJ gene family is involved in the defense mechanisms of plants in wheat [12]. However, there is a paucity of information regarding the DnaJ gene family in Citrus.

In the present study, the DnaJ family members were identified throughout the Citrus genome. Gene structures, conserved domains, phylogenetic relationships, chromosomal locations, and expression patterns of all 86 *DnaJ* genes in Citrus were also analyzed. According to the results of *ClDJC24* expression analysis, the role of *ClDJC24* was explored further. Subsequently, the functions of the *DnaJ* genes were elucidated through the evaluation of gene silencing effects in Citrus plants during HLB infection. This study provided novel insights for resistant gene resources and provided useful clues for further functional characterization in Citrus.

2. Results

2.1. Genome-Wide Identification of DnaJ Genes from Citrus

According to the characteristics of the DnaJ domain (Pfam: PF00226), all 86 DnaJ protein sequences were identified from *Citrus clementina*. As illustrated in Table 1, the number of amino acids (AAs), molecular weight (Mw), and theoretical pI varied from 101 (Ciclev10026825m.1) to 2592 (Ciclev10007224m.1), 12.00 (Ciclev10010382m.1) to283.90 (Ciclev10007224m.1) kDa, and 4.64 (Ciclev10016883m.1) to 10.07 (Ciclev10003089m.1), respectively (Table 2). Additionally, the subcellular localization of all members was predicted using WoLF PSORT, and the results indicated that most members (such as Ciclev10029600m.1 (*ClDJC24*) are localized in the nucleus (nucl), with a few localized in the chloroplast (chlo) and cyto (Table 1).

Table 1. Physicochemical analysis of *DnaJ* genes in Citrus.

Annotation Number	Gene Name	AAs	Mw (kDa)	pI	Subcellular Localization	Arabidopsis Ortholog	Clade
Ciclev10030731m.1	ClDJC41-1	803	90.54	8.82	nucl	DJC41	I
Ciclev10030729m.1	ClDJC41-2	806	89.42	8.61	nucl	DJC41	I
Ciclev10004367m.1	ClDJC41-3	779	86.57	8.88	nucl	DJC41	I
Ciclev10000081m.1	ClDJC44	1142	127.65	8.78	nucl	DJC44	I
Ciclev10001879m.1	ClDJC45	317	35.15	8.54	cyto	DJC45	I
Ciclev10000551m.1	ClDJC46	646	73.81	9.27	nucl	DJC46	I
Ciclev10004241m.1	ClDJC67	1000	113.3	5.43	nucl	DJC67	I
Ciclev10014483m.1	ClDJC84 [$]	689	76.14	8.8	nucl	DJC84 [$]	I
Ciclev10028303m.1	ClDJLD13 [$]	490	54.74	8.7	nucl	DJLD13 [$]	I
Ciclev10029411m.1	ClDJLD9 [$]-1	167	18.84	5.4	cyto	DJLD9 [$]	I
Ciclev10029403m.1	ClDJLD9 [$]-2	169	19.19	5.64	cyto	DJLD9 [$]	I
Ciclev10017157m.1	ClDJLD9 [$]-3	131	14.9	9.45	nucl	DJLD9 [$]	I
Ciclev10029096m.1	ClDJLD9 [$]-4	259	28.71	8.91	nucl	DJLD9 [$]	I
Ciclev10020330m.1	ClDJA1-1	418	46.94	6.65	nucl	DJA1	II
Ciclev10028531m.1	ClDJA1-2	416	46.16	6.06	nucl	DJA1	II
Ciclev10015385m.1	ClDJA1-3	416	46.46	6.01	nucl	DJA1	II
Ciclev10031530m.1	ClDJA3	448	49.38	9.11	nucl	DJA3	II
Ciclev10008033m.1	ClDJA4	513	56.45	8.9	chlo	DJA4	II
Ciclev10025609m.1	ClDJA6	443	47.83	9.41	chlo	DJA6	II
Ciclev10025441m.1	ClDJA7	499	53.56	8.89	chlo	DJA7	II
Ciclev10020947m.1	ClDJA8	345	39.08	5.93	cyto	DJA8	II
Ciclev10016063m.1	ClDJB1-1	305	34.34	8.91	cyto	DJB1	II
Ciclev10012139m.1	ClDJB1-2	339	37.39	9.31	cyto	DJB1	II
Ciclev10015887m.1	ClDJB1-3	331	37.21	7.61	nucl	DJB1	II
Ciclev10008858m.1	ClDJB10	337	36.91	7.69	nucl	DJB10	II
Ciclev10015820m.1	ClDJB3	344	37.54	8.97	cyto	DJB3	II
Ciclev10032269m.1	ClDJB6	295	32.81	9.22	cyto	DJB6	II
Ciclev10001008m.1	ClDJC20	478	53.19	8.96	chlo	DJC20	II
Ciclev10020899m.1	ClDJC51	350	39.13	8.98	nucl	DJC51	II
Ciclev10003893m.1	ClDJC79	385	43.69	5.73	cyto	DJC79	III
Ciclev10028788m.1	ClDJC80	340	38.25	7.71	cyto	DJC80	III
Ciclev10001330m.1	ClDJC81	410	45.14	5.5	cyto	DJC81	III
Ciclev10026825m.1	ClDJC23	101	11.35	8.64	nucl	DJC23	IV
Ciclev10029600m.1	ClDJC24	119	13.13	9.06	nucl	DJC24	IV
Ciclev10011261m.1	ClDJC2	646	73.62	8.55	nucl	DJC2	IX
Ciclev10008750m.1	ClDJB14	358	40.48	8.75	nucl	DJB14	V
Ciclev10005299m.1	ClDJC35	352	41.26	8.54	nucl	DJC35	V
Ciclev10005578m.1	ClDJC30	282	32.34	8.95	nucl	DJC30	VI
Ciclev10033027m.1	ClDJC33	139	16.3	4.96	nucl	DJC33	VII
Ciclev10002824m.1	ClDJC50	133	16	5.18	nucl	DJC50	VII
Ciclev10016883m.1	ClDJC28	181	21.11	4.64	chlo	DJC28	VIII
Ciclev10026366m.1	ClDJC29	242	26.78	5.04	chlo	DJC29	VIII

Table 1. *Cont.*

Annotation Number	Gene Name	AAs	Mw (kDa)	pI	Subcellular Localization	Arabidopsis Ortholog	Clade
Ciclev10002746m.1	ClDJC47-1	159	17.86	4.98	nucl	DJC47	VIII
Ciclev10022709m.1	ClDJC47-2	148	16.62	4.8	chlo	DJC47	VIII
Ciclev10030891m.1	ClDJC17	671	74.96	5.7	plas	DJC17	X
Ciclev10004408m.1	ClDJB16	741	82.34	9.16	chlo	DJB16	XI
Ciclev10019555m.1	ClDJC36/ClDJC37	555	63.4	8.78	nucl	DJC36/DJC37	XI
Ciclev10015402m.1	ClDJB11	414	45.84	6.18	nucl	DJB11	XII
Ciclev10031692m.1	ClDJB12	410	46.33	8.72	nucl	DJB12	XII
Ciclev10011610m.1	ClDJC16	484	54.19	6.66	cyto	DJC16	XIII
Ciclev10014072m.1	ClDJC31	1214	133.52	5.66	chlo	DJC31	XIII
Ciclev10004152m.1	ClDJC62	1408	153.92	6.23	nucl	DJC62	XIII
Ciclev10012465m.1	ClDJC61	270	31.88	8.28	chlo	DJC61	XIV
Ciclev10011366m.1	ClDJB9	577	63.66	9.36	nucl	DJB9	Others
Ciclev10003422m.1	ClDJC10	182	20.53	4.49	nucl	DJC10	Others
Ciclev10016093m.1	ClDJC19	300	35.17	9.31	chlo	DJC19	Others
Ciclev10026659m.1	ClDJC22	163	18.43	5.62	chlo	DJC22	Others
Ciclev10019624m.1	ClDJC25	540	60.21	8.96	chlo	DJC25	Others
Ciclev10006058m.1	ClDJC26-1	170	19.16	8.74	chlo	DJC26	Others
Ciclev10022177m.1	ClDJC26-2	218	24.76	10.03	nucl	DJC26	Others
Ciclev10007224m.1	ClDJC27/ClDJC74	2592	283.9	5.91	nucl	DJC27/DJC74	Others
Ciclev10029128m.1	ClDJC32	250	29.6	9.21	nucl	DJC32	Others
Ciclev10030221m.1	ClDJC34	412	45.96	9.3	nucl	DJC34	Others
Ciclev10000724m.1	ClDJC57	567	64.79	8.7	nucl	DJC57	Others
Ciclev10002163m.1	ClDJC65	268	31.78	9.71	nucl	DJC65	Others
Ciclev10003690m.1	ClDJC66-1	127	14.44	8.92	nucl	DJC66	Others
Ciclev10009810m.1	ClDJC66-2	147	16.06	9.34	chlo	DJC66	Others
Ciclev10002683m.1	ClDJC66-3	170	18.9	9.8	chlo	DJC66	Others
Ciclev10003089m.1	ClDJC66-4	147	16.62	10.07	chlo	DJC66	Others
Ciclev10010382m.1	ClDJC66-5	104	12	6.27	nucl	DJC66	Others
Ciclev10006484m.1	ClDJC66-6	166	18.42	6.18	cyto	DJC66	Others
Ciclev10021147m.1	ClDJC69	325	36.21	10.05	chlo	DJC69	Others
Ciclev10012897m.1	ClDJC70	181	20.51	9.54	nucl	DJC70	Others
Ciclev10027520m.1	ClDJC71	528	60.83	8.22	nucl	DJC71	Others
Ciclev10026217m.1	ClDJC72	285	32.93	9.56	nucl	DJC72	Others
Ciclev10028974m.1	ClDJC73	289	33.14	5.27	nucl	DJC73	Others
Ciclev10002209m.1	ClDJC75	259	29.66	6.02	chlo	DJC75	Others
Ciclev10016190m.1	ClDJC76-1	281	31.72	9.24	chlo	DJC76	Others
Ciclev10031498m.1	ClDJC76-2	456	50.3	9.02	chlo	DJC76	Others
Ciclev10017707m.1	ClDJC77	340	38.33	6.27	chlo	DJC77	Others
Ciclev10002309m.1	ClDJC78	241	28.82	9.55	nucl	DJC78	Others
Ciclev10002196m.1	ClDJC82	262	29.42	7.01	chlo	DJC82	Others
Ciclev10006857m.1	ClDJC83 $	287	33.73	9.57	nucl	DJC83 $	Others
Ciclev10019373m.1	ClDJC9	604	69.09	5.49	nucl	DJC9	Others
Ciclev10009451m.1	ClDJLD4	215	24.23	5.85	chlo	DJLD4	Others
Ciclev10019092m.1	ClDJLD8	703	79.48	5.42	chlo	DJLD8	Others

Note: $ newly found J-proteins in *Arabidopsis* according to the study by Zhang et al.

Table 2. The primer sequences used for RT-PCR.

Genes	Forward Primers	Reverse Primers
Ciclev10002196m.1	TATTCATGTGTAAAAACAGCCCCTA	GTGCGTCCAAGATACTCTCAAGTAA
Ciclev10002746m.1	ATGAGAACAAGAGGTCAATGTACGA	TATTCATCATGGAGACCATCTCTTG
Ciclev10029600m.1	GGAAATGTCGGCAAATGAGTTTATC	TTCTGAGGAAAGCGAAGAAGAAGA
Ciclev10011366m.1	TTCAAGTTCATTTGGTTTTGGTCTA	TAAATTTATGGCTCTGATGTTTCCA
Ciclev10008750m.1	AAGAGGGGTGAATTTTTATGTGAAG	AATGGGGAGTCTCCTTTATAAAACC
Ciclev10030731m.1	GAGTACCTCCGAACTTATCTCAACA	AGTGTATGTGTTTCCTGTCATGGTA
Ciclev10016190m.1	AAGAAATCAAAGAGGCTTACAGGAA	TGGAAGCATCATAATCTTTTCTGAG

Table 2. *Cont.*

Genes	Forward Primers	Reverse Primers
Ciclev10001008m.1	ACATTGCCTCAATCGATACTTCATA	ATACACCTCCTAGCCAAAAACATTC
Ciclev10015385m.1	CATAAATGATGAAGGAATGCCTATG	CTTCACACTCATCCAGTTCCATATC
ClDJC24-dsRNA	CGGCTACAACAGCTTGTTCA	GCCGTTGAGTCCTTGTCTTC
CrNPR1	TGAGGGATTATGAAGTTGGGTTTG	CTTGTCCAGAATGTCTAGGAGGTG
CrPR1	ACAGTGAGACAATTGCATGGAG	AGCACAACCTAAACGCACAG
CrPR2	CTCTACGCTCCGAATCAAGCAACT	CGGCGATGTAGCGAAACTTG
CrGAPA	TACACTGGCGACCAAAGGCTACTC	GCGGCGTTCACCTCTTCTGC
Actin	TCTGTACTCAACCGTCTTCC	ATCACAACCACTCTTCATCTC

2.2. Chromosomal Distribution of DnaJ Genes from Citrus

Chromosomal distribution and the tandem duplications of the *DnaJ* genes from Citrus were analyzed (Figure 1). As illustrated in Figure 1, analysis of the 86 *DnaJ* genes from citrus genome chromosomal locations showed that they were unevenly anchored to nine chromosomes (Chr) Chr1-Chr9 by MapChart1.0 software. Most *DnaJ* genes were from Chr5 (15 DnaJ genes) of citrus, followed by Chr2 (10 DnaJ genes). In contrast, six genes were found on Chr6. These results indicate that *DnaJ* genes from Citrus are distributed unevenly throughout the genome. Furthermore, according to the analysis, four genes were clustered into two tandem duplication events (Ciclev10015820m.1/Ciclev10015887m.1 and Ciclev10016093m.1/Ciclev10017157m.1) on Chr2 (Figure 1). And Chr8 contained a tandem duplication event of Ciclev10029096m.1 and Ciclev10029411m.1. These results suggest that some of the *DnaJ* genes from citrus might be produced by gene duplication, which undoubtedly promotes their evolution.

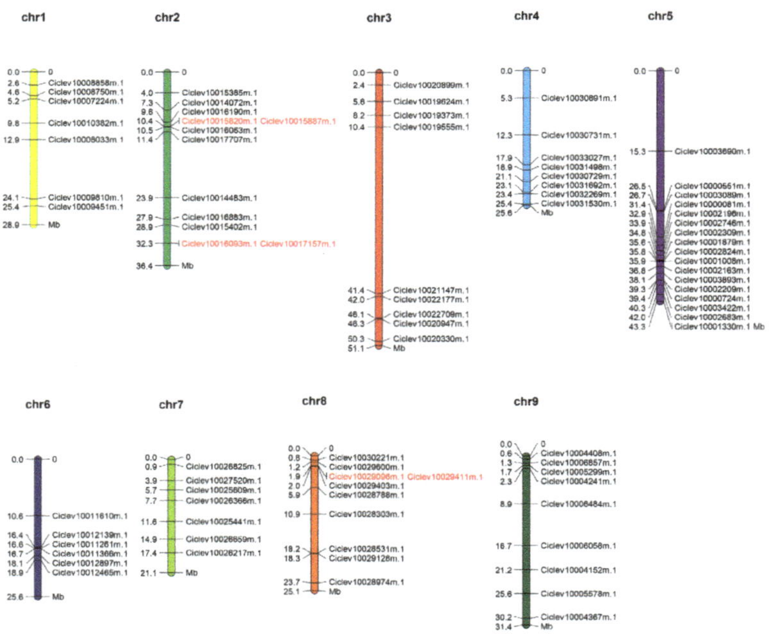

Figure 1. Chromosomal locations of *DnaJ* genes in Citrus. The chromosomal position of each *DnaJ* gene was mapped according to the genome of Citrus. The number at the top of each chromosome represents the chromosome number, and the scale unit is megabases (Mb). The gene duplication events are marked in red.

2.3. Gene Structures and Motif Analysis of DnaJ Genes from Citrus

A multiple sequence alignment was performed on the protein sequences derived from 86 *DnaJ* genes in citrus. Subsequently, a phylogenetic tree was constructed using the neighbor-joining (NJ) method in MEGA-X with 1000 bootstrap replicates (Figure 2A). As shown in Figure 2B,C, the conserved motifs and gene structures of 86 *DnaJ* genes were visualized by TBtools. The motif architectures motif 1 and motif 2 were the two most frequently presented motifs, covering almost all *DnaJ* gene families according to frequencies of occurrence (Figure 2B). Motif 9 was a specific motif that was solely found in Ciclev10003089m.1. Additionally, Motif 4 was a specific motif in Ciclev10002309m.1, Ciclev10012465m.1, Ciclev10007224m.1, and Ciclev10026217m.1. Ciclev10020330m.1 contained the largest number of motifs, including in motif 1, motif 2, motif 3, motif 6, motif 8, motif 9, and motif 10. To gain more insight into the evolutionary relationships within the Citrus *DnaJ* genes, we examined the exon and intron structures of all *DnaJ* gene members. In Figure 2C, among the 86 *DnaJ* genes, 13 had no introns, whereas the other *DnaJ* genes possessed 1 to 21 exons (9 with one exon, 12 with two exons, 2 with three exons, 7 with four exons, 6 with five exons, 9 with six exons, 2 with ten exons 2 with sixteen exons, 1 with twenty exons and 1 with twenty-one exons) (Figure 2C). These results indicated varying degrees of evolutionary conservation and divergence between the *DnaJ* genes from Citrus.

2.4. Characterization of Citrus HLB-Related DnaJ Genes

To investigate the evolutionary characteristics of the DnaJ genes family, we conducted the phylogenetic and clades analyses of DnaJ protein sequences from Arabidopsis and Citrus according to the classification criteria of Zhang et al. in Arabidopsis and Brassica [15]. Figure S2 shows that the phylogenetic tree divided 86 DnaJ proteins into 15 major clades (I-XV). The gene names of the 86 DnaJ proteins in Citrus, Arabidopsis ortholog, and the clades analysis are presented in Table 1. As shown in Figure 3, the clades are indicated by different frame colors. The study by Zhang et al. indicates that J-proteins in *Arabidopsis thaliana* and *brassica oleracea* are grouped into 15 major clades (I–XV), often separated by low-confidence branches that may represent unrelated single-gene clades. Clades I and II, each containing more than 10 members in Arabidopsis and B. oleracea, are classified as multigene clades. Clades III–XV contain 2–7 members and are termed oligogene clades. The remaining clades, with a single member in at least one of the two species, are designated as monogene clades [15]. In Figure 3, the largest number of *DnaJ* genes was observed in clade II with 18 *DnaJ* genes, followed by subgroup I (13) and clade VIII (4). However, clades VI, IX, X, XIII, and XIV only contained 1 DnaJ member from Citrus. In this study, candidate gene Ciclev10029600m.1 (*ClDJC24*) was notably divided into clade IV. The above results indicated the distinct phylogenetic characteristics of *DnaJ* genes in Citrus. Functional analysis of these genes might provide more information for the evolution of the DnaJ gene family.

To identify *DnaJ* genes related to citrus HLB, transcriptome data from control and HLB-infected citrus were obtained from earlier experiments generated in this lab [30]. The heatmap was created based on fragments per kilobase of transcript per million fragments mapped (FPKM) values from no HLB-infected and HLB-infected samples (Table S3). In the transcriptome data, sample A (A-1, A-2, A-3): post-treatment sample (new tissues), 90 days after FOS and CAC combination therapy; Sample B (B-1, B-2, B-3): pre-treatment sample. A-1, A-2, A-3, and B-1, B-2, B-3 represented no HLB-infected and HLB-infected with three biological replicates in this study, respectively. As shown in Figure 3, based on transcriptome data, nine genes were screened and indicated with asterisks.

Figure 2. The gene structure analysis of *DnaJ* gene family from citrus. (**A**) Phylogenetic relationship of *DnaJ* genes from citrus. (**B**) Conserved motifs analysis of the DnaJ from Citrus proteins. All motifs were identified by the MEME database with the complete amino acid sequences of the *DnaJs*. The detailed information for each motif is provided in Figure S1. (**C**) Gene structures of Citrus DnaJ members. Yellow boxes and gray lines indicate exons and introns, respectively, and the UTR is marked in green.

2.5. Cis-Acting Elements in DnaJ Genes Promoters from Citrus

To understand the functional properties of *DnaJ* gene family members, we predicted cis-acting elements in the promoter regions using the PlantCARE database. In Figure 4, cis-elements of the nine *DnaJ* genes are illustrated to further examine the regulatory mechanisms of *DnaJ* genes related to Citrus HLB. The findings disclosed a variety of *cis*-acting elements, encompassing those associated with defense and stress responsiveness, low-temperature responsiveness, drought inducibility, and hormone responsiveness. As illustrated in Figure 4, defense and stress response (TC-rich, GTTTTCTTAC) were discovered in five genes, whereas they were not detected in the other three *DnaJ* genes. In addition, four contained TCA-element (CCATCTTTTT), which is a salicylic acid (SA) responsive element (Figure 4). Ciclev10029600m.1 (*ClDJC24*) included MeJA responsiveness elements (TGACG-motif, TGACG and CGTCA), abscisic acid responsiveness (ABRE, ACGTG), and defense and stress responsiveness elements (TC-rich repeats, GTTTTCTTAC) (Figure 4). These results suggest that *DnaJ* genes play important roles in Citrus defense and stress and hormone responses.

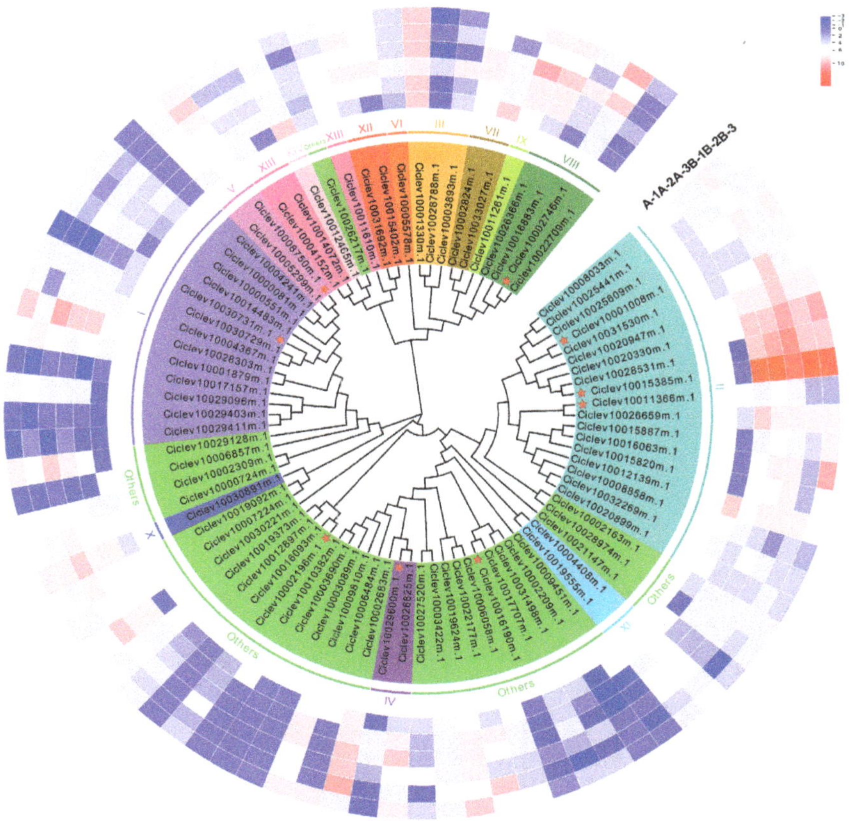

Figure 3. Phylogenetic analysis and expression patterns analysis of DnaJ proteins in Citrus with HLB. The clades are indicated by different frame colors. Amino acid sequences alignment of 86 DnaJ proteins in citrus. The heat map was generated using TBtools. The bar at the right of the heat map represents relative expression values. In the transcriptome data, sample A (A-1, A-2, A-3): post-treatment sample (new tissues), 90 days after FOS and CAC combination therapy; Sample B (B-1, B-2, B-3): pre-treatment sample. A-1, A-2, A-3, and B-1, B-2, B-3 represented no HLB-infected and HLB-infected with three biological replicates, respectively. Red star indicates the nine genes identified based on transcriptome data.

2.6. Relative Expressions of Citrus HLB-Related 9 DnaJ Genes

qRT-PCR was carried out for us to further verify the expression patterns of these nine genes, and the data are presented in Figure 5. The primers for qRT-PCR are listed in Table 1. As shown in Figure 5, compared to group 1 (uninfected leaves), five *DnaJ* genes (Ciclev10002746m.1 (*ClDJC47-1*), Ciclev10029600m.1 (*ClDJC24*), Ciclev10011366m.1 (*ClDJB9*), Ciclev10008750m.1 (*ClDJB14*) and Ciclev10030731m.1 (*ClDJC41-1*)) were significantly decreased in group 2 and group 3, while they were increased in group 4. Furthermore, compared to group 1, the other four genes (Ciclev10002196m.1 (*ClDJC82*), Ciclev10016190m.1 (*ClDJC76-1*), Ciclev10001008m.1 (*ClDJC20*) and Ciclev10015385m.1 (*ClDJA1-3*)) were increased in group 2, whereas they were decreased in group 3. However, expression of the *ClDJC24* revealed identical patterns based on qRT-PCR data and the RNA-Seq. In conclusion, the qRT-PCR data showed that expression patterns of most genes were consistent between RNA-seq and qRT-PCR assays (Figure 5), which warranted the reliability of the RNA-seq data. Therefore, the *ClDJC24* gene would be used as a candidate gene for subsequent functional validation experiments.

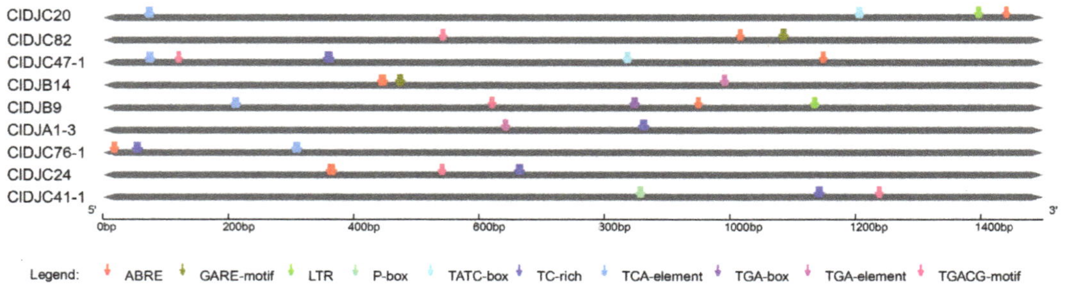

Figure 4. Cis-element analysis of 9 selected *DnaJ* genes from the upstream 1500 bp sequence to the transcription start site. Note: auxin responsive element (TGA-element, AACGAC; TGA-box, TGACGTAA), defense and stress responsiveness (TC-rich repeats, GTTTTCTTAC), gibberellin responsive element (GARE-motif, TCTGTTG; P-box, CCTTTTG; TATC-box, TATCCCA), low-temperature responsiveness (LTR, CCGAAA), MeJA responsiveness (TGACG-motif, TGACG; CGTCA), salicylic acid responsiveness (TCA-element, CCATCTTTTT), abscisic acid responsiveness (ABRE, ACGTG).

2.7. Synthesis and Characterization of Gold Nanoparticles–Polyethylenimine (AuNPs-PEI)

First, 1 mM gold nanoparticles–polyethylenimine (AuNPs-PEI) was synthesized as previously described [31,32]. In this subject, the AuNPs-PEI adopted the PEI reduction method [32–34]. The theoretical size range for nanoparticles that can enter cells is 6–200 nm [35]. As illustrated in Figure 6A, the prepared gold glue solution exhibited a ruby-red color. The results of particle size measurement showed that the particle size of AuNPs-PEI is mainly distributed in the 15–18 nm range (Figure 6B). In Figure 6C, the zeta potential of AuNPs-PEI was 33 mV. Furthermore, the binding assay of AuNPs-PEI and dsRNA was performed. Second, AuNPs-PEI and dsRNA in different volume ratios (15:1, 10:1, 5:1, 2:1) were incubated for 30 min at 25 °C, 50 rpm. After incubation, the binding samples were electrophoresed with 1.2% agarose. As shown in Figure 6D, the result showed that the best bind of AuNPs-PEI and dsRNA increases with decreasing AuNPs volume. According to the result, the 10:1 ratio was chosen for further analysis (Figure 6D). These results indicate that AuNPs-PEI and dsRNA maintain good dispersibility and can easily enter cells.

2.8. ClDJC24-dsRNA Silencing Affects the Expression Patterns of Immune Genes in Response to Citrus HLB

To investigate the biological function of *ClDJC24* in response to citrus HLB, wounded HLB-infected citrus leaves were treated with ClDJC24-dsRNA and AuNPs-PEI + CLDJC24-dsRNA, water and AuNPs-PEI + water as controls [36]. As illustrated in Figure 7A, ClDJC24 expression was reduced following treatment with ClDJC24-dsRNA compared to water-treated control leaves, indicating successful silencing of *ClDJC24*. However, compared to the control leaves treated with AuNPs-PEI + water, the expression of dsRNA was significantly decreased following treatment with AuNPs-PEI + CLDJC24-dsRNA (Figure 7A). We measured the CLas titers of HLB-infected Citrus leaves treated with ClDJC24-dsRNA and AuNPs-PEI + ClDJC24-dsRNA in 3 d and 7 d. The results showed that the Ct values of CLas were not significantly changed in 3 d and 7 d relative to controls. Furthermore, the expression patterns of immune genes *CrNPR1*, *CrPR1*, and *CrPR2* were analyzed by qRT-PCR in response to Citrus HLB (Figure 7C–E). The results showed that *CrNPR1* expression was upregulated in HLB-infected citrus leaves treated with AuNPs-PEI + ClDJC24-dsRNA when compared to AuNPs-PEI + water in 3 d (Figure 6C). The expression of *CrPR1* was increased in the leaves treated with both AuNPs-PEI + ClDJC24-dsRNA and ClDJC24-dsRNA as compared to controls in 7 d (Figure 6D). However, *CrPR2* was significantly expressed in 3 d. In addition, *CrGAPA*, a secretory protein in citrus HLB, was increased in HLB-infected Citrus leaves treated with AuNPs-PEI + ClDJC24-dsRNA in 3 d relative to AuNPs-PEI + water (Figure 7F). The results showed that *ClDJC24* can inhibit immune gene expression of *CrNPR1*, *CrPR1*, *CrPR1*, and *CrGAPA* in 3 d.

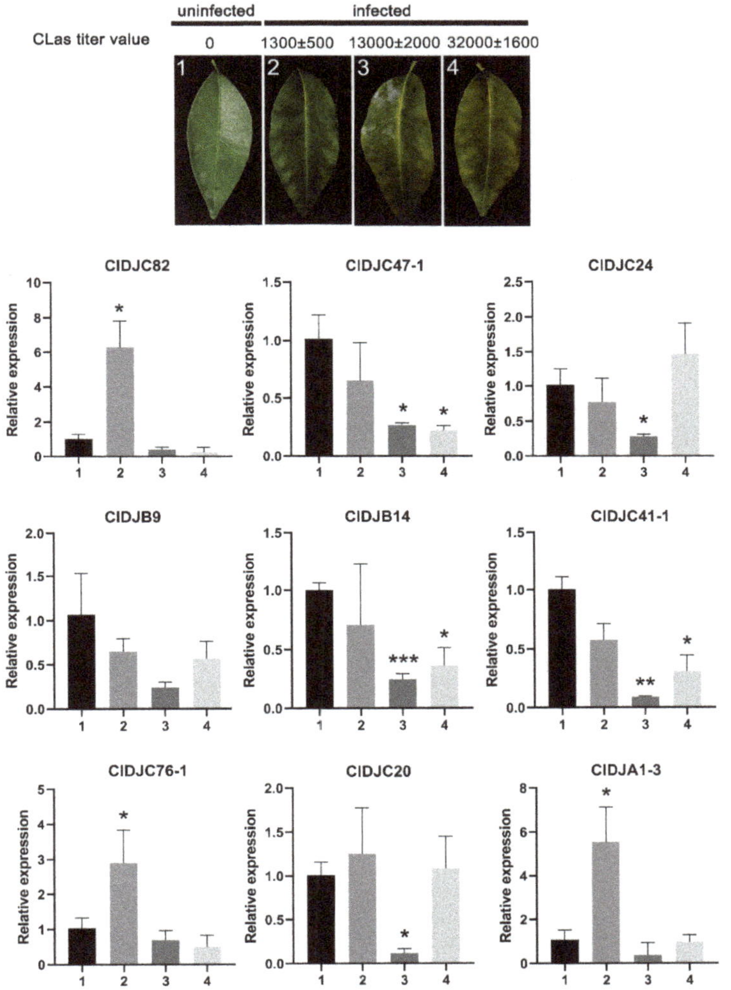

Figure 5. Relative expression of 9 selected *DnaJ* genes in groups 1, 2, 3, and 4. CLas titer refers to the bacterial load in 100 ng of genomic DNA. Based on the symptoms and CLas content, the leaves were divided into groups 1 to 4. Group 1 (CLas titer value is 0) uninfected citrus leaves were used as a control, group 2 (CLas titer value is 1300 ± 500), group 3 (CLas titer value is 13,000 ± 2000), group 4 (CLas titer value is 32,000 ± 1600). qRT-PCR data were normalized using *Actin* gene. qRT-qPCR was performed in three biological replicates in groups 1, 2, 3, and 4. Bars represent the standard deviation (±SD) calculated for three biological replicates. * represents $p < 0.05$, ** represents $p < 0.01$, *** represents $p < 0.001$ (Student's *t*-test).

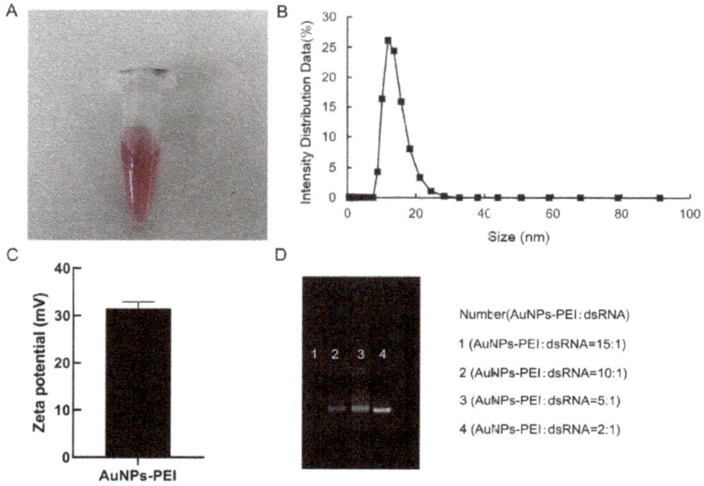

Figure 6. AuNPs-PEI synthesis and characterization. (**A**) The color of AuNPs-PEI solution. (**B**) The particle size of AuNPs-PEI solution and (**C**) Zeta potential of AuNPs-PEI solution. (**D**) Agarose gel electrophoresis of the mixture of AuNPs-PEI solution and dsRNA. A total volume of 16 μL of AuNPs-PEI solution and dsRNA (1 μL) was incubated in different volume ratios (15:1, 10:1, 5:1, 2:1) for 30 min at 28 °C. The mixture was run on 1.2% agarose gel electrophoresis. All experiments were performed with more than three biological replicates.

To determine the subcellular localization of *ClDJC24*, we designed a ClDJC24-GFP construct using a pUBQ10-GFP vector and transiently expressed it using a nuclear-localized marker transgenic *Nicotiana benthamiana* (NLS-mCherry-OE). As illustrated in Figure 7G, confocal fluorescence imaging of ClDJC24-GFP in NLS-mCherry-OE leaves cells demonstrated the nuclear and cytoplasm localization of ClDJC24-GFP and co-localizes with a nuclear marker. Additionally, ClDJC24-GFP is distributed throughout the entire cell, including the cytoplasm, similar to the localization of the empty vector control. This observation pertains to NLS-mCherry-OE leaves harvested three days post-infiltration.

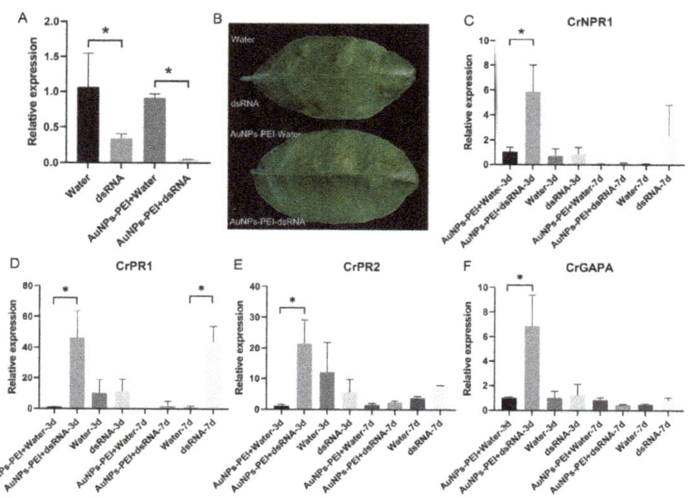

Figure 7. *Cont.*

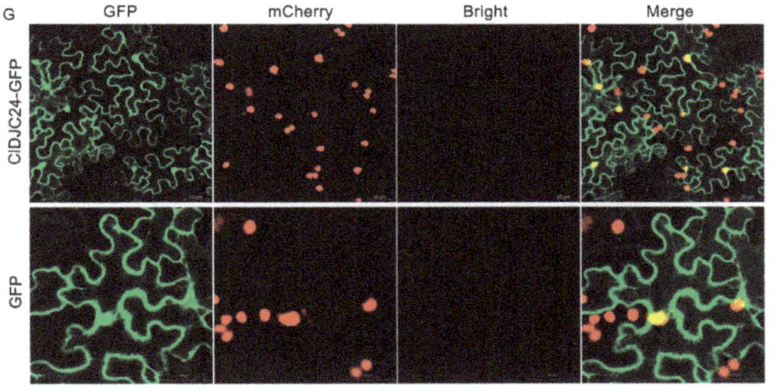

Figure 7. ClDJC24-dsRNA silencing affects the expression patterns of immune genes in response to Citrus HLB. (**A**) The expression level of *ClDJC24* in HLB-infected citrus leaves in 3 d treated with ClDJC24-dsRNA and AuNPs-PEI + ClDJC24-dsRNA, water, and AuNPs-PEI + water was analyzed by qRT-PCR. (**B**) The disease symptoms of dsRNA and AuNPs-PEI + dsRNA, water, and AuNPs-PEI + water treated HLB-infected citrus leaves. (**C–F**) represented the expression patterns of immune genes (*CrNPR1, CrPR1, CrPR2, CrGAPA*) in response to Citrus HLB. Data are presented as means ± SDs (n = 3). "*" symbol above the columns indicates a significant difference at $p < 0.05$. (**G**) Subcellular localization of *ClDJC24* uses a nuclear-localized marker transgenic *Nicotiana benthamiana* (NLS-mCherry-OE). The GFP empty vector was used as a control. Photographs were taken at 3 d post-infiltration. Scale bar = 50 μm.

3. Discussion

Huanglongbing (HLB), caused by the bacterium *Candidatus* Liberibacter asiaticus (CLas), represents one of the most devastating diseases affecting citrus crops worldwide [27,28]. Despite extensive research, no effective strategies have been identified for curing citrus HLB, and no susceptibility genes associated with HLB have been discovered in citrus species. This absence complicates the application of advanced gene-editing technologies, such as CRISPR-Cas9, to enhance citrus resistance to HLB [37–39]. The DnaJ protein family is known to play crucial roles in regulating various physiological processes and is implicated in diverse pathological pathways [40]. However, there is limited research on the mechanisms of interaction between the DnaJ family and plant pathogenic bacteria in citrus species. In this study, we conducted a comprehensive analysis of DnaJ family members in citrus, including their gene structures, conserved domains, phylogenetic relationships, chromosomal locations, and expression patterns in response to HLB infection. We also explored the effects of ClDJC24-dsRNA silencing on the expression patterns of immune-related genes in response to citrus HLB. Previous studies have shown that 117, 115, and 88 *DnaJ* homologs were identified in Arabidopsis [15], rice [16], and wheat [6], respectively. We identified 86 DnaJ family members throughout the Citrus genome. In this study, we identified 86 DnaJ family members across the citrus genome. The lower number of DnaJ members in citrus, compared to Arabidopsis (117) and rice (115), suggests gene loss during evolutionary processes and genome ploidy differences among these species. The isoelectric point (pI) of DnaJs protein sequences ranged from 4.64 (Ciclev10016883m.1) to 10.07 (Ciclev10003089m.1) (Table 1), with 62% of the family members exhibiting a pI greater than 7, a trend consistent with findings in *Capsicum annuum L.* and *Tolypocladium guangdongense* [41,42]. This pI distribution might indicate functional diversification of DnaJ proteins under varying physiological conditions and stress responses in citrus.

Additionally, our analysis revealed an uneven chromosomal distribution of the 86 *DnaJ* genes within the citrus genome. Notably, chromosomes 2 and 8 each contained three tandem duplication events, indicating that the evolution of DnaJ genes has been significantly influenced by gene duplication processes [43,44]. Such duplication events are known to contribute to genetic diversity and functional innovation within gene families. Furthermore, in this study, the analysis of gene structures showed that the number and distribution of introns and exons were highly conservative, and most *DnaJ* gene members contain 4–6 introns. The intron–exon structure offers significant insights into the systematic and comprehensive understanding of the conservation and evolutionary characteristics of gene families [45]. It serves as an important clue for evolutionary studies and functional analyses. This structurally similar evolutionary branch indicates that these genes may corroborate this finding of physiological functions [46].

In this study, we conducted an analysis of the cis-acting elements within the promoters of DnaJ genes in citrus species (Table S2). The results revealed that the *ClDJC24* gene in citrus possesses a significant number of cis-regulatory elements associated with MeJA responsiveness elements (TGACG-motif, TGACG and CGTCA-motif, CGTCA), abscisic acid responsiveness (ABRE, ACGTG) and defense and stress responsiveness elements (TC-rich repeats, GTTTTCTTAC). These findings suggest a potential role for the ClDJC24 gene in regulating various biological activities and responses to environmental stressors. The promoter is a crucial element of genes that regulates the initiation timing and expression levels of gene transcription. The analysis of promoters constitutes the foundation and prerequisite for research on gene expression regulation [47].

These findings suggest that DnaJ genes may play a significant role in regulating various biological activities within citrus plants. Furthermore, based on previously obtained transcriptome data related to Huanglongbing (HLB) from our laboratory [30], we identified nine candidate DnaJ genes for further investigation. Notably, the expression patterns of the DnaJ genes Ciclev10002746m.1 (ClDJC47-1), Ciclev10029600m.1 (ClDJC24), Ciclev10011366m.1 (ClDJB9), Ciclev10008750m.1 (ClDJB14), and Ciclev10030731m.1 (ClDJC41-1) exhibited a consistent decrease in expression levels as the severity of HLB infection escalated, as evidenced by quantitative real-time PCR (qRT-PCR) analysis. This downregulation of DnaJ gene expression in response to HLB infection highlights their potential involvement in the plant's defense mechanisms against this devastating disease. The correlation between the severity of infection and the expression of these genes underscores the importance of investigating their functional roles in the context of citrus immunity. Future research should focus on elucidating the specific contributions of these DnaJ genes to the overall stress response and exploring their potential applications in enhancing disease resistance in citrus through targeted gene-editing approaches.

To investigate the biological function of the ClDJC24 gene, we examined its interactions with other HSP70 proteins and evaluated the effects of gene silencing in citrus during HLB infection using yeast two-hybrid technology and AuNPs-PEI + dsRNA delivery systems. Our yeast two-hybrid assay results indicated that ClDJC24 does not interact with HSP70 proteins (Figure S3). As a member of the HSP40 protein subfamily, DnaJ genes are characterized by the presence of a conserved DnaJ domain, which typically facilitates interaction with HSP70 proteins [3,7,48,49]. However, our results are inconsistent with previous research findings, suggesting that certain HSP70 family members may have evolved distinct functions in citrus. Additionally, we also synthesized 1 mM gold nanoparticles polyethylenimine (AuNPs-PEI) as previously described [31,32]. The binding assay of AuNPs-PEI and dsRNA revealed that a 10:1 volume ratio effectively silences the target gene (Figure 6D). Previous studies from our laboratory demonstrated that gene silencing was effective at a 9:1 ratio [31,32]. Consistent with these findings, we opted to use the 10:1 volume ratio for further analyses.

In examining the effects of ClDJC24 silencing in HLB-infected citrus leaves, we observed that the cycle threshold (Ct) values of CLas did not significantly change in 3, 7, and 14 d post-treatment compared to the controls. This lack of change in CLas levels suggests that ClDJC24 may not directly influence CLas proliferation under the conditions tested. However, we did find that the expression patterns of immune-related genes *CrNPR1*, *CrPR1*, and *CrGAPA* were inhibited by *ClDJC24* silencing, indicating a potential role for *ClDJC24* in negatively regulating disease resistance. This aligns with previous findings that *SDE420* can inhibit host immunity by interacting with cytoplasmic glyceraldehyde-3-

phosphate dehydrogenase (GAPDH), which in turn regulates autophagy [50]. Additionally, *AtGAPA* has been identified as a defense-related target protein of SDEs that inhibits host defenses in *Arabidopsis thaliana* [51]. In addition, subcellular localization of *DnaJ* genes in Citrus was predicted in chloroplast (chlo), nucleus (nucl), and cytoplasm (cyto). Subcellular localization results indicated that the *ClDJC24* was expressed in the nucleus nuclear and cytoplasm (Figure 7G). This result is consistent with the localization of the homologous *DnaJ* gene in *Arabidopsis thaliana* [52]. This nuclear localization suggests that ClDJC24 may play a significant role in the regulation of gene expression and cellular responses to stress, further emphasizing the need for additional studies to elucidate its precise functions in the context of HLB and citrus immunity.

In conclusion, we conducted a comprehensive genome-wide identification and characterization of the DnaJ family genes in Citrus. This study encompassed the analysis of phylogenetic relationships, gene structures, chromosomal locations, and expression patterns for a total of 86 *DnaJ* genes. Based on the results of the expression analysis of *ClDJC24*, the functional role of *ClDJC24* was further investigated. *ClDJC24* gene may be a negative regulator of disease resistance in Citrus for its silencing effect. Yeast two-hybrid (Y2H) assay results demonstrated that the ClDJC24 protein does not interact with HSP70 proteins. This study offers valuable insights for further functional characterization and establishes a theoretical basis for the development of disease-resistant Citrus cultivars.

4. Materials and Methods

4.1. Identification of DnaJ Genes in Citrus

A Hidden Markov Model (HMM) profile of the DnaJ domain (Pfam: PF00226) was downloaded from the Pfam database (http://pfam.xfam.org, accessed 11 April 2024) and was adopted for the identification of *DnaJ* genes in the Citrus genome (https://www.citrusgenomedb.org/, accessed 11 April 2024). The threshold was 0.01, and the parameter was set to the default value. To confirm the presence of the DnaJ domain, a batch search of the sequences of all the obtained *DnaJ* genes in the Citrus was conducted using the online databases SMART (http://smart.embl.de/smart/, accessed 15 April 2024), NCBI Conserved Domain Search databases (NCBI CDD) (https://www.ncbi.nlm.nih.gov/cdd/, accessed 15 April 2024) and Pfam (http://pfam.xfam.org/, accessed 15 April 2024). Redundant sequences were manually deleted. Furthermore, the amino acid numbers (AAs), molecular weights (Mw), and isoelectric points (pI) of the identified *DnaJ* genes were obtained using the ExPasy website (http://web.expasy.org/protparam/, accessed 5 May 2024). Additionally, using the online WoLF PSORT (https://www.genscript.com/wolf-psort.html, accessed 8 June 2024) predicted subcellular localization.

4.2. Chromosomal Locations and Gene Structure Analysis and Cis-Element Analysis of DnaJ Genes in Citrus

According to the locations of the identified *DnaJ* genes in Citrus in the database, they were obtained from annotated gff3 files. All the *DnaJ* genes were mapped to the banana chromosomes with MapChart software according to their positions in the database [53]. An online program of the gene structure display server (GSDS2.0) (https://gsds.gao-lab.org/index.php, accessed 10 June 2024) was applied to draw the exon/intron organization of each *DnaJ* gene in Citrus by comparing the cDNAs with their corresponding full-length sequences [54]. The MEME tool (http://meme-suite.org/tools/meme, accessed 15 June 2024) was used to identify conserved motifs of 86 *DnaJ* gene proteins [55]. Parameters applied were as follows: the number of motifs searched was set as 10, and the limits of motif widths were between 6 and 50 residues [56]. For cis-acting element analysis, genomic DNA sequences in the upstream 1500 bp region of all *DnaJ* genes were obtained, and the sequences were identified using the PlantCARE database (http://bioinformatics.psb.ugent.be/webtools/plantcare/html/, accessed 5 October 2024) [57].

4.3. Phylogenetic Analysis and Heat Map Construction of DnaJ Genes from Citrus

The phylogenetic analysis of the DnaJ gene family in Citrus was performed using MEGA X with 1000 bootstrap replicates. The phylogenetic tree was constructed employing the neighbor-joining (NJ) method and pairwise gap deletion mode [56]. The domain information was retrieved from the SMART database (http://smart.emblheidelberg.de/, accessed 2 August 2024). To further insights into the expression patterns of *DnaJ* genes, FPKM counts were regarded as thresholds to define significant differences in gene expression. *DnaJ* gene expression levels were analyzed in relation to the expression patterns in HLB. RNA-Seq data were visualized with heat maps using TBtools [58]. In the transcriptome data, sample A (A-1, A-2, A-3): post-treatment sample (new tissues), 90 days after FOS and CAC combination therapy; Sample B (B-1, B-2, B-3): pre-treatment sample. A-1, A-2, A-3, and B-1, B-2, B-3 represented no HLB-infected and HLB-infected with three biological replicates, respectively. The treatment method for Group A is as follows: the roots of the citrus trees were first soaked in 1 L of Fosthiazate solution at a concentration of 480 μg/mL, applied as a soil drench (75% Fosthiazate from Hebei Sannong Agricultural Chemical Co., Ltd., Shijiazhuang, China). After 3 to 5 days, we applied 1 L of 200 μg/mL CAC to each citrus sample in the greenhouse through root drenches (the Fosthiazate or CAC dosage was increased to 3 L per tree in the orchard). The samples were collected 90 days post-treatment.

4.4. RNA Extraction and Quantitative Reverse Transcription qRT-PCR

Total RNA was extracted with the CTAB method [51,59]. Briefly, qRT-PCR was performed on a StepOne real-time PCR system with ChamQ Universal SYBR qPCR Master Mix (Vazyme Biotech, Nanjing, China) following the manufacturer's instructions. The total RNA was treated to eliminate any potential contamination with DNA using gDNA Eraser (TaKaRa, Dalian, Japan). Subsequently, DNA-free total RNA was reverse transcribed into cDNA with a reverse transcription kit (TaKaRa, Dalian, Japan), following the manufacturer's instructions. Real-time Quantitative Polymerase Chain Reaction (qRT-PCR) was performed as described by Li et al. [56]. Primer pairs of 9 *DnaJ* genes were designed using the online software Primer 3 (http://primer3.ut.ee/, accessed 8 April 2024). The primers are listed in Table 2. Gene relative expression levels were normalized to that of the reference gene Citrus *Actin* [60]. The expression levels were calculated using the $2^{-\Delta\Delta Ct}$ method [61]. There were three independent biological replicates of each sample type, and the data were shown as the mean ± standard deviation.

4.5. Subcellular Localization

ClDJC24 genes were cloned from cDNA generated from citrus leaves following the manufacturer's instructions (Max Master Mix, Vazyme Biotech, Nanjing, China). The amplified fragments were ligated into the pUBQ10-GFP empty vector using the In-Fusion Cloning Kit (Vazyme Biotech, Nanjing, China). The primers used for plasmid construction are listed in Table S1. Successfully constructed ClDJC24-GFP vectors were transformed into Agrobacterium tumefaciens strain GV3101. Agrobacterium-mediated transient expression in *Nicotiana benthamiana* leaves for 72 h was carried out as previously described in Huang's methods [36,62]. Using LSM880 confocal microscope (Carl Zeiss, Oberkochen, Germany) subcellular localization of ClDJC24-GFP was observed. The excitation wavelength for GFP was 488 nm, and the detection wavelength ranged from 500 to 550 nm [51].

4.6. Yeast Two-Hybrid (Y2H) Assays

The yeast two-hybrid method was performed according to the protocol described by Huang et al. [36,62]. The Y2H system was used to identify protein–protein interactions in vitro. The coding sequence (CDS) of *ClDJC24* was cloned in frame into the bait vector pGBKT7, while the CDSs of HSP90 and HSP70 were cloned in frame into the prey vector pGADT7, which all used the In-Fusion Cloning Kit (Vazyme Biotech, Nanjing, China). The primers used for plasmid construction are listed in Table S1. Following the manufacturer's protocol, the Y2H Gold yeast strain was cotransformed with the resulting bait and prey vectors to identify positive interactions (Clontech, Mountain View, CA, USA). The co-transformants pGBKT7-53 + pGADT7-T and pGBKT7-lam + pGADT7-T serve as the positive and negative controls, respectively. The positive interactors were screened on SD/–leucine–tryptophan–adenine–histidine (SD/-LWAH) medium supplemented with X-α-gal at 30 °C in the dark for approximately 3 days. Meanwhile, SD/–leucine–tryptophan (SD/-LW) medium was used to screen the positive transformants. X-α-gal was used to confirm positive interactions by turning the colonies blue. The experiment's analysis involved observing whether yeast colonies grew on SD/-LWAH and SD/-LW media and whether colonies on SD/-LWAH medium with added X-α-gal turned blue. If the colonies grew on SD/-LW and turned blue on SD/-LWAH after approximately 3 days, the results indicated positive interactions. The plates were then photographed using a Nikon camera.

4.7. AuNPs-PEI Synthesis

Prepare 14 mL PEI (branched polyethylenimine, 25KD, Shyuanye, Shanghai, China) solution (final concentration 0.2 mM) and 1 mL chloroauric acid ($HAuCl_4$, Aladdin, Shanghai, China) solution (final concentration 2 mM) with double distilled water (ddH_2O), respectively. Added PEI solution to $HAuCl_4$ solution and slowly heated to 85 °C until the solution was ruby red in color. Continued to stir until the solution cooled to room temperature to obtain AuNPs-PEI [32]. Centrifuged at 70,403× *g* for 15 min to remove the free PEI from the supernatant and resuspended the pellet in 200 mL ddH_2O. Particle size and zeta potential detection 800 mL of AuNPs-PEI or AuNPs-PEI-pNPR1-GFP ten-fold dilution were tested by Zetasizer Nano ZS90 (Malvern Instruments, Malvern, PA, USA) [35,36].

4.8. Gene Silencing

The dsRNAs, approximately 300 bp in size and complementary to *ClDJC24*, were synthesized using the T7 RNAi Transcription Kit following the manufacturer's protocol (Vazyme Biotech, Nanjing, China). The PCR technique was utilized to add the T7 promoter sequence to both ends of the RNA interference (RNAi) target fragments. Details of the primers used for PCR are provided in Table 1. The sequences containing the T7 promoters were purified and used as templates for dsRNA amplification. Subsequent infection assays were conducted to establish the effects of knocking out target genes, following the method described by Huang [36,62]. HLB-infected citrus leaves were detached and uniformly wounded using the tip of a pipette head. Subsequently, 10 μL of dsRNA and AuNPs-PEI solution at a concentration of 500 ng/μL were applied to the abaxial surface of the leaves with 0.02% Silwet L-77 (Solarbio Science & Technology, Beijing, China). As a control, corresponding sites on the same leaves were treated with 10 μL of water and AuNPs-PEI solution containing 0.02% Silwet L-77. Silwet L-77, a surfactant, was used to enhance the adsorption and spread of dsRNA or water on the Citrus leaves [36]. Ten HLB-infected Citrus leaves were used for each treatment. Each experiment was independently repeated three times. The leaves were also harvested for qRT-PCR to verify gene silencing in 3 d, 7 d, and 14 d.

4.9. Detection of Citrus Huanglongbing (HLB) Bacterial Titer

Using copy number as the indicator: the Ct value is first obtained using plant genomic DNA as the template, and the copy number is then calculated based on the generated standard curve. A higher copy number indicates a higher bacterial titer. Using genomic DNA from Catharanthus roseus infected with Huanglongbing (HLB) as a template, specific primers HLBas and HLBr were used to amplify the 16S rDNA sequence of CLas. The 16S rDNA sequence was verified by 1.5% agarose gel electrophoresis, and the product was recovered and ligated into the pMD19-T vector to construct the recombinant plasmid pMD19-T-16S rDNA (p16S rDNA), which was then transformed into *E. coli* DH5α. Single colonies were selected for colony identification, and if correctly identified, the culture was expanded, and plasmids were extracted and quantified. The plasmids were then diluted with DEPC water to 10^{-1}, 10^{-2}, 10^{-3}, 10^{-4}, 10^{-5}, 10^{-6}, 10^{-7}, and 10^{-8} gradients.

Using the diluted plasmid p16S rDNA as a template, absolute quantitative PCR was performed with a 2× TaqMan Fast qPCR kit (B639275, Sangon Biotech (Shanghai) Co., Ltd China). The primer and probe sequences are HLBas (TCGAGCGCGTATGCAATACG), HLBr (GCGTTATCCCGTAGAAAAAGGTAG), and HLBp (AGACGGGTGAGTAACGCG). The following program: 95 °C for 5 min; 95 °C for 30 s; 60 °C for 30 s; 40 cycles. Each sample was run with at least three technical replicates, and a standard curve was plotted based on the copy number of CLas and Ct values. The formula for calculating the copy number is as follows: $A = (M \times N) \div (L \times D)$. A: CLas copy number; M: mass concentration (g/mL) = DNA (ng/mL) $\times 10^{-6}$; N: Avogadro's constant (6.022×10^{23} molecules/mol); L: DNA length, which is the number of base pairs in the recombinant plasmid containing the target gene, in kb; D: Conversion factor (6.6×10^{5} g/mol/kb).

4.10. Limitations of the Study Related to the Species Source of the Data

The database used for this transcriptome sequencing is *Citrus clementina*, while the expression analysis, including qRT-PCR and the expression levels of immune genes (*CrNPR1*, *CrPR1*, *CrPR2*, *CrGAPA*), is based on *Citrus reticulata*. Since there was no database available for *Citrus reticulata* at the time, we chose a phylogenetically close database for the transcriptome sequencing. In this study, to ensure consistency with the transcriptome database, we selected the *Citrus clementina* database for analysis.

Supplementary Materials: The following supporting information can be downloaded at: https://www.mdpi.com/article/10.3390/ijms252211967/s1.

Author Contributions: W.C. and Y.Z. conceived the study, participated in its design and coordination, performed the experiments, and collected, analyzed, and deposited the data. H.H., X.W., B.Z., Y.M., L.W. and H.C. performed the experiments and qRT-PCR analysis. Y.T., W.C. and X.D. prepared the draft and proofread the final draft. Y.Z. provided guidelines and revised the manuscript. All authors have read and agreed to the published version of the manuscript.

Funding: This research was supported by the National Natural Science Foundation of China (31570256), the open competition program of the top nine critical priorities of Agricultural Science and Technology Innovation for the 14th Five-year Plan of Guangdong Province (2023SDZG06), Guangzhou Basic and Applied Basic Project (2023A04J0139) and the Key-Area Research and Development Program of Guangdong Province (2018B020202009).

Data Availability Statement: Data contained within the article.

Conflicts of Interest: The authors declared that the research was conducted in the absence of any commercial or financial relationships that could be construed as a potential conflict of interest.

References

1. Tutar, L.; Tutar, Y. Heat shock proteins; an overview. *Curr. Pharm. Biotechnol.* **2010**, *11*, 216–222. [CrossRef] [PubMed]
2. Muthusamy, S.K.; Dalal, M.; Chinnusamy, V.; Bansal, K.C. Genome-wide identification and analysis of biotic and abiotic stress regulation of small heat shock protein (HSP20) family genes in bread wheat. *J. Plant Physiol.* **2017**, *211*, 100–113. [CrossRef] [PubMed]
3. Pulido, P.; Leister, D. Novel dnaj-related proteins in *Arabidopsis thaliana*. *New Phytol.* **2018**, *217*, 480–490. [CrossRef] [PubMed]
4. Walsh, P.; Bursać, D.; Law, Y.C.; Cyr, D.; Lithgow, T. The j-protein family: Modulating protein assembly, disassembly and translocation. *EMBO Rep.* **2004**, *5*, 567–571. [CrossRef] [PubMed]
5. Dang, W.; Zhang, M.; Sun, L. Edwardsiella tarda dnaj is a virulence-associated molecular chaperone with immunoprotective potential. *Fish Shellfish Immunol.* **2011**, *31*, 182–188. [CrossRef]
6. Solana, J.C.; Bernardo, L.; Moreno, J.; Aguado, B.; Requena, J.M. The astonishing large family of HSP40/Dnaj proteins existing in leishmania. *Genes* **2022**, *13*, 742. [CrossRef]
7. Qiu, X.B.; Shao, Y.M.; Miao, S.; Wang, L. The diversity of the Dnaj/HSP40 family, the crucial partners for HSP70 chaperones. *Cell. Mol. Life Sci.* **2006**, *63*, 2560–2570. [CrossRef]
8. Takaya, A.; Tomoyasu, T.; Matsui, H.; Yamamoto, T. The Dnak/Dnaj chaperone machinery of salmonella enterica serovar typhimurium is essential for invasion of epithelial cells and survival within macrophages, leading to systemic infection. *Infect. Immun.* **2004**, *72*, 1364–1373. [CrossRef]

9. Liu, J.Z.; Whitham, S.A. Overexpression of a soybean nuclear localized type-iii dnaj domain-containing HSP40 reveals its roles in cell death and disease resistance. *Plant J. Cell Mol. Biol.* **2013**, *74*, 110–121. [CrossRef]
10. Wang, G.; Zhou, S.; Luo, Y.; Ma, C.; Gong, Y.; Zhou, Y.; Gao, S.; Huang, Z.; Yan, L.; Hu, Y.; et al. The heat shock protein 40 lednaj regulates stress resistance and indole-3-acetic acid biosynthesis in lentinula edodes. *Fungal Genet. Biol.* **2018**, *118*, 37–44. [CrossRef]
11. Kampinga, H.H.; Craig, E.A. The HSP70 chaperone machinery: J proteins as drivers of functional specificity. *Nat. Rev. Mol. Cell Biol.* **2010**, *11*, 579–592. [CrossRef] [PubMed]
12. Liu, T.-T.; Xu, M.-Z.; Gao, S.-Q.; Zhang, Y.; Hu, Y.; Jin, P.; Cai, L.-N.; Cheng, Y.; Chen, J.-P.; Yang, J.; et al. Genome-wide identification and analysis of the regulation wheat Dnaj family genes following wheat yellow mosaic virus infection. *J. Integr. Agric.* **2022**, *21*, 153–169. [CrossRef]
13. Clerico, E.M.; Meng, W.; Pozhidaeva, A.; Bhasne, K.; Petridis, C.; Gierasch, L.M. HSP70 molecular chaperones: Multifunctional allosteric holding and unfolding machines. *Biochem. J.* **2019**, *476*, 1653–1677. [CrossRef]
14. Cao, Y.Q.; Yuan, L.; Zhao, Q.; Yuan, J.L.; Miao, C.; Chang, Y.F.; Wen, X.T.; Wu, R.; Huang, X.B.; Wen, Y.P.; et al. HSP40 protein dnajb6 interacts with viral NS3 and inhibits the replication of the japanese encephalitis virus. *Int. J. Mol. Sci.* **2019**, *20*, 5719. [CrossRef] [PubMed]
15. Zhang, B.; Qiu, H.L.; Qu, D.H.; Ruan, Y.; Chen, D.H. Phylogeny-dominant classification of j-proteins in *Arabidopsis thaliana* and *Brassica oleracea*. *Genome* **2018**, *61*, 405–415. [CrossRef] [PubMed]
16. Luo, Y.; Fang, B.; Wang, W.; Yang, Y.; Rao, L.; Zhang, C. Genome-wide analysis of the rice j-protein family: Identification, genomic organization, and expression profiles under multiple stresses. *3 Biotech* **2019**, *9*, 358. [CrossRef]
17. Knox, C.; Luke, G.A.; Blatch, G.L.; Pesce, E.-R. Heat shock protein 40 (HSP40) plays a key role in the virus life cycle. *Virus Res.* **2011**, *160*, 15–24. [CrossRef]
18. Bekh-Ochir, D.; Shimada, S.; Yamagami, A.; Kanda, S.; Ogawa, K.; Nakazawa, M.; Matsui, M.; Sakuta, M.; Osada, H.; Asami, T.; et al. A novel mitochondrial Dnaj/HSP40 family protein bil2 promotes plant growth and resistance against environmental stress in brassinosteroid signaling. *Planta* **2013**, *237*, 1509–1525. [CrossRef]
19. Sporn, Z.A.; Hines, J.K. Hsp40 function in yeast prion propagation: Amyloid diversity necessitates chaperone functional complexity. *Prion* **2015**, *9*, 80–89. [CrossRef]
20. Park, M.Y.; Kim, S.Y. The Arabidopsis j protein ATJ1 is essential for seedling growth, flowering time control and aba response. *Plant Cell Physiol.* **2014**, *55*, 2152–2163. [CrossRef]
21. Wang, G.; Cai, G.; Kong, F.; Deng, Y.; Ma, N.; Meng, Q. Overexpression of tomato chloroplast-targeted Dnaj protein enhances tolerance to drought stress and resistance to pseudomonas solanacearum in transgenic tobacco. *Plant Physiol. Biochem.* **2014**, *82*, 95–104. [CrossRef] [PubMed]
22. Hwang, D.J.; Tumer, N.E.; Wilson, T.M. Chaperone protein grpe and the groel/groes complex promote the correct folding of tobacco mosaic virus coat protein for ribonucleocapsid assembly in vivo. *Arch. Virol.* **1998**, *143*, 2203–2214. [CrossRef] [PubMed]
23. Hofius, D.; Maier, A.T.; Dietrich, C.; Jungkunz, I.; Börnke, F.; Maiss, E.; Sonnewald, U. Capsid protein-mediated recruitment of host Dnaj-like proteins is required for potato virus y infection in tobacco plants. *J. Virol.* **2007**, *81*, 11870–11880. [CrossRef] [PubMed]
24. Shimizu, T.; Yoshii, A.; Sakurai, K.; Hamada, K.; Yamaji, Y.; Suzuki, M.; Namba, S.; Hibi, T. Identification of a novel tobacco Dnaj-like protein that interacts with the movement protein of tobacco mosaic virus. *Arch. Virol.* **2009**, *154*, 959–967. [CrossRef]
25. Zong, T.; Yin, J.; Jin, T.; Wang, L.; Luo, M.; Li, K.; Zhi, H. A Dnaj protein that interacts with soybean mosaic virus coat protein serves as a key susceptibility factor for viral infection. *Virus Res.* **2020**, *281*, 197870. [CrossRef]
26. Urano, E.; Morikawa, Y.; Komano, J. Novel role of HSP40/Dnaj in the regulation of hiv-1 replication. *J. Acquir. Immune Defic. Syndr. (1999)* **2013**, *64*, 154–162. [CrossRef]
27. Tipu, M.M.H.; Masud, M.M.; Jahan, R.; Baroi, A.; Hoque, A. Identification of citrus greening based on visual symptoms: A grower's diagnostic toolkit. *Heliyon* **2021**, *7*, e08387. [CrossRef]
28. Merli, M.L.; Padgett-Pagliai, K.A.; Cuaycal, A.E.; Garcia, L.; Marano, M.R.; Lorca, G.L.; Gonzalez, C.F. '*Candidatus* liberibacter asiaticus' multimeric lotp mediates citrus sinensis defense response activation. *Front. Microbiol.* **2021**, *12*, 661547. [CrossRef] [PubMed]
29. Jacob, P.; Hirt, H.; Bendahmane, A. The heat-shock protein/chaperone network and multiple stress resistance. *Plant Biotechnol. J.* **2017**, *15*, 405–414. [CrossRef]
30. Duan, J.; Li, X.; Zhang, J.; Cheng, B.; Liu, S.; Li, H.; Zhou, Q.; Chen, W. Cocktail therapy of fosthiazate and cupric-ammoniun complex for citrus huanglongbing. *Front. Plant Sci.* **2021**, *12*, 643971. [CrossRef]
31. Lei, W.X.; An, Z.S.; Zhang, B.H.; Wu, Q.; Gong, W.J.; Li, J.M.; Chen, W.L. Construction of gold-sirna (NPR1) nanoparticles for effective and quick silencing of npr1 in *Arabidopsis thaliana*. *RSC Adv.* **2020**, *10*, 19300–19308. [CrossRef] [PubMed]
32. Zhang, J.; Lei, W.; Meng, Y.; Zhou, C.; Zhang, B.; Yuan, J.; Wang, M.; Xu, D.; Meng, X.; Chen, W. Expression of pei-coated gold nanoparticles carrying exogenous gene in periwinkle mesophyll cells and its practice in huanglongbing research. *iScience* **2022**, *25*, 104479. [CrossRef]
33. Graczyk, A.; Pawlowska, R.; Jedrzejczyk, D.; Chworos, A. Gold nanoparticles in conjunction with nucleic acids as a modern molecular system for cellular delivery. *Molecules* **2020**, *25*, 204. [CrossRef] [PubMed]

34. Li, M.; Guo, Z.; Yang, H.; Liu, Y.; Tong, Y. Electrochemical sensor based on polyethyleneimine-aunps- anthraquinone-2-carboxylic acid nanocomposite for cysteine detection. *Int. J. Electrochem. Sci.* **2019**, *14*, 943–956. [CrossRef]
35. Brondani, D.; de Souza, B.; Souza, B.S.; Neves, A.; Vieira, I.C. Pei-coated gold nanoparticles decorated with laccase: A new platform for direct electrochemistry of enzymes and biosensing applications. *Biosens. Bioelectron.* **2013**, *42*, 242–247. [CrossRef]
36. Huang, H.; Tian, Y.; Huo, Y.; Liu, Y.; Yang, W.; Li, Y.; Zhuo, M.; Xiang, D.; Li, C.; Yi, G.; et al. The autophagy-related *musa acuminata* protein MaATG8F interacts with maatg4b, regulating banana disease resistance to *Fusarium oxysporum* f. Sp. *Cubense* tropical race 4. *J. Fungi* **2024**, *10*, 91. [CrossRef]
37. Zaynab, M.; Sharif, Y.; Fatima, M.; Afzal, M.Z.; Aslam, M.M.; Raza, M.F.; Anwar, M.; Raza, M.A.; Sajjad, N.; Yang, X.; et al. Crispr/cas9 to generate plant immunity against pathogen. *Microb. Pathog.* **2020**, *141*, 103996. [CrossRef] [PubMed]
38. Chen, K.; Wang, Y.; Zhang, R.; Zhang, H.; Gao, C. Crispr/cas genome editing and precision plant breeding in agriculture. *Annu. Rev. Plant Biol.* **2019**, *70*, 667–697. [CrossRef]
39. Samanta, M.K.; Dey, A.; Gayen, S. Crispr/cas9: An advanced tool for editing plant genomes. *Transgenic Res.* **2016**, *25*, 561–573. [CrossRef]
40. Verchot, J. Cellular chaperones and folding enzymes are vital contributors to membrane bound replication and movement complexes during plant rna virus infection. *Front. Plant Sci.* **2012**, *3*, 275. [CrossRef]
41. Wang, G.; Li, M.; Cheng, H.; Zhang, C.; Deng, W.; Li, T. Expression profiling of cordyceps dnaj protein family in tolypocladium guangdongense during developmental and temperature stress processes. *Gene* **2020**, *743*, 144563. [CrossRef] [PubMed]
42. Fan, F.F.; Liu, F.; Yang, X.; Wan, H.; Kang, Y. Global analysis of expression profile of members of DnaJ gene families involved in capsaicinoids synthesis in pepper (*Capsicum annuum* L). *BMC Plant Biol.* **2020**, *20*, 326. [CrossRef] [PubMed]
43. Abdullah; Faraji, S.; Mehmood, F.; Malik, H.M.T.; Poczai, P.J.A. The GASA gene family in theobroma cacao: Genome wide identification and expression analyses. *BioRxiv* **2021**, *11*, 1425. [CrossRef]
44. Musavizadeh, Z.; Najafi-Zarrini, H.; Kazemitabar, S.K.; Hashemi, S.H.; Faraji, S.; Barcaccia, G.; Heidari, P. Genome-wide analysis of potassium channel genes in rice: Expression of the osakt and oskat genes under salt stress. *Genes* **2021**, *12*, 784. [CrossRef]
45. Li, Y.; Li, W.; Hu, D.; Shen, P.; Zhang, G.; Zhu, Y. Comparative analysis of the metabolome and transcriptome between green and albino zones of variegated leaves from hydrangea macrophylla 'maculata' infected by hydrangea ringspot virus. *Plant Physiol. Biochem.* **2020**, *157*, 195–210. [CrossRef]
46. Wang, T.; Liu, M.; Wu, Y.; Tian, Y.; Han, Y.; Liu, C.; Hao, J.; Fan, S. Genome-wide identification and expression analysis of mapk gene family in lettuce (*lactuca sativa* L.) and functional analysis of lsmapk4 in high- temperature-induced bolting. *Int. J. Mol. Sci.* **2022**, *23*, 11129. [CrossRef]
47. Valassakis, C.; Livanos, P.; Minopetrou, M.; Haralampidis, K.; Roussis, A. Promoter analysis and functional implications of the selenium binding protein (SBP) gene family in *Arabidopsis thaliana*. *J. Plant Physiol.* **2018**, *224–225*, 19–29. [CrossRef]
48. Craig, E.A.; Huang, P.; Aron, R.; Andrew, A. The diverse roles of j-proteins, the obligate HSP70 co-chaperone. *Rev. Physiol. Biochem. Pharmacol.* **2006**, *156*, 1–21. [CrossRef]
49. Miernyk, J.A. The j-domain proteins of *Arabidopsis thaliana*: An unexpectedly large and diverse family of chaperones. *Cell Stress Chaperones* **2001**, *6*, 209–218. [CrossRef]
50. Shi, J.; Gong, Y.; Shi, H.; Ma, X.; Zhu, Y.; Yang, F.; Wang, D.; Fu, Y.; Lin, Y.; Yang, N.; et al. '*Candidatus* liberibacter asiaticus' secretory protein sde3 inhibits host autophagy to promote huanglongbing disease in citrus. *Autophagy* **2023**, *19*, 2558–2574. [CrossRef]
51. Li, X.; Guo, Z.; Zhou, Y.; Zhang, B.; Ruan, H.; Chen, W. Three new discovery effector proteins from *Candidatus* liberibacter asiaticus psy62 inhibit plant defense through interaction with ATCAT3 and ATGAPA. *Plant Cell Rep.* **2024**, *43*, 130. [CrossRef] [PubMed]
52. Chen, K.M.; Holmström, M.; Raksajit, W.; Suorsa, M.; Piippo, M.; Aro, E.M. Small chloroplast-targeted Dnaj proteins are involved in optimization of photosynthetic reactions in *Arabidopsis thaliana*. *BMC Plant Biol.* **2010**, *10*, 43. [CrossRef] [PubMed]
53. Xie, T.; Chen, C.; Li, C.; Liu, J.; Liu, C.; He, Y. Genome-wide investigation of WRKY gene family in pineapple: Evolution and expression profiles during development and stress. *BMC Genom.* **2018**, *19*, 490. [CrossRef] [PubMed]
54. Hu, B.; Jin, J.; Guo, A.Y.; Zhang, H.; Luo, J.; Gao, G. GSDS 2.0: An upgraded gene feature visualization server. *Bioinformatics* **2015**, *31*, 1296–1297. [CrossRef]
55. Bailey, T.L.; Boden, M.; Buske, F.A.; Frith, M.; Grant, C.E.; Clementi, L.; Ren, J.; Li, W.W.; Noble, W.S. Meme suite: Tools for motif discovery and searching. *Nucleic Acids Res.* **2009**, *37*, W202–W208. [CrossRef]
56. Li, B.; Fan, R.; Yang, Q.; Hu, C.; Sheng, O.; Deng, G.; Dong, T.; Li, C.; Peng, X.; Bi, F.; et al. Genome-wide identification and characterization of the NAC transcription factor family in *musa acuminata* and expression analysis during fruit ripening. *Int. J. Mol. Sci.* **2020**, *21*, 634. [CrossRef]
57. Lescot, M.; Déhais, P.; Thijs, G.; Marchal, K.; Moreau, Y.; Van de Peer, Y.; Rouzé, P.; Rombauts, S. Plantcare, a database of plant cis-acting regulatory elements and a portal to tools for in silico analysis of promoter sequences. *Nucleic Acids Res.* **2002**, *30*, 325–327. [CrossRef]
58. Chen, C.; Chen, H.; Zhang, Y.; Thomas, H.R.; Frank, M.H.; He, Y.; Xia, R. TBtools: An integrative toolkit developed for interactive analyses of big biological data. *Mol. Plant* **2020**, *13*, 1194–1202. [CrossRef]
59. Mushtaq, S.; Tayyeb, A.; Firdaus, E.B. A comparison of total RNA extraction methods for RT-PCR based differential expression of genes from trichoderma atrobrunneum. *J. Microbiol. Methods* **2022**, *200*, 106535. [CrossRef]

60. Zhao, H.; Sun, R.; Albrecht, U.; Padmanabhan, C.; Wang, A.; Coffey, M.D.; Girke, T.; Wang, Z.; Close, T.J.; Roose, M.; et al. Small RNA profiling reveals phosphorus deficiency as a contributing factor in symptom expression for citrus huanglongbing disease. *Mol. Plant* **2013**, *6*, 301–310. [CrossRef]
61. Takeuchi, T.; Yamazaki, Y.; Katoh-Fukui, Y.; Tsuchiya, R.; Kondo, S.; Motoyama, J.; Higashinakagawa, T. Gene trap capture of a novel mouse gene, jumonji, required for neural tube formation. *Genes Dev.* **1995**, *9*, 1211–1222. [CrossRef] [PubMed]
62. Huang, H.; Zhang, X.; Zhang, Y.; Yi, G.; Xie, J.; Viljoen, A.; Wang, W.; Mostert, D.; Fu, G.; Peng, C.; et al. Fovecm33, a gpi-anchored protein, regulates vegetative growth and virulence in *Fusarium oxysporum* f. Sp. *Cubense* tropical race 4. *Fungal Biol.* **2022**, *126*, 213–223. [CrossRef] [PubMed]

International Journal of
Molecular Sciences

MDPI

Correction

Correction: Tian et al. Genome-Wide Identification of the DnaJ Gene Family in Citrus and Functional Characterization of *ClDJC24* in Response to Citrus Huanglongbing. *Int. J. Mol. Sci.* 2024, *25*, 11967

Yuzhen Tian [1,†], Xizi Wang [1,†], Huoqing Huang [2], Xin Deng [3,4], Baihong Zhang [5], Yixuan Meng [1], Libo Wu [1], Hang Chen [1], Yun Zhong [2,*] and Wenli Chen [1,*]

1 MOE Key Laboratory of Laser Life Science & Institute of Laser Life Science, Guangdong Provincial Key Laboratory of Laser Life Science, Guangzhou Key Laboratory of Spectral Analysis and Functional Probes, College of Biophotonics, School of Optoelectronic Science and Engineering, South China Normal University, Guangzhou 510631, China; tianyzh@126.com (Y.T.); wangxizi1103@163.com (X.W.); 2021023392@m.scnu.com (Y.M.); 2023023373@m.scnu.edu.cn (L.W.); ccc2020xxsc@126.com (H.C.)

2 Institute of Fruit Tree Research, Guangdong Academy of Agricultural Sciences, Key Laboratory of South Sub-Tropical Fruit Biology and Genetic Research Utilization, Ministry of Agriculture and Rural Affairs, Guangdong Provincial Key Laboratory of Science and Technology Research on Fruit Tree, Guangzhou 510640, China; hqhuang07@163.com

3 Department of Biomedical Sciences, City University of Hong Kong, Kowloon Tong, Hong Kong SAR, China; xindeng@cityu.edu.hk

4 Shenzhen Research Institute, City University of Hong Kong, Shenzhen 518057, China

5 Institute of Nanfan & Seed Industry, Guangdong Academy of Science, Guangzhou 510640, China; zhangbyhome@126.com

* Correspondence: zhongyun@gdaas.cn (Y.Z.); chenwl@scnu.edu.cn (W.C.)

† These authors contributed equally to this work.

In the original publication [1], there was an error in Figure 7G as published. The incorrect image was uploaded due to a review oversight. The corrected Figure 7G appears below. The authors state that the scientific conclusions remain unaffected. This correction has been approved by the Academic Editor. The original publication has also been updated.

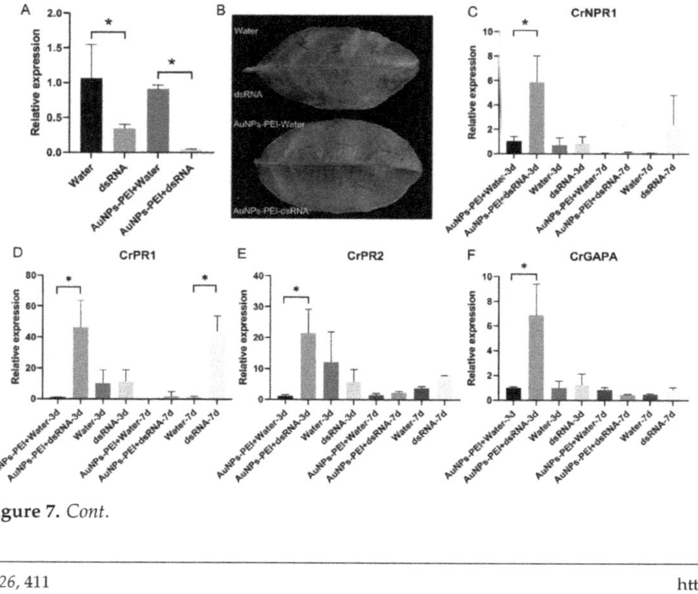

Figure 7. *Cont.*

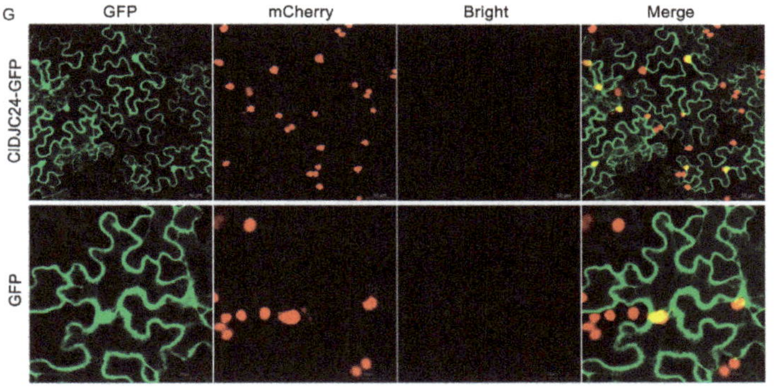

Figure 7. ClDJC24-dsRNA silencing affects the expression patterns of immune genes in response to Citrus HLB. (**A**) The expression level of *ClDJC24* in HLB-infected citrus leaves in 3 d treated with ClDJC24-dsRNA and AuNPs-PEI + ClDJC24-dsRNA, water, and AuNPs-PEI + water was analyzed by qRT-PCR. (**B**) The disease symptoms of dsRNA and AuNPs-PEI + dsRNA, water, and AuNPs-PEI + water treated HLB-infected citrus leaves. (**C–F**) represented the expression patterns of immune genes (*CrNPR1, CrPR1, CrPR2, CrGAPA*) in response to Citrus HLB. Data are presented as means ± SDs ($n = 3$). "*" symbol above the columns indicates a significant difference at $p < 0.05$. (**G**) Subcellular localization of *ClDJC24* uses a nuclear-localized marker transgenic *Nicotiana benthamiana* (NLS-mCherry-OE). The GFP empty vector was used as a control. Photographs were taken at 3 d post-infiltration. Scale bar = 50 μm.

Reference

1. Tian, Y.; Wang, X.; Huang, H.; Deng, X.; Zhang, B.; Meng, Y.; Wu, L.; Chen, H.; Zhong, Y.; Chen, W. Genome-Wide Identification of the DnaJ Gene Family in Citrus and Functional Characterization of *ClDJC24* in Response to Citrus Huanglongbing. *Int. J. Mol. Sci.* **2024**, *25*, 11967. [CrossRef]

International Journal of
Molecular Sciences

MDPI

Article

Identification of the Brassinazole-Resistant (*BZR*) Gene Family in Wheat (*Triticum aestivum* L.) and the Molecular Cloning and Functional Characterization of *TaBZR2.1*

Yan Zhang [1,†], Jingzi Qin [1,2,†], Jinna Hou [1], Congcong Liu [1], Shenghui Geng [1], Maomao Qin [1], Wenxu Li [1], Ziju Dai [1], Zhengqing Wu [1], Zhensheng Lei [1,2,3,*] and Zhengfu Zhou [1,3,*]

1 Henan Institute of Crop Molecular Breeding, Henan Academy of Agricultural Sciences, Zhengzhou 450002, China; zhangyan123@hnagri.org.cn (Y.Z.); qinjingzi@hnagri.org.cn (J.Q.); houjinna@hnagri.org.cn (J.H.); liucongcong@hnagri.org.cn (C.L.); gengshenghui@hnagri.org.cn (S.G.); qinmaomao@hnagri.org.cn (M.Q.); liwenxu@hnagri.org.cn (W.L.); daiziju@hnagri.org.cn (Z.D.); wuzhengqing@hnagri.org.cn (Z.W.)

2 National Key Laboratory of Wheat and Maize Crop Science, College of Agronomy, Henan Agricultural University, Zhengzhou 450002, China

3 The Shennong Laboratory, Zhengzhou 450002, China

* Correspondence: leizhensheng@hnagri.org.cn (Z.L.); zhouzf215@hnagri.org.cn (Z.Z.)

† These authors contributed equally to this work.

Abstract: Brassinazole-resistant (BZR) transcription factors are important transcription factors in Brassinosteroid (BR)-responsive gene expression. However, limited knowledge exists regarding the *BZR* genes in wheat and a limited number of *BZR* family genes have been previously reported in wheat. In this study, the synteny analyses of the *TaBZR* genes suggested that gene duplication events have played an essential role in the *TaBZR* family during evolution. The results of RT-qPCR and transcriptome data analyses exhibited remarkable expression patterns in the *BZR* genes in different tissues and under different treatments. The yeast two-hybrid (Y2H) screen result showed that the TaBZR2.1 protein interacts with Argonaute 4 (AGO4). Taken together, our results not only provide us a basis for understanding the molecular characteristics and expression patterns of the *TaBZR* family genes but also offered the functional characterization of *TaBZR2.1* in wheat.

Keywords: wheat; transcription factor; *BZR*; expression profiles

1. Introduction

Brassinosteroid (BR) is a kind of plant-specific steroidal hormone that plays a crucial role in many plant biological process, such as regulating plant growth, development, and the stress response [1]. It is reported that mutant plants with defects in BR biosynthesis or BR signal components display characteristic phenotypes such as dwarfism [2], constitutive photomorphogenesis in darkness [3], altered fertility [4], changes in seed size and number [5–7], delayed flowering, and defects in xylem differentiation [8]. Unlike animal steroid hormones, plant BRs are perceived by receptors that are located on the cell membrane [9]. Upon receiving the BRs' signal, the cell membrane receptor kinase BRASSINOSTEROID INSENSITIVE1 (BRI1) is activated and can phosphorylate a negative regulator, BRI1 kinase inhibitor 1 (BKI1), inducing its dissociation from the cell membrane [10]. Many plant transcription factors are located downstream of BR signaling events, including key factors such as BRASSINAZOLE-RESISTANT 1 (BZR) [11]. Acting as positive regulators in BR signaling, *Arabidopsis* BES1 and BZR factors are two BES/BZR members of the *BZR* family [12].

BZR acts as an important regulator in maintaining BR homeostasis for plant growth and development [13]. It mediates BR-induced gene expression and suppresses BR biosynthesis through feedback mechanisms [11]. BZR transcription factors induce the BR-induced genes [11,14–16]. In *Arabidopsis*, BZR can be phosphorylated by activating BIN2 and interacting with 14-3-3s to regulate BR signaling, which retained BES1/BZR in the cytosol to prevent BR signaling [17,18]. Dephosphorylation of BZR by PP2A allows it to transfer to the nucleus to regulate BR-induced gene expression [19]. BZR also mediates the ovule and seed number in *Arabidopsis* [5]. It has been reported that BR-activated BZR directly regulates the genes involved in seed size control such as ZmBES1/BZR-5, which positively regulates kernel size by inhibiting *AP2/EREBP* gene transcription [6,7]. Overexpression of *ZmBZR* transgenic *Arabidopsis* results in large organs and seeds [20].

Studies have firmly established that plant development and growth are synergistically regulated by environmental and hormonal signals, with light and BR serving as the two central stimuli regulating plant morphogenesis. It is reported that certain photoreceptors mediate brassinosteroid signaling by interacting with BZR. For instance, by interacting with the photoreceptor phytochrome B (phyB), BZR can repress BR signaling in *Arabidopsis* [21]; by interacting with CRY1, BZR and BIN2 can repress BR signaling [22]; and the interaction of PIF4 and BZR controls a hub of transcription networks, integrating various endogenous and environmental signals [23]. The interaction of BZR and the HY5 mediates the BR and light signaling pathways to regulate the cotyledon opening in seedling photomorphogenesis in *Arabidopsis* [24]. Additionally, BZR also plays a vital role in the response to nitrogen starvation in tomatoes [25].

Furthermore, certain abiotic stresses can induce the expression of *BZRs*, conferring tolerance to stresses. For example, a *TaBZR* gene could be significantly upregulated by salt treatment in wheat, improving salt tolerance through activating the genes involved in ROS scavenging and ABA biosynthesis [26]. Tomato BZR can regulate heat stress responses by ROS signaling [27]. Ectopic expression of a maize *BZR* gene in *Arabidopsis* confers tolerance to osmotic stress [28]. Through CBF-dependent pathways, BZR positively modulates plant freezing tolerance in *Arabidopsis* [29]. Ectopic expression of the maize *ZmBZR1* gene in rice and *Arabidopsis* negatively regulated drought tolerance [30].

The BZR transcription factors have been identified in numerous species [31–33]. Wheat is a globally significant cereal crop globally, providing approximately 20% of the total calories consumed by humans [34]. However, compared to other crop species, our understanding of the BZR transcription factors in wheat remains limited. We do not know whether the wheat *BZR* gene also responds to plant hormones, participates in the stress response, or has a role in seed development. This comparative analysis on gene/protein structure, evolution history, and expression pattern will help to reveal the molecular function of *BZR* family genes and provide gene resources for wheat breeding. Therefore, this study characterized 20 members of the wheat *BZR* gene family and assessed their expression profiles. Additionally, we further characterized the interacting protein of TaBZR2.1 using Y2H assays and performed subcellular localization assays. The findings from this investigation may facilitate further investigation of the molecular functions of the *BZR* family genes in wheat.

2. Results

2.1. Identification of TaBZR Genes in Wheat

The synteny analyses and chromosomal localization of *TaBZR* genes showed that 19 out of the 20 *TaBZR* genes were distributed in 14 of the chromosomes (Figure 1), and the 1 remaining *TaBZR* gene was mapped onto an unattributed scaffold. With the exception of the *BZR* genes on 4B and 4D, the distribution and number of *TaBZR* genes from the A/B/D subgenomes were similar. Among them, *TaBZR4.4* had four transcripts, *TaBZR6* had three transcripts, and *TaBZR4.1* had two transcripts. The CDS and polypeptide length, MW, transcript number, pI, and the predicted subcellular localization were analyzed and shown in Table S1.

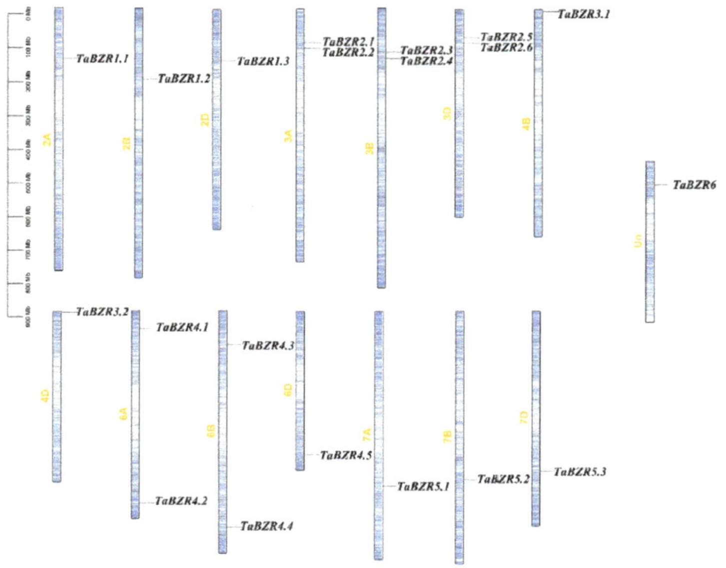

Figure 1. The chromosome distribution of the *TaBZR* genes.

2.2. The Exon–Intron Structure of TaBZR Genes

The gene structures of 20 wheat *BZRs* were investigated in this study (Figure 2). The results revealed that the intron and exon numbers of the *TaBZRs* were different from *TaBZR1.1* to *TaBZR6*. The numbers of exons differed from two to ten, and most of the *TaBZR* genes contained the minimum exon number, i.e., two. Meanwhile, *TaBZR4.1* had ten, and *TaBZR3.1/3.2/4.3/6* contained eight or nine exons. It has been reported that the *TaBZR* genes in tomatoes, Chinese cabbage, and apples do not contain the upstream and downstream structures [35]. Meanwhile, all of the *TaBZR* genes in wheat contained these structures, and all of them had exons and introns.

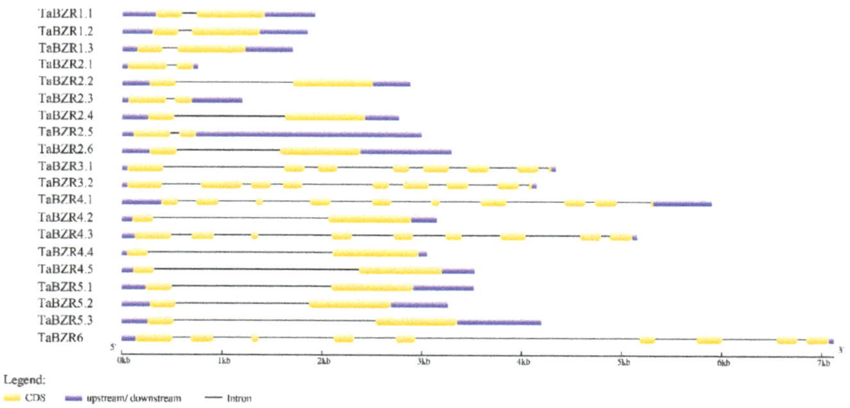

Figure 2. The genomic structure of *TaBZR* genes. The blue boxes in the schematic diagram represent the upstream/downstream sequences; the exon sequences and the introns are represented by yellow boxes and black lines, respectively.

2.3. Analysis of Three-Dimensional Modeling

To further investigate the protein structural effects of the BZRs in wheat, a three-dimensional (3D) model of BZR proteins was constructed on the SWISS-MODEL website, and the optimal model was selected according to Global Model Quality Estimation (GMQE) (Figure 3). The results showed that their structural integrity was preserved throughout their evolution, which is crucial for their function. A phylogenetic analysis of wheat BZR genes was conducted in a previous study, and our results strongly support the reliability of their phylogenetic analysis of the BZR genes [36].

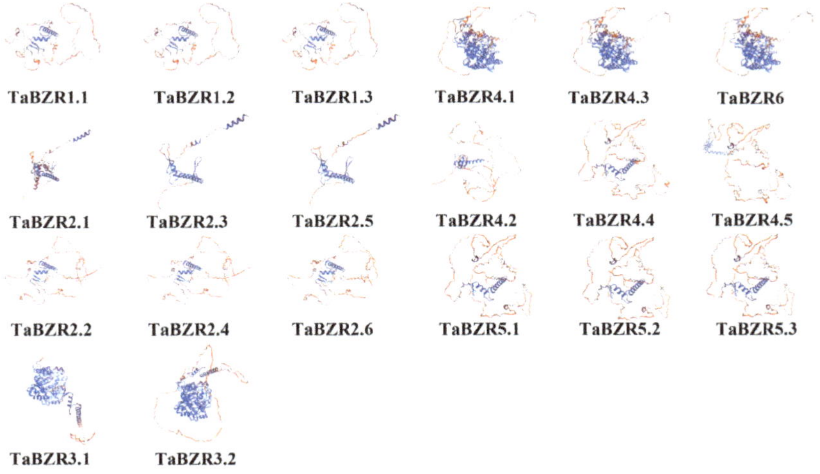

Figure 3. The tertiary structure prediction of the BZR protein in wheat. The tertiary structures of the TaBZR proteins were generated using the SWISS-MODEL. Twenty BZR proteins were modeled based on GMQE.

2.4. Synteny Analyses of TaBZR Genes

Gene duplication events are known as an important source of gene complexity diversity [37]. Synteny analyses of the *TaBZR* genes were performed to better understand *TaBZR* gene family gene duplication and expansion events in wheat (Figure 4). Gene duplication acts a major factor leading to expansion of the gene family. Current hexaploid wheat was produced as a result of two natural hybridization events between three diploid species. In theory, every wheat gene often has three homologous genes, and they are caused by polyploidization [38]. Figure 4 showed that the *TaBZRs* were distributed on the 14 wheat chromosomes. Fourteen *TaBZR* genes (*TaBZR-1.1/1.2/1.3*, *TaBZR2.1/TaBZR2.2/TaBZR2.3*, *TaBZR3.1/TaBZR-3.2*, *TaBZR4.2/TaBZR4.4/TaBZR4.5*, and *TaBZR5.1/TaBZR5.2/TaBZR5.3*) were clustered into eleven tandem duplication event regions. These results proved that gene duplication events are an important driving force during *TaBZR* gene family evolution.

2.5. Cis-Regulatory Elements in the Promoter Region of TaBZRs

The promoter region of many of the genes contained many cis-acting elements. The cis-elements in the promoter regions of the *TaBZR* genes were identified. To understand the stress-responsive regulatory role of the *BZR* genes in wheat, promoter cis-acting elements of *TaBZR* genes were further analyzed (Figure 5). There were many cis-elements in the 20 *TaBZR* genes' promoter regions, such as anoxic-specific inducibility elements, anaerobic induction elements, low-temperature-responsive elements, and some plant-hormone-responsive elements, such as MeJA, ABA, GA, and SA response elements. Some of the *TaBZR* gene promoters contain elements associated with the seed-specific regulation element, light-responsive element, and meristem expression. Additionally, MYB TF binding sites involved in light and drought are also found in the promoter regions.

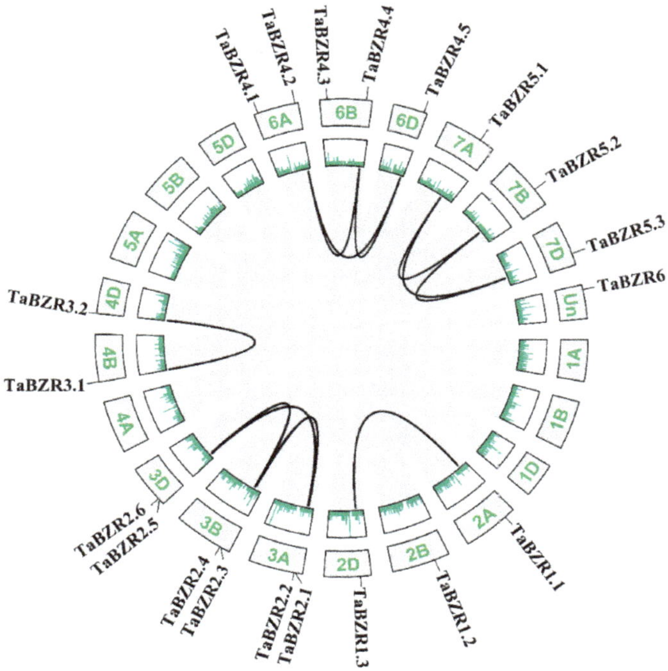

Figure 4. The interchromosomal relationships of the *TaBZR* gene family in wheat. In the middle of the figure, the black lines indicate the *TaBZR* gene pairs, and the gray lines represent all of the synteny blocks in the wheat genome. The green bars on each chromosome represent gene density. The number of each chromosome is indicated in green.

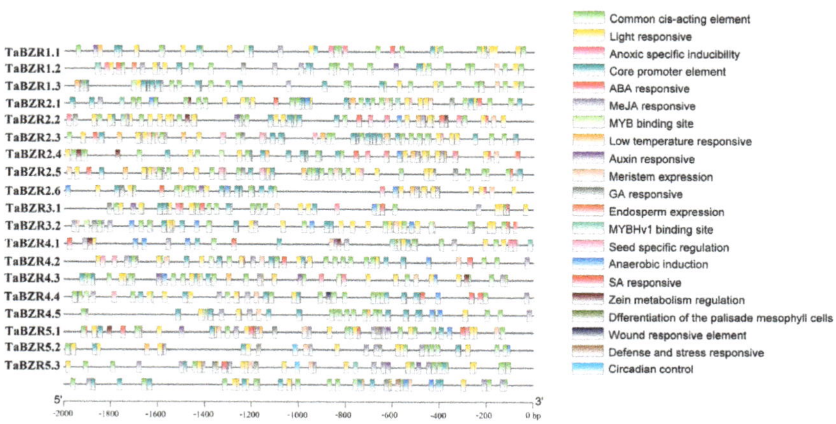

Figure 5. The cis-elements in the promoter regions of the *TaBZR* genes.

2.6. Conserved Motifs of TaBZRs

The protein motifs of the TaBZR protein were analyzed utilizing MEME online tools (Figure 6). We found that motif 1 was conserved in most of the TaBZR transcripts, suggesting that this region was important for BZR protein functions. Some of the motifs were found only in a few of the TaBZR subfamilies. For example, motifs 2, 3, 4, 5, 7, 8, and 10 were specific to the subfamily group II. The results of the conserved motif analysis of TaBZR were generally consistent with those of the TaBZR phylogenetic analysis.

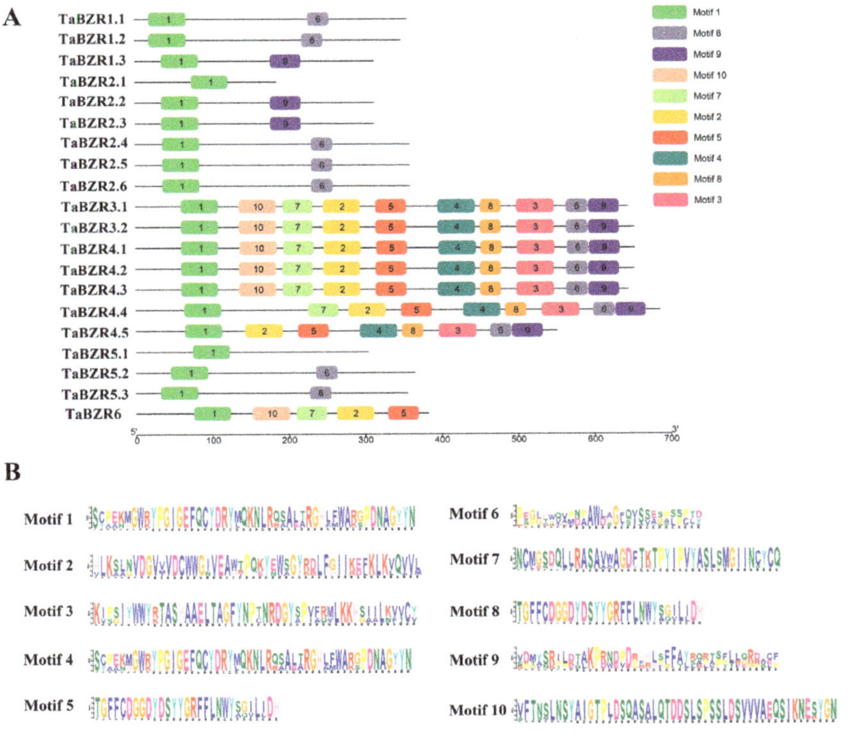

Figure 6. Distribution of the conserved motifs of the TaBZRs. (**A**). Conserved motif analysis of TaBZR was performed in this study. The different colored boxes numbered 1–10 indicate different motifs. The annotations of the motifs are listed on the right. (**B**). The conserved amino acid sequences in each motif.

2.7. Expression Profiles of Wheat BZR Genes

Published data were downloaded and analyzed in this study to gain insight into the spatial and temporal expression profiles of wheat *BZR* genes, and the expression patterns of all 20 *BZR* genes were derived from different development stages of the wheat tissues that were investigated (Figure 7). The result revealed that some of the *TaBZR* genes showed preferential expression in some detected tissues. For example, two genes in anther (*TaBZR3.1/3.2*) and four of the genes in lemma (*TaBZR2.2/2.4/2.6/3.2*) showed the highest expression levels. In different development stages, some of the *TaBZR* genes showed significant expression. For example, the *TaBZR2.2/2.4/2.6* genes were gradually increased along with the root development. Moreover, among the 20 *BZR* genes, *TaBZR2.1/2.3/4.2/4.4/4.5* were barely expressed in all of the detected samples, which may have been due to them having spatial or temporal expression patterns that may not have been examined in our study. Overall, the expression profiles of the *TaBZRs* varied considerably. The differential expression patterns of the *TaBZR* genes could provide crucial clues for further investigating their biological function, implying that these genes may play different roles in wheat.

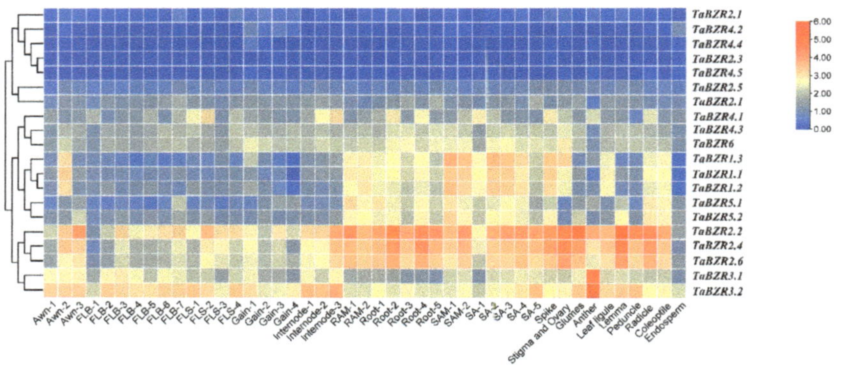

Figure 7. The wheat *BZR* gene expression profiles. Hierarchical clustering of the wheat *BZR* gene expression profiles in 45 of the samples, including different tissues and development stages. The numbers in the schematic diagram represent the development stages of the same tissue.

Certain *BZR* genes were randomly selected from 20 *TaBZR* genes to further investigate the expression of *TaBZR* genes under various different treatments. A qRT-PCR assay was performed to analyze the expression levels under different treatments (Figure 8). Many of the *TaBZR* genes were markedly induced under various treatments. For example, under cold stress, the expression levels of *TaBZR* increased, and those of *TaBZR1.1/1.2/1.3* and *TaBZR2.1/2.3* initially increased and then later decreased. In the presence of heat and ABA treatments, the *TaBZR1.1/1.2/1.3* expression levels exhibited an intensive increase then decrease, while *TaBZR2.1/2.3* and *TaBZR5.1/5.3* were reduced. *TaBZR1.1/1.2/1.3* and *TaBZR2.1/2.2/2.3* were reduced under NaCl and mannitol treatments first, but they then increased, peaking after 1 or 2 h of treatment. In contrast, *TaBZR6* showed a continuous downregulation under cold, NaCl, ABA, mannitol, and 24-Epibrassinolide treatments. The expression levels of *TaBZR1.1/1.2/1.3* and *TaBZR2.1/2.2/2.3* were markedly increased under 24-BR treatments, and they peaked after 2 h or 0.5 h of treatment, implying that they play a crucial role in the BR signal pathway.

2.8. Subcellular Localization Analysis of TaBZR2.1 and Its Tissue-Specific Expression Analysis in Wheat

Since the *TaBZR2.1* gene responded to almost all of the treatments in the expression profile analysis of wheat *BZRs*, and the expression levels of *TaBZR2.1* initially increased in the early treatment stage and then later decreased, we chose *TaBZR2.1* for further analysis. The TaBZR2.1-GFP fusion vector, which was driven by the CAMV35S, was transformed into wheat protoplasts and expressed for 24 h to determine the subcellular localization of TaBZR2.1. The fluorescence signals were detected using confocal microscopy. The pCAMBIA2300-35S-EGFP vector and the pCAMBIA1300:35S:AtHY5-mCherry were used as the negative control and as a nuclear marker, respectively [39]. As shown in Figure 9A, the CaMV35S:GFP signal was observed in the whole cell, whereas the signal of TaBZR2.1-GFP was overlapped with 35S:AtHY5-mCherry. This result implies that the TaBZR2.1 protein was confirmed as a nuclear protein. In addition, the *TaBZR2.1* tissue-specific expression profiles showed that *TaBZR2.1* gene had a high expression in grain (Figure 9B).

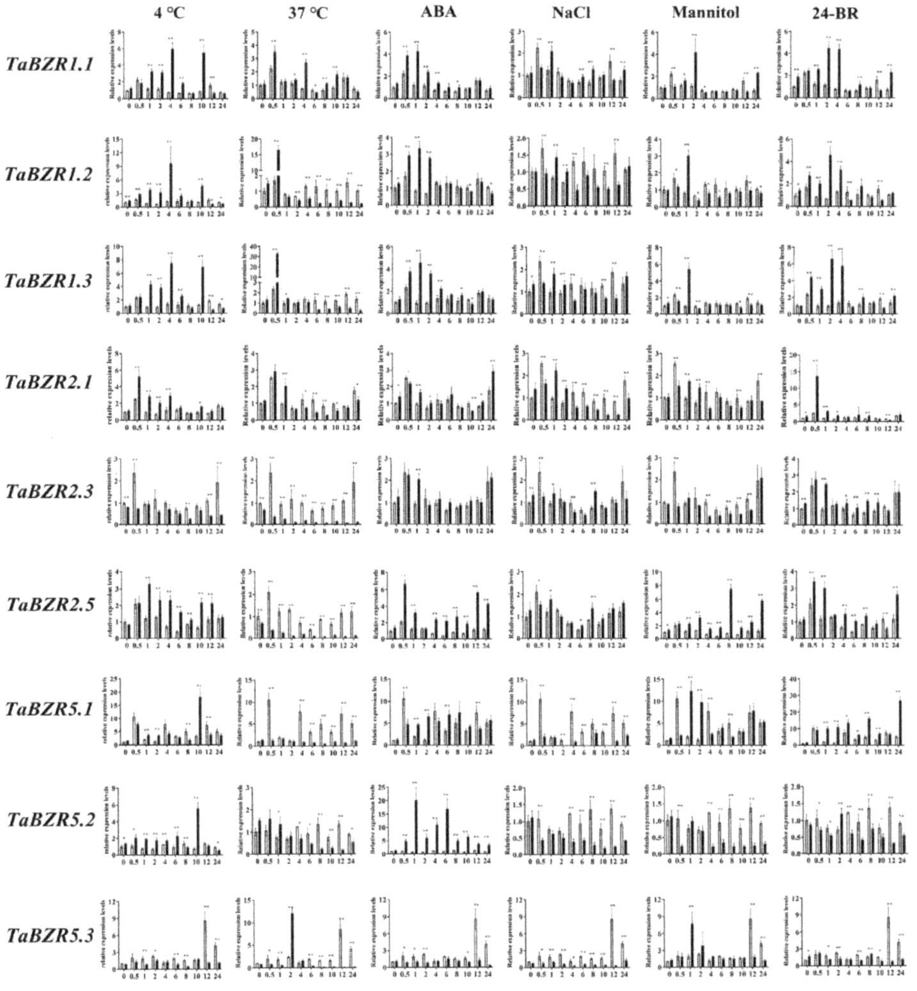

Figure 8. Expression analysis of the *TaBZR* genes in response to different treatments by qRT-PCR. Seeds of wheat cultivar Zhengmai366 (ZM366) were germinated for 3 days in the dark and then transferred to Hoagland liquid solution. The nutrient solution was changed every three days. At the trefoil stage (about three weeks old), seedlings were transferred to Hoagland liquid nutrient solution with ABA (200 mM), NaCl (200 mM), 20% PEG, and epi-BR (1µM) for ABA, NaCl, PEG, and epi-BR treatment, and seedlings were transferred to chambers at 37 °C or 4 °C to initiate heat and cold stress. The data were normalized with *TaACTIN* and *TaGAPDH*. The white and black columns in the diagrams represent the control and treatment groups, respectively. *, $p < 0.05$. **, $p < 0.01$.

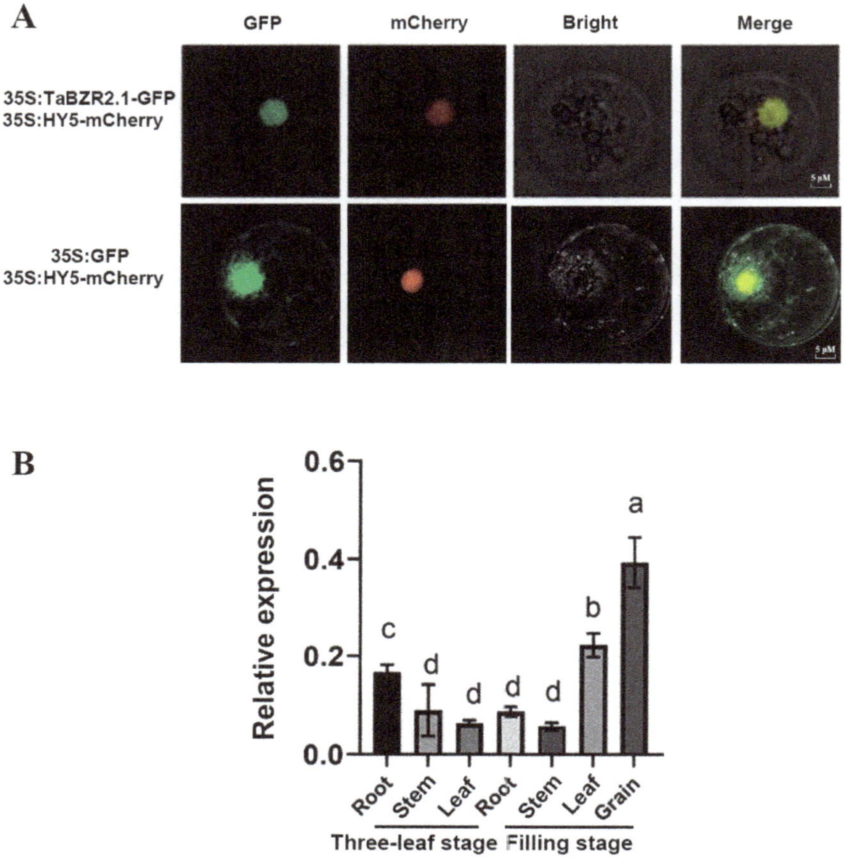

Figure 9. Subcellular localization of the TaBZR2.1-GFP fusion protein in the protoplast and TaBZR2.1 tissue-specific expressions. (**A**). TaBZR2.1-GFP protein driven by the 35S promoter were transiently expressed in protoplast cells of wheat, and they were observed using a confocal microscope. The GFP signals are represented by a green color; the red color represents the mCherry signals. Scale bars = 5 μM. (**B**). TaBZR2.1 tissue-specific expression profiles. Samples of the three-leaf and filling stages were collected, respectively, and the transcription levels of *TaBZR2.1* were measured using RT-qPCR assays, which were normalized with *TaACTIN* and *TaGAPDH*. The letters indicated significant at $p < 0.05$. Data are the mean ± SD (n = 3).

2.9. Overexpression of TaBZR2.1 in Arabidopsis Decreases Brassinazole Resistance

The full-length *TaBZR2.1* was transferred into the *Arabidopsis* wild-type Col-0 to further investigate the molecular function of *TaBZR2.1*. A total of 20 T_1 transgenic plants were obtained, and we selected three T_3 homozygous lines for further investigation (Figure S1). Transgenic *Arabidopsis* lines were exposed to different treatments, including ABA, NaCl, mannitol, 24-Epibrassinolide, and brassinazole, to further investigate the role of the *TaBZR2.1* gene in stresses. As shown in Figure S2, on 1/2 MS medium, the overexpression lines showed no significant difference compared to WT. Meanwhile, on 1/2 MS medium with 1 μM of brassinazole, the overexpression lines exhibited a shorter root length and a smaller leaf area; that is, the transgenic lines were more sensitive to the brassinazole treatments compared to WT. Additionally, the expression of some of the stress-induced genes (including *AtHsp17.6A*, *AtHsp17.6B*, *AtHsp17.6C*, *AtHsp17.8*, *SOS1*, and *CAT2*) were quantified using RT-qPCR. The results showed that four of the small Hsps were downregulated in seedlings of the control group, while they were upregulated in the brassinazole treatment groups; the expression of *SOS1* and *CAT2* was upregulated both in the control and brassinazole treatments (Figure 10). Overall, the overexpressed *TaBZR2.1* plants displayed a greater sensitivity to brassinazole tolerance than the Col-0 plants, suggesting that the ectopic expression of *TaBZR2.1* brassinazole treatment negatively affects the root length and leaf area in *Arabidopsis*. It was also found that the *TaBZR2.1* overexpression lines were more sensitive to brassinazole-induced stress.

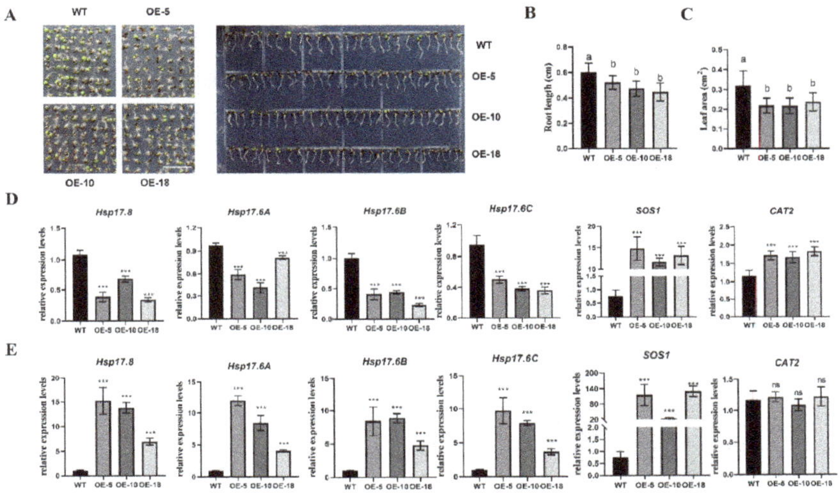

Figure 10. Overexpression of the *TaBZR2.1* gene in *Arabidopsis* negatively regulated the brassinazole-induced stress tolerance. (**A**). Seedling photographs of the 6-day-old Col-0 and TaBZR2.1-overexpressing *Arabidopsis* grown on 1/2 MS with brassinazole (1 μM). (**B**,**C**). The root length and leaf area of the seedlings in (**A**). The letters indicated significance at $p < 0.05$ (n = 30). (**D**,**E**). The expression levels of *Hsp17.8*, *Hsp17.6A*, *Hsp17.B*, *Hsp17.C*, *SOS1*, and *CAT2* in the seedlings under control (**D**) and brassinazole (**E**) treatment. ***, $p < 0.001$. ns, not significant.

2.10. TaBZR2.1 Physically Interacts with Wheat AGO4

A Y2H library was constructed, and the quality of the library was qualified (Figure 11A,B). A Y2H screening was performed to elucidate the molecular mechanism by which TaBZR2.1 modulates, and a large number of proteins were enriched (Table S2). We were particularly interested in one protein, AGO4. The full-length CDS of AGO4 was cloned and it interacted with TaBZR2.1 in yeast (Figure 11C) Then, a split-LUC assay was performed to further confirm the interaction. We generated constructs encoding the nLuc and TaBZR2.1 fusion protein (TaBZR2.1-nLuc) and the cLuc and AGO fusion protein (cLuc-AGO4). Then, these constructs were infiltrated into tobacco leaves. Likewise, in the split-LUC assay, only constructs harboring TaBZR2.1 and AGO4 fused to the two halves of the luciferase protein showed strong LUC signals in the *N. benthamiana* leaves (Figure 11D), indicating TaBZR2.1 interacts with AGO4 in planta. Taken together, these results showed that TaBZR2.1 interacts with AGO4 in vivo and in vitro.

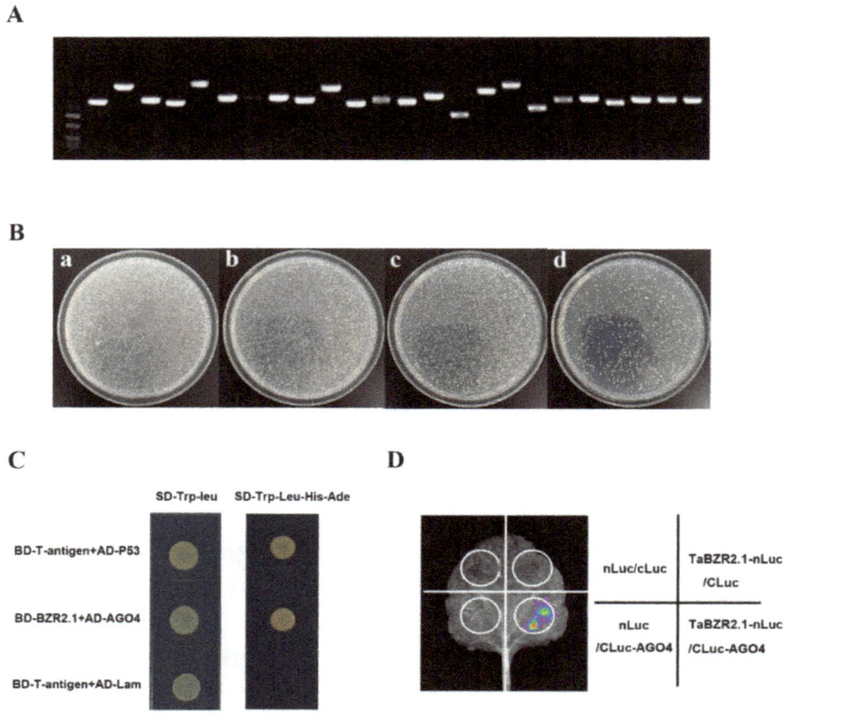

Figure 11. TaBZR2.1 physically interacting with AGO4. (**A**). Gel electrophoresis results for the identification of the insert fragments from the yeast library. (**B**). Yeast library titration. In the experiment, 100 μL of the 1/10, 1/100, 1/1000, and 1/10,000 dilutions and 100 μL of the yeast library were plated on SD/-Trp medium. Dilution factor = 10^{-1} (**a**), 10^{-2} (**b**), 10^{-3} (**c**), and 10^{-4} (**d**). (**C**). Yeast two-hybrid assay of TaBZR2.1 interacting with AGO4. BD-TaBZR2.1 (bait) and AD-AGO4 (prey) plasmids were transformed into the yeast (Y2H-gold) competent cell, as indicated and grown on the selection medium. (**D**). Split-LUC assay of TaBZR2.1 interacting with AGO4 in the tobacco leaves.

3. Discussion

3.1. Identification and Characteristic Analysis of Wheat BZR Genes

In this study, 20 *BZR* family genes in wheat were identified, which were named *TaBZR1.1* through *TaBZR6* on the basis of wheat gene symbolization guidelines (Figure 1). The genes in wheat usually have three homologs in three subgenomes (A, B, and D) because common wheat has endured two naturally interspecific hybridization events. However, in this study, no *BZR* gene was found on the 4A chromosome. The absence of the *BZR* gene on 4A could be the results of gene modification or the recombination events during the evolutionary process of the wheat [40,41]. Meanwhile, the synteny analyses showed that the *BZR* family expansion in wheat was generated by gene duplication events (Figure 4). In addition, three-dimensional modeling of TaBZRs was performed. Moreover, we found that the TaBZR proteins from the same subgenome showed similar models (Figure 3), and they were in agreement with previous findings of the phylogenetic and synteny analysis findings [42].

BZR genes can be induced by certain phytohormones and various stresses [43–45]. Previous study shows that maize *BZR* gene promoters contains light- and stress-responsive elements [33], while *Arabidopsis* and soybean *BZR* gene promoters revealed the presence of development-related and stress-, hormone-, and light-responsiveness-related cis-acting elements [46]. The cis-regulatory elements in the promoter region of *TaBZR* were analyzed

(Figure 5). Certain *TaBZR* promoters included the GA-, ABA-, MeJA-, and SA-responsive elements, suggesting that some *TaBZR* genes might be regulated by phytohormone and stress conditions. It was observed in a previous study that some BZRs are key to regulating organ and seed sizes [5,23]. Moreover, some of the elements of seed-specific regulation and meristem expression elements have also been found in some *TaBZR* promoters. The results imply that the *TaBZR* genes might be involved in plant growth and development, the stress response, and phytohormone metabolism networks. Previous studies showed that the *BZR* genes can be induced by ABA, cold, drought, and salt stress [31]. And the expression of *ZmBZR* genes is responsive to ABA and light [33]. Similarly, in view of the crucial role played in the stress response, the expression levels of the *TaBZR* genes under cold (4 °C) and hot (37 °C) conditions, as well as under ABA, NaCl, mannitol, and 24-Epibrassinolide, were investigated in this study (Figure 8). The expression levels of most of the *TaBZR* genes dramatically changed when subjected to stress treatments.

Subcellular localization analysis is a valuable tool for gaining insights into the cellular activities of proteins [42], and the nuclear localization of proteins is a common feature of transcription factors [23]. Subcellular localization results showed that the TaBZR2.1 protein was localized in the nucleus (Figure 9A).

3.2. Molecular Functional Analysis of TaBZR2.1

BZR genes have been comprehensively investigated in rice and *Arabidopsis*, and they have been implicated in numerous plant processes. There is substantial evidence indicating that *BZR* genes play crucial roles in the regulation of plant growth, development, and response to environmental stress [30,31,34]. Recent studies have demonstrated that wheat *TaBZR2* provides resistance to wheat stripe rust [47]. Overexpression of *TaBZR1* in wheat increases grain size. TaBZR1 mediates wheat chilling tolerance by directly binding to the *TaSAMT1* promoter to activate its expression [48].

BZR genes respond to abiotic stress and plant hormone treatment. For instance, the overexpression of the tomato *BZR1D* gene in *Arabidopsis* positively regulates BR and salt tolerance [36]. By promoting ABA biosynthesis and ROS scavenging, the *TaBZR1* gene can enhance wheat salt tolerance. To dissect the molecular function of *TaBZR2.1*, *Arabidopsis* overexpression lines were generated using Agrobacterium-mediated transformation (Figure S1). We identified that the wheat *TaBZR2.1* gene is involved in the brassinazole stress response. In *Arabidopsis*, the ectopic expression of *TaBZR2.1* indicated the shoot root length and small leaf area under brassinazole treatment, and the *TaBZR2.1* overexpression lines were found to be more sensitive to brassinazole-induced stress (Figure 10). Overall, *TaBZR2.1* played a negative role in brassinazole stress response.

AGO proteins are present in various organisms and exhibit expression across a diverse array of tissues [49,50]. They are central components of the RNA-induced silencing complex (RISC) and play an important role in plant development and growth [51]. For example, *Arabidopsis* AGO1 regulates leaf and floral stem cell development by binding to miR172 and miR165/166 [3]. In rice, OsAGO17 regulates grain weight and seed size through miR172 [47]. Additionally, OsAGO18 can bind to miR168, thus participating in antiviral rice defense pathways [52]. In *Arabidopsis* and rice, the AGO1 gene is required for normal plant development [53]. In this study, the Y2H assay was used to identify TaBZR2.1-interacting proteins. Our findings indicated that TaBZR2.1 directly interacts with AGO4, as predicted, suggesting that BZR-mediated BR signaling may serve as an important strategy in miRNA-mediated gene regulation (Figure 11C,D).

Overall, our findings provide valuable clues about the *TaBZR* family genes and their functional response to a variety of hormones and stresses in the different plant developmental processes. However, the molecular functions of *TaBZR* genes need to be further investigated.

4. Materials and Methods

4.1. Identification of BZRs in Wheat

The sequences of DNA (FASTA), proteins (FASTA), and the GFF3 files of wheat were downloaded from EnsemblPlants. The hidden Markov model of the BES1/BZR plant transcription factor N-terminal (Pfam login number: IPR008540) was obtained from the Pfam database and BLAST tool was used to identify the homologous proteins in wheat, maize, rice, and *Arabdopsis*. The obtained protein sequences were further analyzed by the NCBI Conserved Domain Database, and the redundant sequences were removed. The 20 *BZR* genes were named in accordance with gene symbolization guidelines.

4.2. Chromosomal Localization and Collinearity of the TaBZR Genes

The *TaBZRs* were mapped onto the chromosomes by analyzing the wheat GFF3 files. We utilized the TBtools software (V1.098, Guangzhou, China) [54] and MCScanX toolkit program (Athens, OH, USA) to visualize the chromosomal localizations and gene duplications.

4.3. Phylogenetic Analysis of BZR Proteins

The protein sequences of the BZR family in wheat and other selected species were downloaded and aligned via MUSCLE [55] using the default parameters to assess the evolutionary relationships of *BZR* in plants. Using MEGA-X software (V11.0., Philadelphia, PA, USA), a neighbor-joining phylogenetic tree was generated. Evolview software (V4.0, Beijing, China) was used to further modify the tree.

4.4. Subcellular Localization of TaBZR in Wheat Protoplasts

The coding sequence (CDS) of *TaBZR2.1* (gene ID: TraesCS3A02G123500) was PCR-amplified from the cDNA of the Zhengmai 366 genetic background of wheat using specific primers and KOD OneTM PCR Master Mix (TOYOBO, KMM-101, Osaka, Japan), and it was cloned into the pAN580-GFP vector. The HY5 protein of *Arabidopsis* was used in this study as a nuclear marker [27]. These vectors were then transformed into wheat protoplasts. The steps of the wheat protoplast extraction and transformation were conducted as has been described in a previous study [56]. The GFP fluorescence signals of the protoplast were observed under a confocal microscope (Zeiss, Jena, Germany).

4.5. Plant Materials and Stress Treatments

Healthy seeds of wheat cultivar Zhengmai366 (ZM366) were germinated in distilled water for 3 days in the dark. The germinated seeds were then transferred into a Hoagland liquid nutrient solution. At the trefoil stage, the seedlings were transferred to nutrient solution with ABA (final concentration 200 mM), NaCl (final concentration 200 mM), 20% PEG, and epi-BR (final concentration 1 μM) for the ABA, NaCl, PEG, and epi-BR treatments, respectively. Meanwhile, trefoil-stage wheat was transferred into chambers at either 37 °C or 4 °C to initiate heat and cold stress. The solution was changed every three days. Leaf samples were collected at different time points after treatment.

The full-length open reading frame (ORF) of the *TaBZR2.1* gene was cloned into the Kpn I/Sal I-digested 35S-eGFP vector to generate transgenic plants overexpressing *TaBZR2.1*. These constructs were then transformed into *Arabdopsis* (Col-0) via the *Agrobacterium* dipping flower method. The genomic DNA and RNA of the T1 plant were extracted to verify the presence of overexpression genes, and the expression levels of the *TaBZR2.1* gene and the specific primer pairs were used for PCR amplification (Table S3). Seeds of the T1 transgenic lines were sterilized and sown on a 1/2 MS medium containing kanamycin (final concentration 50 μg/mL) for further investigation.

4.6. Gene Expression Pattern Analysis

The RNA of samples collected in this study was extracted using an RNAiso Plus (No: 9108, Takara, Osaka, Japan). The quantity and quality of the extracted RNA from samples were examined using a Qubit 4.0 Fluorometer (Invitrogen, Carlsbad, CA, USA). The RNA were then reverse-transcribed to cDNA for real-time qPCR analysis. The expression levels of *TaBZR2.1* genes were normalized against those of *TaACTIN* (gene ID: TraesCS5B02G124100) and *TaGAPDH* (gene ID: TraesCS6B02G243700). The transcriptomic data from different tissues were downloaded from the Wheat eFP Browser. A hierarchical cluster map was obtained by the Tbtools software (V1.098, Guangzhou, China) [54].

4.7. Yeast Two-Hybrid Assay

A Y2H experiment was performed in accordance with the Matchmaker Gold Yeast Two-Hybrid System (https://www.takarabiomed.com.cn/, accessed on 20 November 2024). To generate the pGBKT7-TaBZR2.1 bait vector, the CDS of *TaBZR2.1* was amplified from the cDNA of wheat and cloned into the digested pGBKT7 vector. The CDS of *AGO4* (gene ID: TraesCS3A02G188400) was amplified and ligated into the pGADT7 plasmid in order to generate the prey vector, producing pGADT7-AGO4. Then, the prey vectors were co-transformed with bait vectors. The transformants were plated on DDO (SD/-Leu/-Trp) and QDO medium (SD/-Ade/-His/-Leu/-Trp) with ABA (125 ng/mL).

4.8. Split Luciferase Complementation Assay

cLuc-AGO4 and TaBZR2.1-nLuc vectors were used in the analysis. Constructs were electroporated into the agrobacterium competent cell (GV3101) and then injected into *N. benthamiana* leaves. After 48 h of infiltration, tobacco leaves were sprayed with D-luciferin (final concentration 1 mM), and then removed from light for 5 min. The signals were detected using a Night SHADE LB 985 system (Berthold, Stuttgart, Germany).

4.9. Conserved Protein Motifs and Structure of the TaBZRs

We used the MEME software (V5.5.7, Reno, NV, USA) to analyze the conservative motifs of the TaBZRs, and the results were visualized using the TBtools software (V1.098, Guangzhou, China) [54]. The exon–intron structures of the *TaBZR* genes were analyzed using GSDS 2.0.

4.10. Statistical Analysis

The IBM SPSS Statistics 22 software (Chicago, IL, USA) was used in this study to analyze the data using one-way ANOVA.

5. Conclusions

A total of 20 *TaBZR* gene members were identified in wheat in this study, and comprehensive bioinformatic analyses were performed for them. The promoter regions of the *TaBZR* genes contained elements associated with stress responses and the phytohormone response. Meanwhile, we confirmed that the TaBZR2.1 protein was located in the nucleus. In addition, we observed TaBZR2.1 physically interacting with AGO4 in vivo and in vitro. The discoveries in this study offer valuable clues for further investigation of *BZR* genes in wheat.

Supplementary Materials: The following supporting information can be downloaded at https://www.mdpi.com/article/10.3390/ijms252312545/s1.

Author Contributions: Y.Z. and Z.Z. conceived and designed the experiments; Y.Z. and J.Q. performed the experiment; Y.Z., J.Q. and C.L. analyzed the data; Y.Z. wrote the manuscript; Y.Z., J.H., S.G., M.Q., W.L., Z.D., Z.W. and Z.L. completed the writing—review and editing. All authors have read and agreed to the published version of the manuscript.

Funding: This research was supported by the Henan scientific and technological research projects (No. 242102110294), Outstanding Youth Technology Fund of the Henan Academy of Agricultural Sciences (No. 2024YQ07), and the Postdoctoral research project of Henan Province (No. HN2024191).

Institutional Review Board Statement: Not applicable.

Informed Consent Statement: Not applicable.

Data Availability Statement: No new data were created.

Acknowledgments: The authors are grateful to all lab members for their useful suggestions, support, and encouragement.

Conflicts of Interest: The authors declare no conflicts of interest.

References

1. Tong, H.; Chu, C. Functional Specificities of Brassinosteroid and Potential Utilization for Crop Improvement. *Trends Plant Sci.* **2018**, *23*, 1016–1028. [CrossRef] [PubMed]
2. Clouse, S.D.; Sasse, J.M. BRASSINOSTEROIDS: Essential Regulators of Plant Growth and Development. *Annu. Rev. Plant Physiol. Plant Mol. Biol.* **1998**, *49*, 427–451. [CrossRef] [PubMed]
3. Li, J.; Nagpal, P.; Vitart, V.; McMorris, T.C.; Chory, J. A Role for Brassinosteroids in Light-Dependent Development of Arabidopsis. *Science* **1996**, *272*, 398–401. [CrossRef] [PubMed]
4. Huang, Y.; Liu, L.; Chai, M.; Su, H.; Ma, S.; Liu, K.; Tian, Y.; Cao, Z.; Xi, X.; Zhu, W.; et al. Epigenetic regulation of female germline development through ERECTA signaling pathway. *New Phytol.* **2023**, *240*, 1015–1033. [CrossRef] [PubMed]
5. Huang, H.-Y.; Jiang, W.-B.; Hu, Y.-W.; Wu, P.; Zhu, J.-Y.; Liang, W.-Q.; Wang, Z.-Y.; Lin, W.-H. BR Signal Influences Arabidopsis Ovule and Seed Number through Regulating Related Genes Expression by BZR1. *Mol. Plant* **2013**, *6*, 456–469. [CrossRef]
6. Jiang, W.-B.; Huang, H.-Y.; Hu, Y.-W.; Zhu, S.-W.; Wang, Z.-Y.; Lin, W.-H. Brassinosteroid Regulates Seed Size and Shape in Arabidopsis. *Plant Physiol.* **2013**, *162*, 1965–1977. [CrossRef]
7. Sun, F.; Ding, L.; Feng, W.; Cao, Y.; Lu, F.; Yang, Q.; Li, W.; Lu, Y.; Shabek, N.; Fu, F.; et al. Maize transcription factor ZmBES1/BZR1-5 positively regulates kernel size. *J. Exp. Bot.* **2020**, *72*, 1714–1726. [CrossRef]
8. Saito, M.; Kondo, Y.; Fukuda, H. BES1 and BZR1 Redundantly Promote Phloem and Xylem Differentiation. *Plant Cell Physiol.* **2018**, *59*, 590–600. [CrossRef]
9. Thummel, C.S.; Chory, J. Steroid signaling in plants and insects—Common themes, different pathways. *Genes Dev.* **2002**, *16*, 3113–3129. [CrossRef]
10. Wang, X.; Chory, J. Brassinosteroids regulate dissociation of BKI1, a negative regulator of BRI1 signaling, from the plasma membrane. *Science* **2006**, *313*, 1118–1122. [CrossRef]
11. Wang, Z.Y.; Nakano, T.; Gendron, J.; He, J.; Chen, M.; Vafeados, D.; Yang, Y.; Fujioka, S.; Yoshida, S.; Asami, T.; et al. Nuclear-localized BZR1 mediates brassinosteroid-induced growth and feedback suppression of brassinosteroid biosynthesis. *Dev. Cell* **2002**, *2*, 505–513. [CrossRef] [PubMed]
12. Cui, X.-Y.; Gao, Y.; Guo, J.; Yu, T.-F.; Zheng, W.-J.; Liu, Y.-W.; Chen, J.; Xu, Z.-S.; Ma, Y.-Z. BES/BZR Transcription Factor TaBZR2 Positively Regulates Drought Responses by Activation of *TaGST1*. *Plant Physiol.* **2019**, *180*, 605–620. [CrossRef] [PubMed]
13. He, J.-X.; Gendron, J.M.; Sun, Y.; Gampala, S.S.L.; Gendron, N.; Sun, C.Q.; Wang, Z.-Y. BZR1 Is a Transcriptional Repressor with Dual Roles in Brassinosteroid Homeostasis and Growth Responses. *Science* **2005**, *307*, 1634–1638. [CrossRef] [PubMed]
14. Bai, M.Y.; Zhang, L.Y. FunctionsofOs-BZR1and14-3-3proteinsinbrassinosteroidsignalinginrice. *Proc. Natl. Acad. Sci. USA* **2007**, *104*, 13839–13844. [CrossRef]
15. Sun, Y.; Fan, X.-Y.; Cao, D.-M.; Tang, W.; He, K.; Zhu, J.-Y.; He, J.-X.; Bai, M.-Y.; Zhu, S.; Oh, E.; et al. Integration of Brassinosteroid Signal Transduction with the Transcription Network for Plant Growth Regulation in Arabidopsis. *Dev. Cell* **2010**, *19*, 765–777. [CrossRef]
16. Yu, X.; Li, L.; Zola, J.; Aluru, M.; Ye, H.; Foudree, A.; Guo, H.; Anderson, S.; Aluru, S.; Liu, P.; et al. A brassinosteroid transcriptional network revealed by genome-wide identification of BESI target genes in *Arabidopsis thaliana*. *Plant J.* **2010**, *65*, 634–646. [CrossRef]
17. Chory, J.; Vert, G. Downstream nuclear events in brassinosteroid signalling signalling. *Nature* **2006**, *441*, 96–100.
18. Gampala, S.S.; Kim, T.W.; He, J.X.; Tang, W.; Deng, Z.; Bai, M.Y.; Guan, S.; Lalonde, S.; Sun, Y.; Gendron, J.M.; et al. An essential role for 14-3-3 proteins in brassinosteroid signal transduction in Arabidopsis. *Dev. Cell* **2007**, *13*, 177–189. [CrossRef]
19. Tang, W.; Yuan, M.; Wang, R.; Yang, Y.; Wang, C.; Oses-Prieto, J.A.; Kim, T.-W.; Zhou, H.-W.; Deng, Z.; Gampala, S.S.; et al. PP2A activates brassinosteroid-responsive gene expression and plant growth by dephosphorylating BZR1. *Nat. Cell Biol.* **2011**, *13*, 124–131. [CrossRef]
20. Zhang, X.; Guo, W.; Du, D.; Pu, L.; Zhang, C. Overexpression of a maize BR transcription factor ZmBZR1 in Arabidopsis enlarges organ and seed size of the transgenic plants. *Plant Sci.* **2019**, *292*, 110378 [CrossRef]
21. Dong, H.; Liu, J.; He, G.; Liu, P.; Sun, J. Photoexcited phytochrome B interacts with brassinazole resistant 1 to repress brassinosteroid signaling in Arabidopsis. *J. Integr. Plant Biol.* **2019**, *62*, 652–667. [CrossRef] [PubMed]
22. He, G.; Liu, J.; Dong, H.; Sun, J. The Blue-Light Receptor CRY1 Interacts with BZR1 and BIN2 to Modulate the Phosphorylation and Nuclear Function of BZR1 in Repressing BR Signaling in Arabidopsis. *Mol. Plant* **2019**, *12*, 689–703. [CrossRef] [PubMed]
23. Oh, E.; Zhu, J.-Y.; Wang, Z.-Y. Interaction between BZR1 and PIF4 integrates brassinosteroid and environmental responses. *Nat. Cell Biol.* **2012**, *14*, 802–809. [CrossRef] [PubMed]
24. Li, Q.-F.; He, J.-X. BZR1 Interacts with HY5 to Mediate Brassinosteroid- and Light-Regulated Cotyledon Opening in Arabidopsis in Darkness. *Mol. Plant* **2015**, *9*, 113–125. [CrossRef]
25. Wang, Y.; Cao, J.-J.; Wang, K.-X.; Xia, X.-J.; Shi, K.; Zhou, Y.-H.; Yu, J.-Q.; Zhou, J. BZR1 Mediates Brassinosteroid-Induced Autophagy and Nitrogen Starvation in Tomato. *Plant Physiol.* **2018**, *179*, 671–685. [CrossRef]

26. Yang, R.; Yang, Z.; Xing, M.; Jing, Y.; Zhang, Y.; Zhang, K.; Zhou, Y.; Zhao, H.; Qiao, W.; Sun, J. TaBZR1 enhances wheat salt tolerance via promoting ABA biosynthesis and ROS scavenging. *J. Genet. Genom.* **2023**, *50*, 861–871. [CrossRef]
27. Yin, Y.; Qin, K.; Song, X.; Zhang, Q.; Zhou, Y.; Xia, X.; Yu, J. BZR1 transcription factor regulates heat stress tolerance through FERONIA receptor-like kinase-mediated reactive oxygen species signaling in tomato. *Plant Cell Physiol.* **2018**, *59*, 2239–2254. [CrossRef] [PubMed]
28. Sun, F.; Yu, H.; Qu, J.; Cao, Y.; Ding, L.; Feng, W.; Khalid, M.H.B.; Li, W.; Fu, F. Maize ZmBES1/BZR1-5 decreases ABA sensitivity and confers tolerance to osmotic stress in transgenic Arabidopsis. *Int. J. Mol. Sci.* **2020**, *21*, 996. [CrossRef]
29. Li, H.; Ye, K.; Shi, Y.; Cheng, J.; Zhang, X.; Yang, S. BZR1 Positively Regulates Freezing Tolerance via CBF-Dependent and CBF-Independent Pathways in Arabidopsis. *Mol. Plant* **2017**, *10*, 545–559. [CrossRef]
30. Feng, W.; Zhang, H.; Cao, Y.; Liu, Y.; Zhao, Y.; Sun, F.; Yang, Q.; Zhang, X.; Zhang, Y.; Wang, Y.; et al. Maize ZmBES1/BZR1-1 transcription factor negatively regulates drought tolerance. *Plant Physiol. Biochem.* **2023**, *205*, 108188. [CrossRef]
31. Saha, G.; Park, J.-I.; Jung, H.-J.; Ahmed, N.U.; Kayum, A.; Kang, J.-G.; Nou, I.-S. Molecular characterization of BZR transcription factor family and abiotic stress induced expression profiling in *Brassica rapa*. *Plant Physiol. Biochem.* **2015**, *92*, 92–104. [CrossRef]
32. Wu, P.; Song, X.; Wang, Z.; Duan, W.; Hu, R.; Wang, W.; Li, Y.; Hou, X. Genome-wide analysis of the BES1 transcription factor family in Chinese cabbage (*Brassica rapa* ssp. pekinensis). *Plant Growth Regul.* **2016**, *80*, 291–301. [CrossRef]
33. Yu, H.; Feng, W.; Sun, F.; Zhang, Y.; Qu, J.; Liu, B.; Lu, F.; Yang, L.; Fu, F.; Li, W. Cloning and characterization of BES1/BZR1 transcription factor genes in maize. *Plant Growth Regul.* **2018**, *86*, 235–249. [CrossRef]
34. The International Wheat Genome Sequencing Consortium (IWGSC); Appels, R.; Eversole, K.; Feuillet, C.; Keller, B.; Rogers, J.; Stein, N.; Pozniak, C.J.; Stein, N.; Choulet, F.; et al. Shifting the limits in wheat research and breeding using a fully annotated reference genome. *Science* **2018**, *361*, eaar7191. [CrossRef]
35. Song, X.; Ma, X.; Li, C.; Hu, J.; Yang, Q.; Wang, T.; Wang, L.; Wang, J.; Guo, D.; Ge, W.; et al. Comprehensive analyses of the BES1 gene family in Brassica napus and examination of their evolutionary pattern in representative species. *BMC Genom.* **2018**, *19*, 346. [CrossRef]
36. Jia, C.; Zhao, S.; Bao, T.; Zhao, P.; Peng, K.; Guo, Q.; Gao, X.; Qin, J. Tomato BZR/BES transcription factor SlBZR1 positively regulates BR signaling and salt stress tolerance in tomato and Arabidopsis. *Plant Sci.* **2020**, *302*, 110719. [CrossRef]
37. Fu, Y.; Zhang, Y.; Mason, A.S.; Lin, B.; Zhang, D.; Yu, H.; Fu, D. NBS-Encoding Genes in Brassica napus Evolved Rapidly After Allopolyploidization and Co-localize With Known Disease Resistance Loci. *Front. Plant Sci.* **2019**, *10*, 26. [CrossRef]
38. Feldman, M.; Levy, A. Allopolyploidy—A shaping force in the evolution of wheat genomes. *Cytogenet. Genome Res.* **2005**, *109*, 250–258. [CrossRef]
39. Ang, L.-H.; Chattopadhyay, S.; Wei, N.; Oyama, T.; Okada, K.; Batschauer, A.; Deng, X.-W. Molecular Interaction between COP1 and HY5 Defines a Regulatory Switch for Light Control of Arabidopsis Development. *Mol. Cell* **1998**, *1*, 213–222. [CrossRef]
40. Chen, Z.J. Genetic and Epigenetic Mechanisms for Gene Expression and Phenotypic Variation in Plant Polyploids. *Annu. Rev. Plant. Biol.* **2007**, *58*, 377. [CrossRef]
41. Otto, S.P. The Evolutionary Consequences of Polyploidy. *Cell* **2007**, *131*, 452–462. [CrossRef]
42. Kesawat, M.S.; Kherawat, B.S.; Singh, A.; Dey, P.; Kabi, M.; Debnath, D.; Saha, D.; Khandual, A.; Rout, S.; Manorama; et al. Genome-Wide Identification and Characterization of the Brassinazole-resistant (*BZR*) Gene Family and Its Expression in the Various Developmental Stage and Stress Conditions in Wheat (*Triticum aestivum* L.). *Int. J. Mol. Sci.* **2021**, *22*, 8743. [CrossRef]
43. Yang, Y.; Mao, Y.; Sun, B. Basic performance and future developments of BeiDou global navigation satellite system. *Satell. Navig.* **2020**, *1*, 1. [CrossRef]
44. Su, D.; Xiang, W.; Wen, L.; Lu, W.; Shi, Y.; Liu, Y.; Li, Z. Genome-wide identification, characterization and expression analysis of BES1 gene family in tomato. *BMC Plant Biol.* **2021**, *21*, 161. [CrossRef]
45. Yang, J.; Wu, Y.; Li, L.; Li, C. Comprehensive analysis of the BES1 gene family and its expression under abiotic stress and hormone treatment in *Populus trichocarpa*. *Plant Physiol. Biochem.* **2022**, *173*, 1–13. [CrossRef]
46. Li, Q.; Guo, L.; Wang, H.; Zhang, Y.; Fan, C.; Shen, Y. In silico genome-wide identification and comprehensive characterization of the BES1 gene family in soybean. *Heliyon* **2019**, *5*, e01868. [CrossRef]
47. Bai, X.; Zhan, G.; Tian, S.; Peng, H.; Cui, X.; Islam, A.; Goher, F.; Ma, Y.; Kang, Z.; Xu, Z.-S.; et al. Transcription factor BZR2 activates chitinase *Cht20.2* transcription to confer resistance to wheat stripe rust. *Plant Physiol.* **2021**, *187*, 2749–2762. [CrossRef]
48. Chu, W.; Chang, S.; Lin, J.; Zhang, C.; Li, J.; Liu, X.; Liu, Z.; Liu, D.; Yang, Q.; Zhao, D.; et al. Methyltransferase TaSAMT1 mediates wheat freezing tolerance by integrating brassinosteroid and salicylic acid signaling. *Plant Cell* **2024**, *36*, 2607–2628. [CrossRef]
49. Meng, F.; Jia, H.; Ling, N.; Xue, Y.; Liu, H.; Wang, K.; Yin, J.; Li, Y. Cloning and characterization of two Argonaute genes in wheat (*Triticum aestivum* L.). *BMC Plant Biol.* **2013**, *13*, 18. [CrossRef]
50. Carmell, M.A.; Xuan, Z.; Zhang, M.Q.; Hannon, G.J. The Argonaute family: Tentacles that reach into RNAi, developmental control, stem cell maintenance, and tumorigenesis. *Genes Dev.* **2002**, *16*, 2733–2742. [CrossRef]
51. Zhong, J.; He, W.; Peng, Z.; Zhang, H.; Li, F.; Yao, J. A putative AGO protein, OsAGO17, positively regulates grain size and grain weight through OsmiR397b in rice. *Plant Biotechnol. J.* **2019**, *18*, 916–928. [CrossRef]
52. Wu, J.; Yang, Z.; Wang, Y.; Zheng, L.; Ye, R.; Ji, Y.; Zhao, S.; Ji, S.; Liu, R.; Xu, L.; et al. Viral-inducible Argonaute18 confers broad-spectrum virus resistance in rice by sequestering a host microRNA. *eLife* **2015**, *4*, e05733. [CrossRef] [PubMed]
53. Liu, T.-L.; Li, J.-C.; Gu, J.-T.; Lu, W.-J.; Guo, C.-J.; Li, R.-J.; Xiao, K. Cloning and molecular characterization of TaAGO1, a member of argonaute gene family in wheat (*Triticum aestivum* L.). *Afr. J. Biotechnol.* **2011**, *10*, 13407–13417. [CrossRef]

54. Chen, C.J.; Chen, H.; Zhang, Y.; Thomas, H.R.; Frank, M.H.; He, Y.H.; Xia, R. TBtools: An Integrative Toolkit Developed for Interactive Analyses of Big Biological Data. *Mol. Plant* **2020**, *13*, 1194–1202. [CrossRef]
55. Edgar, R.C. MUSCLE: A multiple sequence alignment method with reduced time and space complexity. *BMC Bioinform.* **2004**, *5*, 113. [CrossRef]
56. Zheng, J.; Liu, H.; Wang, Y.; Wang, L.; Chang, X.; Jing, R.; Hao, C.; Zhang, X. TEF-7A, a transcript elongation factor gene, influences yield-related traits in bread wheat (*Triticum aestivum* L.). *J Exp. Bot.* **2014**, *65*, 5351–5365. [CrossRef]

International Journal of
Molecular Sciences

Review

Adaptation of High-Altitude Plants to Harsh Environments: Application of Phenotypic-Variation-Related Methods and Multi-Omics Techniques

Kai-Lu Zhang [1,2,†], Ya-Nan Leng [1,†], Rui-Rui Hao [1], Wen-Yao Zhang [1], Hong-Fei Li [1], Mo-Xian Chen [1,*] and Fu-Yuan Zhu [1,2,*]

1 The Southern Modern Forestry Collaborative Innovation Center, State Key Laboratory of Tree Genetics and Breeding, Key Laboratory of State Forestry and Grassland Administration on Subtropical Forest Biodiversity Conservation, College of Life Sciences, Nanjing Forestry University, Nanjing 210037, China
2 State Key Laboratory of Desert and Oasis Ecology, Xinjiang Institute of Ecology and Geography, Chinese Academy of Sciences, Urumqi 830011, China
* Correspondence: cmx2009920734@gmail.com (M.-X.C.); fyzhu@njfu.edu.cn (F.-Y.Z.)
† These authors contributed equally to this work.

Abstract: High-altitude plants face extreme environments such as low temperature, low oxygen, low nutrient levels, and strong ultraviolet radiation, causing them to adopt complex adaptation mechanisms. Phenotypic variation is the core manifestation of ecological adaptation and evolution. Many plants have developed a series of adaptive strategies through long-term natural selection and evolution, enabling them to survive and reproduce under such harsh conditions. This article reviews the techniques and methods used in recent years to study the adaptive evolution of high-altitude plants, including transplantation techniques, genomics, transcriptomics, proteomics, and metabolomics techniques, and their applications in high-altitude plant adaptive evolution. Transplantation technology focuses on phenotypic variation, which refers to natural variations in morphological, physiological, and biochemical characteristics, exploring their key roles in nutrient utilization, photosynthesis optimization, and stress-resistance protection. Multiple omics technologies, including genomics, transcriptomics, proteomics, and metabolomics, have revealed genes, regulatory pathways, and metabolic networks associated with phenotypic variations at the genetic and molecular levels. At the same time, the limitations and deficiencies of current technologies used to study plant adaptation to high-altitude environments were discussed. In addition, we propose future improvements to existing technologies and advocate for the integration of different technologies at multiple levels to study the molecular mechanisms of plant adaptation to high-altitude environments, thus providing insights for future research in this field.

Keywords: high-altitude plants; adaptation; transplant experiment; multi-omics techniques; environmental stress

1. Introduction

With an average elevation of about 4000 m, the Qinghai–Tibet Plateau (QTP), commonly referred to as the Third Pole or the Roof of the World, is the biggest and highest plateau on earth [1]. Its geographical location is extremely special, and the environment is extremely diverse. It has typical environmental features such as low temperature, low oxygen levels, and intense ultraviolet radiation. Therefore, it is being used as a natural laboratory to study the adaptation and evolution of species in extreme environments [2].

To withstand the extreme natural environment of the plateau, QTP plants have evolved a number of special adaptive traits, such as distinct morphological traits of the root, stem, and leaf; plant interactions; pollinator interactions; petal color; reproduction; and resource allocation patterns [3]. Cushion plants, as a dominant plant in the QTP, have a compact growth pattern and a shorter plant height [4]. Perennial herbaceous plants found on the QTP have consistent leaf proportions, but their biomass allocation patterns are characterized by an increase in fine roots and a decrease in aboveground reproductive structures. Increasing the surface area of fine roots for absorption and maintaining biomass investment in photosynthetic structures helps high-altitude plants compensate for the carbon gain and reduced nutrient uptake at cold temperatures [5]. According to several studies [6–9], plants on the QTP have evolved specialized adaptive traits in their anti-stress physiological metabolism. For instance, enhancing the concentration of flavonoid compounds aids in resisting ultraviolet radiation, growing ephemeral shoots to completely avoid freezing, and increasing the leaf thickness and stomatal density to adapt to high light intensity and oxidative stress. High-altitude environments are mainly exposed to adversity stresses such as low oxygen, low temperature, and strong UV radiation. Therefore, compared with other extreme environments (e.g., arid or saline environments), the uniqueness of high-altitude adaptation is mainly reflected in low oxygen tolerance, UV resistance, and special metabolic pathways.

High-altitude plants usually have narrow distribution ranges and unique genetic resources, which are essential for maintaining ecosystem stability and function. Identifying their adaptive characteristics can provide a scientific basis for biodiversity conservation in these fragile ecosystems [10,11]. Many high-elevation plants play important roles in water regulation, soil stabilization, and carbon sequestration, services that are critical for downstream ecosystems [12,13]. With global climate change, crops are facing increasing stress from adversity. Plants at high altitude show unique advantages in adversity adaptation, such as resistance to cold, drought, and disease [14]. By studying the adaptive mechanisms of high-altitude plants, we can tap into and utilize these advantageous genetic resources to provide new ideas and approaches for breeding crops for stress tolerance, thus improving crop stress tolerance and yield stability [15–18]. In addition, by understanding how high-altitude plants survive in poor soils and extreme conditions, we can provide inspiration for sustainable agricultural practices in marginal lands [19,20]. Given the vulnerability of high-altitude ecosystems and the severe challenges of global climate change, in-depth research on the adaptive mechanisms of high-altitude plants is particularly important and urgent. This will not only help us to better understand and cope with these challenges but also provide a scientific basis and technical support for ecological conservation and agricultural development. However, the regulatory mechanisms of genes related to cold resistance, UV resistance, and low-oxygen adaptation in high-altitude environments are not yet clear, and the interactions between high-altitude plant community compositions, their distribution with the environment, and the potential of emerging technologies (such as remote sensing and artificial intelligence) in monitoring high-altitude plant growth and health status have not been fully explored, revealing significant gaps in applying genomics and bioinformatics to analyze the adaptation and evolution of high-altitude plants.

Emerging technologies are being used in high-altitude ecosystem studies in addition to conventional techniques such as ecological surveys, physiological experiments (e.g., greenhouse experiments, garden experiments), geochemical analysis, and genetic investigations. This paper reviews the progress made in conventional garden experiments and reciprocal transplant trials while also outlining the application of cutting-edge technologies to comprehensively understand vegetation ecology, species distribution, ecological functions, and adaptation mechanisms in high-altitude regions. The multi-omics approach has played a key role in revealing the mechanisms of energy consumption in organisms and their responses to environmental stress. This paper reviews relevant studies on high-altitude adaptation using omics techniques (Table 1). For high-altitude non-model plants, the de novo assembly technique not only lays the foundation for decoding the genome, but it also provides key data for analyzing the mechanisms of adaptation to extreme environments. By integrating transcriptomics and proteomics, these methods can identify genes that are subject to positive selection pressure under specific environmental conditions, deepening our understanding of how organisms adapt and respond to environmental changes through gene regulatory mechanisms. We discuss the limitations of current techniques for studying plant adaptation to high-altitude environments and make suggestions for improving existing methods.

2. Transplant Experiment

To adapt to environmental changes, populations of plants engage in plastic response, migration, and genetic adaptation through selection [21,22]. The most significant research areas in evolutionary genetics and adaptive evolution are phenotypic plasticity and genetic variability, which play a significant role in defining a species' capacity for adaptation [23]. In short-term climate change, phenotypic plasticity can provide rapid adaptive responses, such as plants responding to temperature changes and water stress by adjusting growth patterns, flowering times, or physiological characteristics. As environmental changes continue, genetic adaptability will gradually accumulate and be fixed in the population, ensuring that plants can cope with the constantly changing climate over multiple generations [24]. Therefore, characterizing the adaptive potential of alpine plant species first involves studying a series of traits associated with plant adaptation and altitude adaptation.

The study of plant phenotypes in the process of adaptive evolution is mainly carried out in transplant experiments (also called field experiments) (Figure 1), which include common garden experiments and reciprocal transplant experiments [25,26]. These transplant experiments are classic tools for studying adaptive differentiation, which aims to describe the differentiation of phenotypes between populations derived from different sizes and geographic scales in order to better understand the phenotypes associated with adaptation and the selection pressures that influence them. The difference is that common garden experiments can compare the growth and other phenotypic traits of plant populations from different environments [26,27] to assess the genetic basis of population adaptation to different environments without considering environmental changes [28–30]. Therefore, common garden experiments have been widely used in forest trees to detect the adaptive signals of traits, including life history, phenology and allometric relationships [31–34]. When there is evidence that environmental differences between populations may drive adaptive genetic differentiation, the genetic mechanisms of genetic differentiation and adaptation can be further investigated by controlling for environmental variability in a common garden [35]. Through a common garden experiment, the study examined the differences in protein accumulation and physiological responses among four populations of *Myrsine coriacea* from different altitudes and assessed whether these differences exhibit genetic fixation [36]. The common garden experiment highlighted that the proteomic and physiological changes in the high-altitude population were due to local adaptation and abiotic selection pressures, even under common garden conditions. These studies showed that common garden experiments play a crucial role in understanding local adaptations to abiotic stress.

The reciprocal transplant experiment can directly detect local adaptation and the factors affecting adaptive differentiation of populations [31]. Through reciprocal transplants, local adaptation of a plant population can be demonstrated if it survives and grows best in its native environment and outperforms other populations of the species that have evolved in different environments [26]. Therefore, the quantification of local adaptation by reciprocal transplant experiments is still a common method to detect local adaptation in natural populations [27,31,37]. For example, in slope vegetation restoration experiments conducted in Jiuzhaigou County, Sichuan Province, to address the harsh environmental conditions in high-altitude areas, the research team employed film mulching technology. This method holds immense potential as it facilitates plant growth and maintains soil fertility. Experimental results revealed that under different coverage rates, film mulching treatment significantly enhanced plant growth and the soil nutrient content while reducing the plant density per unit area. In particular, at a coverage rate of 90%, both plant growth and the soil nutrient content were at their maximum. Although the rapidly changing environmental conditions in high-altitude areas may affect the effectiveness of membrane covering, such experiments offer reliable methods for addressing these challenges [38].

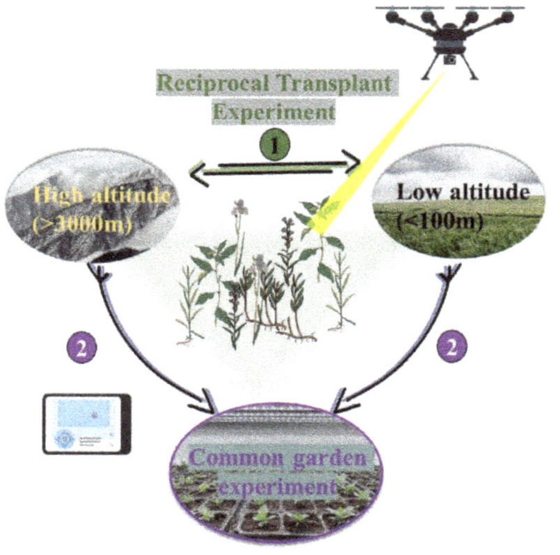

Figure 1. A diagram of the common garden experiment and the reciprocal transplant experiment. The common garden experiment (represented in purple) and the reciprocal transplant experiment (represented in green) are two common methods for studying plant adaptation. In the common garden experiment, seeds or plants from different altitudes are planted under the same standard environmental conditions, such as the same soil type, light exposure, temperature, and humidity. By comparing the growth and performance of plant species from different altitudes under the same environmental conditions, researchers can evaluate the differences in adaptability between different populations. In the reciprocal transplant experiment, seeds or plants from different altitudes are transplanted to altitudes outside of their original source. By comparing the growth and performance at their original source and at the transplanted locations, researchers can assess the adaptability and survival ability of different populations in different environments. Furthermore, precision agricultural technologies such as drones and sensor networks can be used to monitor the growth environment and physiological characteristics of medicinal plants, assisting growers in managing plant cultivation and growth processes more scientifically. The direction of the arrow indicates where the plant was transplanted.

At present, many studies have applied common garden experiments and reciprocal transplant experiments to study the adaptive mechanism of plants in high-altitude areas such as the QTP [39–45]. These studies suggest that inheritance-based trait differentiation and altitude adaptation are common in plants. Many of these traits have moderate to high heritability [30]. Transplant experiments have the advantage of being simple and not requiring a lot of technology, but they may require a lot of time and labor, and the number of tested plants is usually relatively limited. Hyperspectral imaging technology effectively captures subtle changes in plant phenotypes. By using hyperspectral cameras mounted on an unmanned aerial vehicle (UAV) [46], a non-destructive method can be employed to assess the severity of stem rust in different wheat genotypes at very low infection levels, serving as a substitute for trained personnel in large-scale field disease surveillance. Furthermore, since plant phenotypes are dynamic and spectral similarities may exist, most phenotype data across different studies are difficult to compare using standardized methods. Remote sensing technology, based on deep learning models, can distinguish four different sugarcane varieties [47]. These technologies provide new opportunities for research on plant adaptation at high altitudes.

For example, satellite remote sensing can provide wide-area coverage, while UAV remote sensing can provide complementary observations under unfavorable conditions such as cloud cover [48], and ground-based phenotyping detects dynamic plant growth through various sensor technologies [49]. By integrating multiple data sources such as satellite remote sensing, UAV remote sensing, and ground-based observations, the spatial and temporal resolution and accuracy of the data can be improved. In addition, current advanced artificial intelligence techniques can improve data precision and accuracy [50]. For example, deep learning and machine learning algorithms can be used to automate the processing and analysis of large-scale remote sensing data [48,51]. These algorithms can identify plant species-specific features, track changes in plant communities, and even predict future trends in ecological functions.

On the other hand, Artificial Intelligence (AI) is used to analyze plant health in real time (e.g., photosynthesis efficiency, vegetation index) and to predict future trends through modeling, such as the long-term adaptability of plants to environmental changes. However, the application and popularity of these technologies are still far from adequate at high altitudes, and there are still challenges such as data transmission and equipment maintenance. The efficiency of data collection and transmission can be optimized through the development of low-power remote sensing devices and by reliance on satellites or other long-term observatories. The interpretability of the AI algorithms used should be ensured so that ecologists can understand the output of the models and thus make sound ecological judgments. Collaborating with relevant institutions and universities can establish a dedicated monitoring and evaluation platform to provide functions such as data sharing, algorithm optimization, and results presentation. The platform can be open to researchers, administrators, and the public to promote the in-depth development of high-altitude plant adaptation research.

3. Genomics Facilitates the Molecular Basis of Adaptive Evolution

The plant genome governs the intricate responses and adaptive behaviors of plants across diverse environments. Species evolve based on genomic changes. The rapid advancement of high-throughput sequencing technologies has markedly enhanced sequencing throughput and quality while concurrently reducing costs. This progress has greatly accelerated genomic research and functional analysis [50]. Understanding genomic alterations is crucial for elucidating the molecular basis of morphological and physiological enhancements during plant evolution. In other words, plant genomes form the foundation for comprehending the adaptive evolution of plants in the plateau. However, research on plant molecular biology and genomics in the plateau lags behind that of vertebrates and other terrestrial plants. As a result, genome sequencing technology plays a vital role in studying plants in high-altitude regions, and it complements common garden experiments, jointly providing strong support for revealing the molecular mechanisms of plant adaptation to extreme environments.

Advances in genome assembly techniques have led to an increasing number of high-quality reference genomes being published. Since the genome construction of the model plant *Arabidopsis thaliana* [52], with the continuous innovation of high-throughput sequencing technology, more and more high-altitude plant genomes have been reported [51]. In recent years, the extensive collection of plant species and the assembly and resequencing of high-quality whole genomes in the plateau have greatly promoted research on the molecular genetics, phenotypic variation, and adaptive evolution of high-altitude plants [53]. For instance, analysis based on the genome sequence of *Crucihimalaya himalaica* revealed that dramatic changes in gene families in the signs and size of the positive selection of genes that could be candidates for *C. himalaica*'s adaptation to the harsh environments in the QTP [54]. High-altitude plants may have enhanced adaptive capacity through genome duplication, chromosome rearrangements, or segment deletions. Genomic studies can reveal how these structural variations affect a plant's ability to adapt to its environment. For example, the assembly of the genome and the deciphering of whole-genome duplication

(WGD) events of the tetraploid genome of *Rhododendron nivale* subsp. *boreale* have laid the foundation for the evolution and adaptation of *Rhododendrons* and provided valuable resources [55]. In addition, two chromosome-scale genomes of two *Eutrema* species have been assembled, which are from the high-altitude *E. heterophyllum* from the eastern QTP and *E. yunnanense* from lowland [56]. According to the comparative analysis of the two species, the results reveal the expansion of disease-resistance R genes (NBS-LRRs or NLRs) and genes related to DNA damage repair as well as cold-tolerant alpine plants, which show that the genomes of the two *Eutrema* reported are important genetic resources for the adaptative evolution of alpine plants across the world [56]. Moreover, a comparative study of *Prunus* genomes revealed that a striking expansion of the SINE retrotransposons had occurred in the genomes of Tibetan species, which promoted the accumulation of beneficial metabolites of the Tibetan *Prunus* species for adapting to the harsh environment [57]. A comparison of the genomes of several alpine plants as well as their lowland relatives revealed that some genes involved in reproduction and respiration (e.g., *MMD1*, *NBS1*, and *HPR*) may underlie the convergent adaptation of alpine plants to extreme environments [58]. In summary, based on comparative genomics, we can localize regions of selective signals associated with high-altitude acclimatization using methods such as the selective scanning of genomic data from plant populations at different altitudes. For example, certain gene families may exhibit strong positive selection signals, indicating their important roles in high-altitude environments.

The goal of population genetics is to identify evolutionary forces, such as natural selection, gene flow, and demographic fluctuations, that drive plant adaptation to local environments [59]. The increased accessibility of genome-wide data is transitioning population genetics into population genomics, revolutionizing our understanding of local adaptation [60,61]. Population and landscape genomic studies of adaptive evolution typically use sequence data generated by the whole genome, candidate genes, transcriptome, etc. to detect genetic variations in natural populations within an area or an entire species. Genomic signals selected in natural populations can be revealed by various genomic analysis methods, such as FST methods and environmental association analysis, which detect inter-population differentiation or associations between genetic variations and environmental gradients [23,28,62]. Recently, a chromosome-scale genome of *Ilex polyneura* in southwest China was assembled, and by conducting a population genomics study, 34 candidate genes were found to have functions connected to responses to biotic and abiotic stresses under selection at various elevations [63].

However, several unavoidable disadvantages exist when employing population genetics or genomics approaches. For instance, identifying rare alleles crucial for local adaptation can be challenging, especially in complex demographic contexts with strong gene-flow episodes [26]. Recently, landscape genomics has been a useful complementary method for determining the adaptive loci responsible for local adaptation [28,64–66]. By combining approaches from landscape genomics (Bayenv2) and population genomics (BayeScan), two hotspot regions that had strong natural selection signals linked to sun radiation and altitude were identified. These regions contained 14 genes primarily involved in abiotic stress resistance and effective reproduction [59]. High-altitude and alpine plants are ideal materials for studying adaptive evolution. Increasing evidence indicates that population genomics and landscape genomics significantly contribute to the study of plant adaptive evolution [59,67]. Additionally, applying a multiscale landscape genomic approach can detect selection signatures along environmental gradients such as solar radiation, wind exposure, and altitude, indicating that adaptive pressures may be driven by fine-scale topography [67]. Population and landscape genomic approaches study adaptive evolution by detecting genetic variations, which are crucial for understanding how to manage and protect species populations under changing climate conditions.

4. The Role of Multi-Omics in the Adaptive Evolution of Alpine Plants

At present, research on alpine plant adaptation primarily focuses on molecular aspects, revealing some genes, proteins, and metabolites associated with cold resistance, ultraviolet resistance, and hypoxia adaptation. Different omics technologies have their own characteristics in terms of breadth and depth. Based on the four levels of "What possibly happens", "What probably happens", "What happened", and "What actually happens", genomics, transcriptomics, proteomics, and metabolomics can each be used to validate gene expression, protein transcription and modifications, and changes in various metabolic levels. Therefore, establishing an integrated adaptive model can validate and help to understand the adaptation of alpine plants in a multi-level, systematic, and comprehensive manner. By integrating data, it becomes possible to more effectively identify key gene–protein–metabolite interaction pathways and to construct core models of adaptation mechanisms under environmental stress. Moreover, multi-omics integration can mitigate the limitations of single-omics analyses, providing a more comprehensive and accurate perspective. Despite the increasing number of high-altitude plant genomes being reported, a substantial number of plants still lack reference genomes. Consequently, investigating gene expression evolution, gene regulation, and species adaptation through other omics approaches (Figure 2), such as transcriptomics, becomes particularly important.

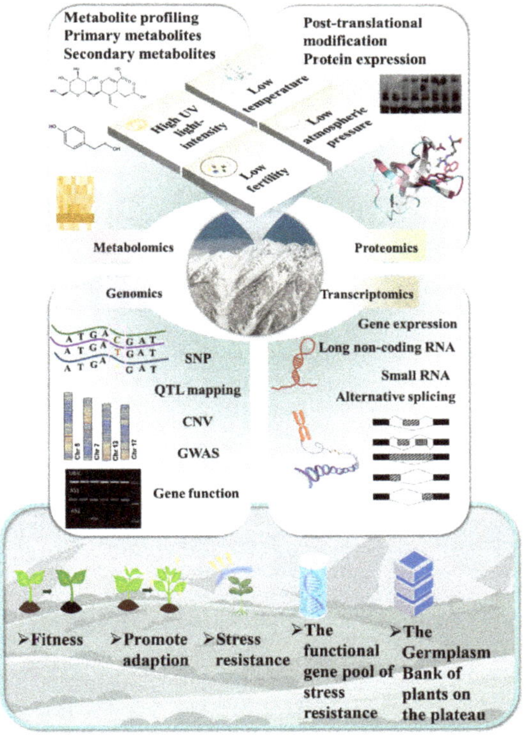

Figure 2. Omics technology research into plant adaptation to high altitude. In high-altitude areas, the environmental conditions, including high UV intensity, low fertility, low atmospheric pressure, and low temperatures, pose significant challenges. To address these challenges, current omics technologies offer multiple research perspectives. Metabolomics, through the analysis of metabolite spectra, primary metabolites, and secondary metabolites, reveals the metabolic adaptation mechanisms of organisms in high-altitude environments. Proteomics focuses on post-translational modifications and

changes in protein expression levels, providing insights into the functional adaptations of proteins to high-altitude environments. Genomics investigates single-nucleotide polymorphisms, quantitative trait loci, copy-number variations, and other genetic aspects, revealing genetic adaptation strategies of organisms to high-altitude environments. Transcriptomics explores gene expression regulation, long non-coding RNAs, small RNAs, and other aspects, elucidating the transcriptional regulatory mechanisms of organisms in high-altitude environments. Future research will focus on enhancing biological adaptability, promoting adaptation processes, expanding the gene pool for stress resistance and stress resistance, and exploring medicinal plant resources. Additionally, efforts will be made to establish a high-altitude plant germplasm repository to support the conservation and utilization of biological resources in high-altitude areas. The experimental figures are quoted from previously published articles [68–71].

Transcriptomics provides valuable insights into the adaptive evolution of high-altitude plants by studying changes in gene expression in plants under different environmental conditions [72]. Differences in adaptation mechanisms can be revealed by comparative transcriptome analysis. It was found that hundreds of positively selected genes (PSGs) were identified in high-elevation *Saussurea* species through comparative transcriptome analysis of five *Saussurea* species from different elevations; among these, *RAD54*, *RUP2*, and *TT5* are involved in DNA repair, the response to UV-B, flavonoid biosynthesis, and the cell response to hypoxia. Some important alpine-species-specific gene families were detected by phylogenetic cluster analysis, which may contribute to the defense against hypothermia and hypoxia in the Tibetan Plateau [73]. The same method was applied to a comparative analysis of two endangered high-alpine herbal plants endemic to China, *Notopterygium incisum* and *Notopterygium franchetii* (Apiaceae), and 18 candidate PSGs, corresponding to the functional groups of RNA splicing, were found that were involved in high-altitude adaptation and binding to RNA-binding protein (RBP) [74,75]. In the application of this technique, a number of precedents were carried out to study non-model plants. For example, high-quality de novo assembly of *Potentilla bifurca* from two altitudes (1725 masl and 3215 masl) was reported for the first time, and pathway analysis revealed a large number of DEGs encoding key enzymes involved in secondary metabolites, including phenylpropane and flavonoids [76].

Transcriptomics can also identify many key genes related to adaptation by analyzing the gene expression profiles of high-altitude plants under specific environmental conditions. For example, for the first time, according to the altitudinal adaptation mechanism of different-color flowers of alpine *Rhododendron* L. along the altitudinal gradient, specific PSGs that may be closely related to their evolution were identified, such as *MEKK1*, *YODA*, *PP2C*, and *SnRK2* in *R. fastigiatum*. These four PSGs are significantly enriched in the plant *MAPK* signaling pathway, while *MEKK1* and *YODA* are closely related to plant stress signal transduction [77]. In addition, based on transcriptome data of the alpine plant *Saussurea obvallata*, genes related to cuticlar waxes (*CER1*, *CER3*, *CER4*, and *MAH1*) and flavonoid biosynthesis pathways (*4CL*, *CHS*, *CH1*, *F3-H*, *TT7*, and *OMT*) were more highly expressed in leafy bracts than in other leaf tissues, and changes of these pathway genes may be important strategies for alpine plants to adapt to the high-altitude environment [58]. These transcriptome data all reveal, at the molecular level, how plants on the Tibetan Plateau survive in extreme environments [73,78–80]. The abundance of reference transcriptome data enables the analysis of specific gene families, candidate genes, and regulatory pathways in the Tibetan Plateau environment, with or without reference genomes. This provides valuable genomic resources and a basis for further studies on the molecular adaptation mechanism of plateau plants. However, transcriptomics is often based on mixed samples (such as entire tissues or organs) and lack details about the different cell types and spatial distribution within cells. Therefore, we propose the use of spatial transcriptome technology to help determine the expression of specific adaptive genes in the leaves, roots, and stems of plants at high altitudes so as to reveal the cooperative adaptation mechanism between different organs of plants [81,82].

Proteomic analysis is a powerful tool for studying the dynamic protein profiles of plants in response to high-altitude stress conditions [83]. Extreme environments alter cellular homeostasis, leading to changes in the final function of plant proteins. Compared with the genome and transcriptome, the proteome is more phenotypic and more directly responsive to environmental stress, which can provide a comprehensive and in-depth perspective on the mechanisms of plant adaptation to alpine environments [84,85]. Proteomic analysis of the accumulation of and changes in various stress-responsive proteins (18.1 class I HSP, HSP 70 family, heat shock factor binding protein, ATP synthase delta chain, and so on), enzymatic antioxidants, and the primary metabolism of *Picrorhiza kurroa* in the Himalayan region revealed their ability to cope with environmental changes along with the altitudinal gradient [86]. But the unique environment at high altitudes provides difficulty for the protein extraction and analysis process of alpine plants, as well as increasing the diversity of protein post-translational modifications, making experiments and data analysis more difficult. In addition, proteomics faces the challenge of detecting low-abundance proteins. In order to overcome these shortcomings, the method of multi-omics integration analysis has been used to make the protein data more reliable. Using comparative proteomics, we studied the dynamic patterns of protein expression in the alpine plant *Lamiophlomis rotata* located at three different altitudes (4350, 4800, and 5200 m) and identified 84 different protein spots. They demonstrated that Hydrogen sulfide (H_2S) can activate antioxidative enzymes and GSNOR (S-nitrosoglutathione reductase), thereby reducing ROS (reactive oxygen species) and RNS (reactive nitrogen species) levels and triggering a range of downstream defense responses, including protein degradation, proline and soluble sugar biosynthesis, and HSP accumulation [87]. Similarly, using comparative proteomics combined with physiological data, it was demonstrated that *Potentilla saundersiana* could enhance tolerance to high-altitude environments through a variety of pathways. For example, it regulates the root structure, leaf phenotype, photosynthetic capacity, and cell-wall structure; increases the contents of proline, soluble sugar, and flavonoids; and enhances the abundance of Autophagy protein 5a and SWI/SNF complex subunit SWI3A in response to environmental stress [88]. Using proteomics, functional proteins can be identified, post-translational modifications can be analyzed, and protein interaction networks can be constructed, providing key clues for the study of plant adaptation at high altitude. These studies not only deepen our understanding of the molecular mechanisms of plant adaptation to extreme environments but also provide a scientific basis for the development of stress-tolerant plants and the response to climate change.

Non-targeted metabolomics makes up for the shortcomings of proteomics in capturing immediate plant responses to environmental changes and comprehensively detecting metabolites. It can quickly reflect the physiological state of plants and provide extensive metabolic information [89,90]. For example, in order to reveal the adaptation strategy of *Picrorhiza kurroa* along the altitude environment, LC-MS-based non-targeted metabolite analysis and targeted analysis of sugars, amino acids, bitter glycosides and their corresponding phenolic acids were performed using leaves, roots and rhizomes. Sugar accumulation is believed to contribute to plant tolerance by acting as osmolytes and stabilizers of proteins and membranes. The accumulation of amino acids under stress may be related to nitrogen storage and serve as precursors for the synthesis of secondary metabolites (SMs). The role of bitter glycosides remains largely unclear, but their differential accumulation could be hypothesized to play a role in plant adaptation, and the results show that the content of picroside (I–IV) vary significantly in an organ-specific manner. The regulation of sugar and amino acid metabolism, remodeling of the root cell wall, ascorbate biosynthesis and cleavage, occurrence of a partial TCA (tricarboxylic acid) cycle, purine catabolism and the salvage route, pyrimidine biosynthesis, alteration of the lipid composition, and inhibition of gibberellins and cytokinin can effectively promote the adaptation of high-altitude *Picrorhiza kurro* [91]. Metabolomics studies revealed that high-altitude environmental conditions were important factors in inducing changes in secondary metabolites in Hemp (*Cannabis sativa* L.) inflorescence. Plants growing at high altitudes showed a higher total amount of terpenes, of which β-myrcene, *trans*-caryophyllene and α-humulene were the main contributors. In addition, the study also showed that the UV-length-exposure environment is conducive to the production of CBDA (cannabidiolic acid) and cannaflavins. Different types of flavonoids are involved in plant protection mechanisms, particularly through their free-radical scavenging activity and their ability to filter short-wavelength UV-B radiation [92]. Using pseudotargeted metabolomics analysis, it was found that in order to adapt to the strong ultraviolet radiation and low-temperature stress conditions in the high-altitude environment, phenylpropanoid biosynthesis, phenylalanine, tyrosine and tryptophan biosynthesis and phenylalanine metabolism related to flavonoid biosynthesis in *Draba oreades* Schrenk plants showed an up-regulated trend [93]. The advantage of metabolomics over other omics is that genomic information is not required. It has been shown that changes in primary metabolites such as sugars and amino acids are adaptive strategies for high-altitude plants, and that reprogramming of these metabolites is responsible for the accumulation of secondary metabolites such as phenolics and flavonoids, which are responsible for the adaptation of plants to high-altitude conditions [14].

The multi-omics analysis more systematically reveals the adaptation mechanism of alpine plants in extreme environments. In studying the adaptation of *Sinopodophyllum hexandrum* to a high-altitude environment, proteomic, transcriptomic, and metabolomic analyses revealed significant increases in the flavonoid, flavonol, and anthocyanin content in plants under conditions of 3300 m altitude and high light intensity. Several differentially expressed proteins (DEPs) and differentially expressed genes (DEGs) involved in the light response and flavonoid biosynthesis were identified at high altitudes, including up-regulation of *PAL, CHS1, IFRL, ANS, MYB4, BHLH137, CYP6, PPO1ERF5, HSP18.1, HSP70, UBC4, ERF5, ERF9, APX3*, and *EX2*. This suggests that high light intensity may enhance the accumulation of flavonoids, thereby improving podophylla's adaptability to high altitude and high light intensity [94].

In summary, multi-omics analysis provides important molecular evidence for studying the adaptability of alpine plants under different altitudinal gradients. These results have laid a solid foundation for revealing the molecular and physiological mechanisms of alpine plant adaptation to stress conditions in the future. In theory, it is entirely possible to transfer the adaptive characteristics unique to high-altitude plants, such as cold resistance, drought resistance, and high UV resistance, to crops using genetic engineering and traditional breeding techniques [95]. These technologies provide scientists with a means of directly manipulating biological genetic information, making it possible to transfer genetic material across species or varieties. However, achieving this goal requires interdisciplinary collaboration, in-depth fundamental research, and strict regulatory measures to ensure safe application of the technology.

5. Conclusions and Perspectives

Overall, we have gained some insights into the molecular and physiological mechanisms of acclimatization in high-altitude plants through the research methods and techniques described above. For example, high-altitude plants enhance low-temperature tolerance by synthesizing antifreeze proteins and by changing the fatty acid composition of cell membranes. These changes help plants maintain cell membrane integrity and stability under low-temperature conditions. In terms of their physiological response, high-elevation plants cope with low-temperature stress by adjusting their growth cycle and increasing biomass accumulation, thereby improving survival and reproduction. In addition, high-altitude plants can protect themselves against intense UV radiation by synthesizing UV-absorbing substances such as flavonoids and phenylpropanoid compounds. These substances absorb and scatter UV rays, thereby protecting plant cells from damage. Physiologically, high-elevation plants reduce UV exposure by increasing the leaf thickness and decreasing the leaf area, thereby increasing their resilience to intense UV radiation. However, high-elevation plants are usually exposed to the superimposed effects of multiple environmental stresses (e.g., low oxygen, low temperature, and intense UV radiation).

However, our understanding of how these stresses interact and how they work together to affect plant acclimatization remains limited. Future studies need to explore in greater depth the interactions among these stresses and their combined effects on plant adaptive networks.

At present, the establishment of genetic transformation systems and plant tissue culture technologies suitable for high-altitude plants is a major technical challenge. At high altitudes, many rare medicinal plants face reproductive problems, such as non-mode propagation, long cycles, conversion technology bottlenecks, and genetic hybridization difficulties [96]. Establishing suitable genetic transformation systems and plant tissue culture techniques is a major technical challenge for cultivating high-altitude plants. Applying the adaptation mechanisms of alpine plants to agricultural breeding, particularly for traits such as cold resistance and UV tolerance, holds great promise. Wild plants, shaped by natural selection, have more diverse and complex adaptation mechanisms. In contrast, crops, which are typically subject to artificial selection and improvement, exhibit strong adaptability and high productivity. Through genetic improvement, combined with multi-omics technologies (such as genomics, transcriptomics, etc.), key genes related to traits like cold resistance and UV tolerance can be identified, providing innovative insights for crop breeding. Today, modern breeding methods, such as gene transformation and gene editing, are evolving rapidly, offering strong support for the development of this field. The most common methods of plant genetic transformation include *Agrobacterium*-mediated transformation and gene gun [97,98], as well as the pollen tube pathway and ultrasonic methods. Recently, the application of nanomaterials in genetic transformation has shown great potential. Common nanomaterials include carbon nanotubes, LDH, gold nanoparticles, carbon dots, and peptide carriers [70,71,99–101]. Nanomaterials can penetrate cell walls without external force and are not restricted by plant species. They can also carry large amounts of nucleic acids and protect them. Although this technology is still in its early stages, with unclear mechanisms and potential bioaccumulation concerns in medicinal plants, its unique advantages in design and modification make it an efficient method for developing stable genetically transformed regenerating plants, especially for high-altitude species. To address the cumbersome tissue culture process in genetic transformation, Dr. Zhu's team developed a cut–dip–budding (CDB) delivery system. The system takes advantage of the plant's "budding" ability, eliminating the need for tissue culture [102]. The system has been tested on *Taraxacum kok-saghyz*, *Coronilla varia*, *Ailanthus altissima*, *Aralia elata*, and *Clerodendrum chinense*, which were previously difficult or impossible to transform. It now enables the successful transformation or gene editing of these species. Although the validation is limited to plants with tillering ability, its application in high-altitude plants is promising [103].

Table 1. Case studies of omics approach for high-altitude plant adaptation.

Plant Species	Altitude	Omics Approach	Effect	Reference
Crucihimalaya himalaica	Seedlings of *C. himalaica* were sampled from an altitude of 4010 m.	Genomic	Gene families showing dramatic changes in size and genes showing signs of positive selection are likely candidates for *C. himalaica*'s adaptation to intense radiation, low temperature, and pathogen–depauperate environments in the QTP. Loss of function at the S-locus, the reason for the transition to self-fertilization of *C. himalaica*, might have enabled its QTP occupation.	[54]
Prunus sp.	A total of 377 accessions of *Prunus* germplasm along altitude gradients ranging from 2067 to 4492 m in the Himalayas were collected and sequenced.	Genomic and metabolomic	A total of 379 metabolites had significant genetic correlations with altitudes; in particular, phenylpropanoids were positively correlated with altitudes. Specific SINE insertions change the expression of altitude-related genes.	[57]

Table 1. *Cont.*

Plant Species	Altitude	Omics Approach	Effect	Reference
Saussurea sp.	Three (*S. pachyneura*, *S. salwinensis*, and *S. velutina*) were collected between 4550 and 4620 m, and the other two (*S. amurensis* and *S. amara*) were collected between 150 and 350 m	Transcriptomic	Gene families specific to alpine species were identified, which involve oxidoreductase activity, pectin metabolism, lipid transport, and polysaccharide metabolism, potentially aiding in the defense against hypoxia and the freezing temperatures of the QTP. Also, hundreds of genes under positive selection were discovered, related to DNA repair, membrane transport, UV-B and hypoxia responses, reproduction, and nutrient metabolism, likely contributing to *Saussurea*'s adaptation to high-altitude environments.	[73]
Potentilla bifurca	Sample selected from two altitude ranges −3215 and 1725 masl	Transcriptomic	Fifty differentially expressed genes (DEG), including peroxidase, superoxide dismutase protein, and the ubiquitin-conjugating enzyme responded to abiotic stresses; a large number of DEGs encode key enzymes involved in secondary metabolites, including phenylpropane and flavonoids; 298 potential genomic SSRs were identified for genetic diversity assessment.	[76]
Rhododendron sp.	Four colored species, *Rhododendron fastigiatum* Franch (4194 m), *Rhododendron lacteum* Franch (3927 m), *Rhododendron facetum* I. B. Balfour & Kingdon Ward (2817 m) and *Rhododendron pachypodum* Balf. f. et W. W. Smith (2413 m), were collected.	Transcriptomic and metabolomic	Genes related to carbohydrates, fatty acids, amino acids and flavonoids biosynthesis play important roles in the altitude adaptability.	[77]
Potentilla saundersiana	Samples from 4350 to 5200 m in altitude.	Proteomic	Proteins involved in antioxidative activity, primary metabolites, epigenetic regulation, and protein post-translational modification play important roles in conferring tolerance to alpine environments.	[88]
Sinopodophyllum hexandrum	Leaves from 3-year-old plants were collected from 3300 and 2300 m.	Proteomic and transcriptomic	Nine DEPs and 41 DEGs were identified as being involved in flavonoid biosynthesis and the light response at 3300 m.	[94]
Herpetospermum pedunculosum	Samples from 2800 m, 3000 m, 3100 m and 3300 m.	Proteomic	High level of expression of some proteins, such as oxygen-evolving enhancer proteins, calreticulins, and S-adenosyl-l-homocysteine hydrolase, might confer greater tolerance in *H. pedunculosum* to the complex environment associated with high altitudes.	[104]
Cannabis sativa	Raw inflorescences material obtained from plants cultivated in 1200 m and 130 m altitude.	Metabolomic	All plants grown at altitude exhibited a higher total amount of terpenes when compared with plains counterparts, with β-Myrcene, trans-Caryophyllene and α-Humulene as the main contributors.	[92]
Draba oreades	Samples were collected at altitudes of 3800 m, 4000 m and 4200 m.	Metabolomic	Phenylalanine, tyrosine, and tryptophan biosynthesis and phenylalanine metabolism related to the biosynthesis of flavonoids were up-regulated in the high-altitude group, and 10 important metabolites were identified as potential biomarkers.	[93]
Cyclocarya paliurus	Mature leaves with the largest leaf area at F4 stage from 280 m and 920 m.	Metabolomic and transcriptomic	High altitude induces more flavonoid accumulation than low altitude, which may be contributed by the up-regulation of genes involved in energy and protein synthesis.	[105]

Author Contributions: F.-Y.Z. and M.-X.C. supervised the research direction and coordinated the writing and revision process. K.-L.Z. and Y.-N.L. were responsible for the preliminary concepts of the review, the literature search, and manuscript writing. R.-R.H., H.-F.L. and W.-Y.Z. contributed significant feedback on the overall structure and made substantial revisions and refinements to various versions of the manuscript. All authors have read and approved the manuscript.

Funding: This work was supported by STI 2030—Major Projects (No. 2023ZD0405602), the Post-graduate Research and Practice Innovation Program of Jiangsu Province (KYCX24_1376), the Basic

Research Program of Jiangsu Province (No. BK20240668), and the Jiangsu Agricultural Science and Technology Innovation Fund (No. CX (21)2023).

Conflicts of Interest: The authors have no conflicts of interest to declare.

References

1. Ruddiman, F.W.; Kutzbach, E.J. Plateau uplift and climatic change. *Sci. Am.* **1991**, *264*, 66–75. [CrossRef]
2. Qiao, Q.; Wang, Q.; Han, X.; Guan, Y.; Sun, H.; Zhong, Y.; Huang, J.; Zhang, T. Transcriptome sequencing of *Crucihimalaya himalaica* (Brassicaceae) reveals how *Arabidopsis* close relative adapt to the Qinghai-Tibet Plateau. *Sci. Rep.* **2016**, *6*, 21729. [CrossRef] [PubMed]
3. Sun, H.; Niu, Y.; Chen, Y.-S.; Song, B.; Liu, C.-Q.; Peng, D.-L.; Chen, J.-G.; Yang, Y. Survival and reproduction of plant species in the Qinghai-Tibet Plateau. *J. Syst. Evol.* **2014**, *52*, 378–396. [CrossRef]
4. Dvorsky, M.; Dolezal, J.; Kopecky, M.; Chlumska, Z.; Janatkova, K.; Altman, J.; de Bello, F.; Rehakova, K. Testing the Stress-Gradient Hypothesis at the Roof of the World: Effects of the Cushion Plant *Thylacospermum caespitosum* on Species Assemblages. *PLoS ONE* **2013**, *8*, e53514. [CrossRef]
5. Ma, W.; Shi, P.; Li, W.; He, Y.; Zhang, X.; Shen, Z.; Chai, S. Changes in individual plant traits and biomass allocation in alpine meadow with elevation variation on the Qinghai-Tibetan Plateau. *Sci. China-Life Sci.* **2010**, *53*, 1142–1151. [CrossRef]
6. Yin, Z.-Y.; Li, M.; Zhang, Y.; Shao, X. Growth-climate relationships along an elevation gradient on a southeast-facing mountain slope in the semi-arid eastern Qaidam Basin, northeastern Tibetan Plateau. *Trees-Struct. Funct.* **2016**, *30*, 1095–1109. [CrossRef]
7. Guo, Q.; Li, H.; Luo, D.; Quan, H.; Bianba, D.; Zhang, W. Comparative drought tolerance of six native deciduous and broad-leaved woody plant species seedlings in the Qinghai-Tibet Plateau. *Acta Physiol. Plant.* **2016**, *38*, 14. [CrossRef]
8. Bu, H.; Du, G.; Chen, X.; Xu, X.; Liu, K.; Wen, S. Community-wide germination strategies in an alpine meadow on the eastern Qinghai-Tibet plateau: Phylogenetic and life-history correlates. *Plant Ecol.* **2008**, *195*, 87–98. [CrossRef]
9. Li, H.Q.; Zhang, F.W.; Li, Y.N.; Cao, G.M.; Zhao, L.; Zhao, X.Q. Seasonal and interannual variations of ecosystem photosynthetic features in an alpine dwarf shrubland on the Qinghai-Tibetan Plateau, China. *Photosynthetica* **2014**, *52*, 321–331. [CrossRef]
10. Alexander, J.M.; Chalmandrier, L.; Lenoir, J.; Burgess, T.I.; Essl, F.; Haider, S.; Kueffer, C.; McDougall, K.; Milbau, A.; Nuñez, M.A.; et al. Lags in the response of mountain plant communities to climate change. *Glob. Chang. Biol.* **2018**, *24*, 563–579. [CrossRef]
11. Rahbek, C.; Borregaard, M.K.; Antonelli, A.; Colwell, R.K.; Holt, B.G.; Nogues-Bravo, D.; Rasmussen, C.M.Ø.; Richardson, K.; Rosing, M.T.; Whittaker, R.J.; et al. Building mountain biodiversity: Geological and evolutionary processes. *Science* **2019**, *365*, 1114–1119. [CrossRef] [PubMed]
12. Albuja-Quintana, M.; Pozo, G.; Gordillo-Romero, M.; Armijos, C.E.; Torres, M.D.L. Genome report: First reference genome of Vaccinium floribundum Kunth, an emblematic Andean species. *G3 Genes | Genomes | Genet.* **2024**, *14*, jkae136. [CrossRef]
13. Wang, Y.; Sun, J.; Lee, T.M. Altitude dependence of alpine grassland ecosystem multifunctionality across the Tibetan Plateau. *J. Environ. Manag.* **2023**, *332*, 117358. [CrossRef] [PubMed]
14. Kumar, V.; Kumar, P.; Bhargava, B.; Sharma, R.; Irfan, M.; Chandora, R. Transcriptomic and Metabolomic Reprogramming to Explore the High-Altitude Adaptation of Medicinal Plants: A Review. *J. Plant Growth Regul.* **2023**, *42*, 7315–7329. [CrossRef]
15. Li, B.-Z.; Liu, R.-N.; Zhang, H.; Tian, Y.-N.; Chen, T.-T.; Li, J.-X.; Jiao, F.-H.; Jia, T.-F.; Li, Y.-X.; Zhang, X.-Y.; et al. ZmMYB56 regulates stomatal closure and drought tolerance in maize seedlings through the transcriptional regulation of ZmTOM7. *New Crops.* **2024**, *1*, 100012. [CrossRef]
16. Tsegaw, M.; Zegeye, W.A.; Jiang, B.; Sun, S.; Yuan, S.; Han, T.; Wu, T. Progress and Prospects of the Molecular Basis of Soybean Cold Tolerance. *Plants* **2023**, *12*, 459. [CrossRef]
17. Vega-Gálvez, A.; Miranda, M.; Vergara, J.; Uribe, E.; Puente, L.; Martínez, E.A. Nutrition facts and functional potential of quinoa (*Chenopodium quinoa* willd.), an ancient Andean grain: A review. *J. Sci. Food Agric.* **2010**, *90*, 2541–2547. [CrossRef]
18. Vélez-Sánchez, J.; Casierra-Posada, F.; Fischer, G. Effect of Regulated Deficit Irrigation (RDI) on the Growth and Development of Pear Fruit (*Pyrus communis* L.), var. Triunfo de Viena. *Sustainability* **2023**, *15*, 13392. [CrossRef]
19. Flores-Marquez, R.; Vera-Vílchez, J.; Verástegui-Martínez, P.; Lastra, S.; Solórzano-Acosta, R. An Evaluation of Dryland Ulluco Cultivation Yields in the Face of Climate Change Scenarios in the Central Andes of Peru by Using the AquaCrop Model. *Sustainability* **2024**, *16*, 5428. [CrossRef]
20. Wu, X.; Xiao, J. Response and Adaptive Mechanism of Flavonoids in Pigmented Potatoes at Different Altitudes. *Plant Cell Physiol.* **2024**, *65*, 1184–1196. [CrossRef]
21. Aitken, S.N.; Yeaman, S.; Holliday, J.A.; Wang, T.; Curtis-McLane, S. Adaptation, migration or extirpation: Climate change outcomes for tree populations. *Evol. Appl.* **2008**, *1*, 95–111. [CrossRef] [PubMed]
22. Jump, A.S.; Penuelas, J. Running to stand still: Adaptation and the response of plants to rapid climate change. *Ecol. Lett.* **2005**, *8*, 1010–1020. [CrossRef]
23. Latreille, A.C.; Pichot, C. Local-scale diversity and adaptation along elevational gradients assessed by reciprocal transplant experiments: Lack of local adaptation in silver fir populations. *Ann. For. Sci.* **2017**, *74*, 77. [CrossRef]

24. Chen, M.X.; Tian, Y.; Zhu, F.Y.; Fan, T.; Yan, H.X.; Sun, P.C.; Li, M.; Hou, X.X.; Lin, P.; Song, Y.C.; et al. Alternative splicing of VRF1 acts as a molecular switch to regulate stress-induced early flowering. *Cell Rep.* **2024**, *11*, 114918. [CrossRef] [PubMed]
25. Hoban, S.; Kelley, J.L.; Lotterhos, K.E.; Antolin, M.F.; Bradburd, G.; Lowry, D.B.; Poss, M.L.; Reed, L.K.; Storfer, A.; Whitlock, M.C. Finding the Genomic Basis of Local Adaptation: Pitfalls, Practical Solutions, and Future Directions. *Am. Nat.* **2016**, *188*, 379–397. [CrossRef] [PubMed]
26. Kawecki, T.J.; Ebert, D. Conceptual issues in local adaptation. *Ecol. Lett.* **2004**, *7*, 1225–1241. [CrossRef]
27. Leimu, R.; Fischer, M. A Meta-Analysis of Local Adaptation in Plants. *PLoS ONE* **2008**, *3*, e4010. [CrossRef]
28. Sork, V.L.; Aitken, S.N.; Dyer, R.J.; Eckert, A.J.; Legendre, P.; Neale, D.B. Putting the landscape into the genomics of trees: Approaches for understanding local adaptation and population responses to changing climate. *Tree Genet. Genomes* **2013**, *9*, 901–911. [CrossRef]
29. Savolainen, O. The Genomic Basis of Local Climatic Adaptation. *Science* **2011**, *334*, 49–50. [CrossRef]
30. Savolainen, V.; Pyhajarvi, T.; Knurr, T. Gene flow and local adaptation in tees. *Annu. Rev. Ecol. Evol. Syst.* **2007**, *38*, 595–619. [CrossRef]
31. de Villemereuil, P.; Gaggiotti, O.E.; Mouterde, M.; Till-Bottraud, I. Common garden experiments in the genomic era: New perspectives and opportunities. *Heredity* **2016**, *116*, 249–254. [CrossRef] [PubMed]
32. McKown, A.D.; Guy, R.D.; Klapste, J.; Geraldes, A.; Friedmann, M.; Cronk, Q.C.B.; El-Kassaby, Y.A.; Mansfield, S.D.; Douglas, C.J. Geographical and environmental gradients shape phenotypic trait variation and genetic structure in *Populus trichocarpa*. *New Phytol.* **2014**, *201*, 1263–1276. [CrossRef] [PubMed]
33. Alberto, F.; Bouffier, L.; Louvet, J.M.; Lamy, J.B.; Delzon, S.; Kremer, A. Adaptive responses for seed and leaf phenology in natural populations of sessile oak along an altitudinal gradient. *J. Evol. Biol.* **2011**, *24*, 1442–1454. [CrossRef] [PubMed]
34. Kawakami, T.; Morgan, T.J.; Nippert, J.B.; Ocheltree, T.W.; Keith, R.; Dhakal, P.; Ungerer, M.C. Natural selection drives clinal life history patterns in the perennial sunflower species, *Helianthus maximiliani*. *Mol. Ecol.* **2011**, *20*, 2318–2328. [CrossRef] [PubMed]
35. Lind, B.M.; Menon, M.; Bolte, C.E.; Faske, T.M.; Eckert, A.J. The genomics of local adaptation in trees: Are we out of the woods yet? *Tree Genet. Genomes* **2018**, *14*, 29. [CrossRef]
36. Paschoa, R.P.D.; Pinto, V.B.; Pereira, J.P.; Cavatte, P.C.; Garbin, M.L.; Godinho, T.; Xavier, L.R.; Carrijo, T.T.; Silveira, V. Proteomic and physiological signatures of altitude adaptation in a *Myrsine coriacea* population under common garden conditions. *J. Proteom.* **2024**, *299*, 105156. [CrossRef]
37. Aitken, S.N.; Whitlock, M.C. Assisted Gene Flow to Facilitate Local Adaptation to Climate Change. *Annu. Rev. Ecol. Evol. Syst.* **2013**, *44*, 367. [CrossRef]
38. Wang, X.; Sun, H.; Tan, C.; Wang, X.; Xia, M. Effects of Film Mulching on Plant Growth and Nutrients in Artificial Soil: A Case Study on High Altitude Slopes. *Sustainability* **2021**, *13*, 11026. [CrossRef]
39. Han, T.-S.; Hu, Z.-Y.; Du, Z.-Q.; Zheng, Q.-J.; Liu, J.; Mitchell-Olds, T.; Xing, Y.-W. Adaptive responses drive the success of polyploid yellowcresses (*Rorippa*, Brassicaceae) in the Hengduan Mountains, a temperate biodiversity hotspot. *Plant Divers.* **2022**, *44*, 455–467. [CrossRef]
40. Lajoie, G.; Vellend, M. Characterizing the contribution of plasticity and genetic differentiation to community-level trait responses to environmental change. *Ecol. Evol.* **2018**, *8*, 3895–3907. [CrossRef]
41. Halbritter, A.H.; Fior, S.; Keller, I.; Billeter, R.; Edwards, P.J.; Holderegger, R.; Karrenberg, S.; Pluess, A.R.; Widmer, A.; Alexander, J.M. Trait differentiation and adaptation of plants along elevation gradients. *J. Evol. Biol.* **2018**, *31*, 784–800. [CrossRef] [PubMed]
42. Tiscar, P.A.; Lucas-Borja, M.E.; Candel-Perez, D. Lack of local adaptation to the establishment conditions limits assisted migration to adapt drought-prone *Pinus nigra* populations to climate change. *For. Ecol. Manag.* **2018**, *409*, 719–728. [CrossRef]
43. Singh, A.; Roy, S. High altitude population of *Arabidopsis thaliana* is more plastic and adaptive under common garden than controlled condition. *BMC Ecol.* **2017**, *17*, 39. [CrossRef] [PubMed]
44. Meng, F.D.; Jiang, L.L.; Zhang, Z.H.; Cui, S.J.; Duan, J.C.; Wang, S.P.; Luo, C.Y.; Wang, Q.; Zhou, Y.; Li, X.E.; et al. Changes in flowering functional group affect responses of-community phenological sequences to temperature change. *Ecology* **2017**, *98*, 734–740. [CrossRef] [PubMed]
45. Zhao, X.; Liu, W.; Zhou, M. Lack of local adaptation of invasive crofton weed (*Ageratina adenophora*) in different climatic areas of Yunnan Province, China. *J. Plant Ecol.* **2013**, *6*, 316–322. [CrossRef]
46. Abdulridha, J.; Min, A.; Rouse, M.N.; Kianian, S.; Isler, V.; Yang, C. Evaluation of Stem Rust Disease in Wheat Fields by Drone Hyperspectral Imaging. *Sensors* **2023**, *23*, 4154. [CrossRef]
47. Kai, P.M.; de Oliveira, B.M.; da Costa, R.M. Deep Learning-Based Method for Classification of Sugarcane Varieties. *Agronomy* **2022**, *12*, 2722. [CrossRef]
48. Ding, R.; Luo, J.; Wang, C.; Yu, L.; Yang, J.; Wang, M.; Zhong, S.; Gu, R. Identifying and mapping individual medicinal plant *Lamiophlomis rotata* at high elevations by using unmanned aerial vehicles and deep learning. *Plant Methods* **2023**, *19*, 38. [CrossRef]
49. Pieruschka, R.; Schurr, U. Plant Phenotyping: Past, Present, and Future. *Plant Phenomics* **2019**, *2019*, 7507131. [CrossRef]
50. Shen, Y.; Zhou, G.; Liang, C.; Tian, Z. Omics-based interdisciplinarity is accelerating plant breeding. *Curr. Opin. Plant Biol.* **2022**, *66*, 102167. [CrossRef]
51. Yang, W.; Feng, H.; Zhang, X.; Zhang, J.; Doonan, J.H.; Batchelor, W.D.; Xiong, L.; Yan, J. Crop Phenomics and High-Throughput Phenotyping: Past Decades, Current Challenges, and Future Perspectives. *Mol. Plant* **2020**, *13*, 28. [CrossRef] [PubMed]

52. Tabata, S.; Kaneko, T.; Nakamura, Y.; Kotani, H.; Kato, T.; Asamizu, E.; Miyajima, N.; Sasamoto, S.; Kimura, T.; Hosouchi, T.; et al. Sequence and analysis of chromosome 5 of the plant *Arabidopsis thaliana*. *Nature* **2000**, *408*, 823–826. [PubMed]
53. Li, Y.; Cao, K.; Li, N.; Zhu, G.; Fang, W.; Chen, C.; Wang, X.; Guo, J.; Wang, Q.; Ding, T.; et al. Genomic analyses provide insights into peach local adaptation and responses to climate change. *Genome Res.* **2021**, *31*, 592–606. [CrossRef] [PubMed]
54. Zhang, T.; Qiao, Q.; Novikova, P.Y.; Wang, Q.; Yue, J.; Guan, Y.; Ming, S.; Liu, T.; De, J.; Liu, Y.; et al. Qiong, Genome of *Crucihimalaya himalaica*, a close relative of *Arabidopsis*, shows ecological adaptation to high altitude. *Proc. Natl. Acad. Sci. USA* **2019**, *116*, 7137–7146. [CrossRef] [PubMed]
55. Lyu, Z.-Y.; Zhou, X.-L.; Wang, S.-Q.; Yang, G.-M.; Sun, W.-G.; Zhang, J.-Y.; Zhang, R.; Shen, S.-K. The first high-altitude autotetraploid haplotype-resolved genome assembled (Rhododendron nivale subsp. boreale) provides new insights into mountaintop adaptation. *GigaScience* **2024**, *13*, giae052. [CrossRef]
56. Guo, X.; Hu, Q.; Hao, G.; Wang, X.; Zhang, D.; Ma, T.; Liu, J. The genomes of two Eutrema species provide insight into plant adaptation to high altitudes. *DNA Res.* **2018**, *25*, 307–315. [CrossRef]
57. Wang, X.; Liu, S.; Zuo, H.; Zheng, W.; Zhang, S.; Huang, Y.; Pingcuo, G.; Ying, H.; Zhao, F.; Li, Y.; et al. Genomic basis of high-altitude adaptation in Tibetan Prunus fruit trees. *Curr. Biol.* **2021**, *31*, 3848–3860. [CrossRef]
58. Zhang, X.; Kuang, T.; Dong, W.; Qian, Z.; Zhang, H.; Landis, J.B.; Feng, T.; Li, L.; Sun, Y.; Huang, J.; et al. Genomic convergence underlying high-altitude adaptation in alpine plants. *J. Integr. Plant Biol.* **2023**, *65*, 1620–1635. [CrossRef]
59. Zheng, C.; Tan, L.; Sang, M.; Ye, M.; Wu, R. Genetic adaptation of Tibetan poplar (*Populus szechuanica* var. *tibetica*) to high altitudes on the Qinghai-Tibetan Plateau. *Ecol. Evol.* **2020**, *10*, 10974–10985. [CrossRef]
60. Weigel, D.; Nordborg, M. Population Genomics for Understanding Adaptation in Wild Plant Species *Annu. Rev. Genet.* **2015**, *492015*, 315–338. [CrossRef]
61. Luikart, G.; England, P.R.; Tallmon, D.; Jordan, S.; Taberlet, P. The power and promise of population genomics: From genotyping to genome typing. *Nat. Rev. Genet.* **2003**, *4*, 981–994. [CrossRef] [PubMed]
62. Tiffin, P.; Ross-Ibarra, J. Advances and limits of using population genetics to understand local adaptation. *Trends Ecol. Evol.* **2014**, *29*, 673–680. [CrossRef] [PubMed]
63. Yao, X.; Lu, Z.; Song, Y.; Hu, X.; Corlett, R.T. A chromosome-scale genome assembly for the holly (*Ilex polyneura*) provides insights into genomic adaptations to elevation in Southwest China. *Hortic. Res.* **2022**, *9*, uhab049. [CrossRef] [PubMed]
64. Manel, S.; Holderegger, R. Ten years of landscape genetics. *Trends Ecol. Evol.* **2013**, *28*, 614–621. [CrossRef] [PubMed]
65. Manel, S.; Joost, S.; Epperson, B.K.; Holderegger, R.; Storfer, A.; Rosenberg, M.S.; Scribner, K.T.; Bonin, A.; Fortin, M.J. Perspectives on the use of landscape genetics to detect genetic adaptive variation in the field. *Mol. Ecol.* **2010**, *19*, 3760–3772. [CrossRef]
66. Holderegger, R.; Buehler, D.; Gugerli, F.; Manel, S. Landscape genetics of plants. *Trends Plant Sci.* **2010**, *15*, 675–683. [CrossRef]
67. Leempoel, K.; Parisod, C.; Geiser, C.; Joost, S. Multiscale landscape genomic models to detect signatures of selection in the alpine plant *Biscutella laevigata*. *Ecol. Evol.* **2018**, *8*, 1794–1806. [CrossRef]
68. Lou, S.; Guo, X.; Liu, L.; Song, Y.; Zhang, L.; Jiang, Y.; Zhang, L.; Sun, P.; Liu, B.; Tong, S.; et al. Allelic shift in cis-elements of the transcription factor RAP2.12 underlies adaptation associated with humidity in Arabidopsis thaliana. *Sci. Adv.* **2022**, *8*, eabn8281. [CrossRef]
69. Song, Y.-C.; Chen, M.-X.; Zhang, K.-L.; Reddy, A.S.N.; Cao, F.-L.; Zhu, F.-Y. QuantAS: A comprehensive pipeline to study alternative splicing by absolute quantification of splice isoforms. *New Phytol.* **2023**, *240*, 928–939. [CrossRef]
70. Lv, Z.; Jiang, R.; Chen, J.; Chen, W. Nanoparticle-mediated gene transformation strategies for plant genetic engineering. *Plant J.* **2020**, *104*, 880–891. [CrossRef]
71. Kwak, S.-Y.; Lew, T.T.S.; Sweeney, C.J.; Koman, V.B.; Wong, M.H.; Bohmert-Tatarev, K.; Snell, K.D.; Seo, J.S.; Chua, N.-H.; Strano, M.S. Chloroplast-selective gene delivery and expression in planta using chitosan-complexed single-walled carbon nanotube carriers. *Nat. Nanotechnol.* **2019**, *14*, 447–455. [CrossRef] [PubMed]
72. Wang, Z.; Gerstein, M.; Snyder, M. RNA-Seq: A revolutionary tool for transcriptomics. *Nat. Rev. Genet.* **2009**, *10*, 57–63. [CrossRef] [PubMed]
73. Zhang, X.; Sun, Y.; Landis, J.B.; Shen, J.; Zhang, H.; Kuang, T.; Sun, W.; Sun, J.; Tiamiyu, B.B.; Deng, T.; et al. Transcriptomes of *Saussurea* (Asteraceae) Provide Insights into High-Altitude Adaptation. *Plants* **2021**, *10*, 1715. [CrossRef]
74. Jia, Y.; Liu, M.-L.; Yue, M.; Zhao, Z.; Zhao, G.-F.; Li, Z.-H. Comparative Transcriptome Analysis Reveals Adaptive Evolution of Notopterygium incisum and Notopterygium franchetii, Two High-Alpine Herbal Species Endemic to China. *Molecules* **2017**, *22*, 1158. [CrossRef] [PubMed]
75. Liu, Y.; Do, S.; Huynh, H.; Li, J.X.; Liu, Y.G.; Du, Z.Y.; Chen, M.X. Importance of pre-mRNA splicing and its study tools in plants. *Adv. Biotechnol.* **2024**, *2*, 4. [CrossRef]
76. Tang, X.; Li, J.; Liu, L.; Jing, H.; Zuo, W.; Zeng, Y. Transcriptome Analysis Provides Insights into Potentilla bifurca Adaptation to High Altitude. *Life* **2022**, *12*, 1337. [CrossRef]
77. Liu, X.-W.; Wang, Y.-H.; Shen, S.-K. Transcriptomic and metabolomic analyses reveal the altitude adaptability and evolution of different-colored flowers in alpine *Rhododendron* species. *Tree Physiol.* **2022**, *42*, 1100–1113. [CrossRef]
78. Zhang, Z.-H.; Chang, X.; Su, D.-Y.; Yao, R.; Liu, X.-D.; Zhu, H.; Liu, G.-X.; Zhong, B.-J. Comprehensive transcriptome analyses of two *Oocystis algae* provide insights into the adaptation to Qinghai-Tibet Plateau. *J. Syst. Evol.* **2021**, *59*, 1209–1219. [CrossRef]

79. Duruflé, H.; Hervé, V.; Ranocha, P.; Balliau, T.; Zivy, M.; Chourré, J.; Clemente, H.S.; Burlat, V.; Albenne, C.; Déjean, S.; et al. Dunand, Cell wall modifications of two *Arabidopsis thaliana* ecotypes, Col and Sha, in response to sub-optimal growth conditions: An integrative study. *Plant Sci.* **2017**, *263*, 183–193. [CrossRef]
80. Zhang, F.; Gao, Q.; Khan, G.; Luo, K.; Chen, S. Comparative transcriptome analysis of aboveground and underground tissues of *Rhodiola algida*, an important ethno-medicinal herb endemic to the Qinghai-Tibetan Plateau. *Gene* **2014**, *553*, 90–97. [CrossRef]
81. Rao, A.; Barkley, D.; França, G.S.; Yanai, I. Exploring tissue architecture using spatial transcriptomics. *Nature* **2021**, *596*, 211–220. [CrossRef] [PubMed]
82. Zhang, B.; Zhang, H.; Xia, Y. Harnessing spatial transcriptomics for advancing plant regeneration research. *Trends Plant Sci.* **2024**, *29*, 718–720. [CrossRef] [PubMed]
83. Kumar, R.; Kumari, M. Adaptive mechanisms of medicinal plants along altitude gradient: Contribution of proteomics. *Biol. Plant.* **2018**, *62*, 630–640. [CrossRef]
84. Hu, J.; Rampitsch, C.; Bykova, N.V. Advances in plant proteomics toward improvement of crop productivity and stress resistance. *Front. Plant Sci.* **2015**, *6*, 209. [CrossRef]
85. Diz, A.P.; Martinez-Fernandez, M.; Rolan-Alvarez, E. Proteomics in evolutionary ecology: Linking the genotype with the phenotype. *Mol. Ecol.* **2012**, *21*, 1060–1080. [CrossRef]
86. Kumari, M.; Kumar, R. Functional trait correlation network and proteomic analysis reveal multifactorial adaptation mechanisms to a climatic gradient associated with high altitude in the Himalayan region. *Plant Cell Environ.* **2024**, *47*, 1556–1574. [CrossRef]
87. Ma, L.; Yang, L.; Zhao, J.; Wei, J.; Kong, X.; Wang, C.; Zhang, X.; Yang, Y.; Hu, X. Comparative proteomic analysis reveals the role of hydrogen sulfide in the adaptation of the alpine plant Lamiophlomis rotata to altitude gradient in the Northern Tibetan Plateau. *Planta* **2015**, *241*, 887–906. [CrossRef]
88. Ma, L.; Sun, X.; Kong, X.; Galvan, J.V.; Li, X.; Yang, S.; Yang, Y.; Yang, Y.; Hu, X. Physiological, biochemical and proteomics analysis reveals the adaptation strategies of the alpine plant Potentilla saundersiana at altitude gradient of the Northwestern Tibetan Plateau. *J. Proteom.* **2015**, *112*, 63–82. [CrossRef]
89. De Vos, R.C.H.; Moco, S.; Lommen, A.; Keurentjes, J.J.B.; Bino, R.J.; Hall, R.D. Untargeted large-scale plant metabolomics using liquid chromatography coupled to mass spectrometry. *Nat. Protoc.* **2007**, *2*, 778–791. [CrossRef]
90. Yuan, H.; Cao, G.; Hou, X.; Huang, M.; Du, P.; Tan, T.; Zhang, Y.; Zhou, H.; Liu, X.; Liu, L.; et al. Development of a widely targeted volatilomics method for profiling volatilomes in plants. *Mol. Plant* **2022**, *15*, 189–202. [CrossRef]
91. Kumari, M.; Joshi, R.; Kumar, R. Metabolic signatures provide novel insights to *Picrorhiza kurroa* adaptation along the altitude in Himalayan region. *Metabolomics* **2020**, *16*, 77. [CrossRef] [PubMed]
92. Giupponi, L.; Leoni, V.; Pavlovic, R.; Giorgi, A. Influence of Altitude on Phytochemical Composition of Hemp Inflorescence: A Metabolomic Approach. *Molecules* **2020**, *25*, 1381. [CrossRef]
93. Lei, L.; Yuan, X.; Fu, K.; Chen, Y.; Lu, Y.; Shou, N.; Wu, D.; Chen, X.; Shi, J.; Zhang, M.; et al. Pseudotargeted metabolomics revealed the adaptive mechanism of *Draba oreades* Schrenk at high altitude. *Front. Plant Sci.* **2022**, *13*, 1052640. [CrossRef] [PubMed]
94. Zhao, Q.; Dong, M.; Li, M.; Jin, L.; Pare, P.W. Light-Induced Flavonoid Biosynthesis in *Sinopodophyllum hexandrum* with High-Altitude Adaptation. *Plants* **2023**, *12*, 575. [CrossRef] [PubMed]
95. He, Z.H.; Zhang, P.; Jia, H.T.; Zhang, S.L.; Nishawy, E.; Sun, X.P.; Dai, M.Q. Regulatory Mechanisms and Breeding Strategies for Crop Drought Resistance. *New Crops.* **2024**, *1*, 100029. [CrossRef]
96. Giménez-Benavides, L.; Escudero, A.; Iriondo, J.M. Reproductive limits of a late-flowering high-mountain Mediterranean plant along an elevational climate gradient. *New Phytol.* **2007**, *173*, 367–382. [CrossRef]
97. Gelvin, S.B. Plant Proteins Involved in *Agrobacterium*-Mediated Genetic Transformation. *Annu. Rev. Phytopathol.* **2010**, *482010*, 45–68. [CrossRef]
98. Rhodes, C.A.; Pierce, D.A.; Mettler, I.J.; Mascarenhas, D.; Detmer, J.J. Genetically transformed maize plants from protoplasts. *Science* **1988**, *240*, 204–207. [CrossRef]
99. Ahmed, S.; Gao, X.; Jahan, M.A.; Adams, M.; Wu, N.; Kovinich, N. Nanoparticle-based genetic transformation of *Cannabis sativa*. *J. Biotechnol.* **2021**, *326*, 48–51. [CrossRef]
100. Demirer, G.S.; Zhang, H.; Matos, J.L.; Goh, N.S.; Cunningham, F.J.; Sung, Y.; Chang, R.; Aditham, A.J.; Chio, L.; Cho, M.-J.; et al. High aspect ratio nanomaterials enable delivery of functional genetic material without DNA integration in mature plants. *Nat. Nanotechnol.* **2019**, *14*, 456–464. [CrossRef]
101. Cunningham, F.J.; Goh, N.S.; Demirer, G.S.; Matos, J.L.; Landry, M.P. Nanoparticle-Mediated Delivery towards Advancing Plant Genetic Engineering. *Trends Biotechnol.* **2018**, *36*, 882–897. [CrossRef] [PubMed]
102. Cao, X.; Xie, H.; Song, M.; Lu, J.; Ma, P.; Huang, B.; Wang, M.; Tian, Y.; Chen, F.; Peng, J.; et al. Cut–dip–budding delivery system enables genetic modifications in plants without tissue culture. *Innovation* **2023**, *4*, 100345. [CrossRef] [PubMed]
103. Lu, J.; Li, S.; Deng, S.; Wang, M.; Wu, Y.; Li, M.; Dong, J.; Lu, S.; Su, C.; Li, G.; et al. A method of genetic transformation and gene editing of succulents without tissue culture. *Plant Biotechnol. J.* **2024**, *22*, 1981–1988. [CrossRef] [PubMed]

104. Kim, H.C.; Peng, M.; Liu, S.; Wang, Y.; Li, Z.; Zhao, S.; Li, S.; Quan, H.; Luo, Q.; Meng, F. Comparative proteomic analysis reveals the adaptation of *Herpetospermum pedunculosum* to an altitudinal gradient in the Tibetan Plateau. *Biochem. Syst. Ecol.* **2018**, *80*, 1–10. [CrossRef]
105. Du, Z.; Lin, W.; Yu, B.; Zhu, J.; Li, J. Integrated Metabolomic and Transcriptomic Analysis of the Flavonoid Accumulation in the Leaves of *Cyclocarya paliurus* at Different Altitudes. *Front. Plant Sci.* **2022**, *12*, 794137. [CrossRef]

International Journal of
Molecular Sciences

MDPI

Article

Genome-Wide Identification of the Lectin Receptor-like Kinase Gene Family in *Avena sativa* and Its Role in Salt Stress Tolerance

Gui Xiong [1,2,3], Dongli Cui [1,2,3], Yaqi Tian [1,2,3], Trude Schwarzacher [4,5], John Seymour Heslop-Harrison [4,5,*] and Qing Liu [1,2,4,*]

1 Laboratory of Plant Resources Conservation and Sustainable Utilization, South China Botanical Garden, Chinese Academy of Sciences, Guangdong Provincial Key Laboratory of Applied Botany, Guangzhou 510650, China; xionggui23@scbg.ac.cn (G.X.); donglicui2018@163.com (D.C.); tianyaqi24@mails.ucas.ac.cn (Y.T.)
2 Key Laboratory of National Forestry and Grassland Administration, Plant Conservation and Utilization in Southern China, Guangzhou 510650, China
3 University of Chinese Academy of Sciences, Beijing 101408, China
4 South China National Botanical Garden, Guangzhou 510650, China; ts32@le.ac.uk
5 Department of Genetics and Genome Biology, Institute for Environmental Futures, University of Leicester, Leicester LE1 7RH, UK
* Correspondence: phh4@le.ac.uk or phh@molcyt.com (J.S.H.-H.); liuqing@scib.ac.cn (Q.L.)

Abstract: Lectin receptor-like kinases (LecRLKs) are membrane-bound receptor genes found in many plant species. They are involved in perceiving stresses and responding to the environment. Oat (*Avena sativa*; $2n = 6x = 42$) are an important food and forage crop with potential in drought, saline, or alkaline soils. Here, we present a comprehensive genome-wide analysis of the *LecRLK* gene family in *A. sativa* and the crop's wild relatives *A. insularis* ($4x$) and *A. longiglumis* ($2x$), unveiling a rich diversity with a total of 390 *LecRLK* genes identified, comprising 219 G-types, 168 L-types, and 3 C-types in oats. Genes were unevenly distributed across the oat chromosomes. GFP constructs show that family members were predominantly located in the plasma membrane. Expression under salt stress demonstrated functional redundancy and differential expression of *LecRLK* gene family members in oats: 173 members of this family were involved in the response to salt stress, and the expression levels of three C-type genes in the root and leaf were significantly increased under salt stress. The results show the diversity, evolutionary dynamics, and functional implications of the *LecRLK* gene family in *A. sativa*, setting a foundation for defining its roles in plant development and stress resilience, and suggesting its potential agricultural application for crop improvement.

Keywords: *Avena sativa*; salt stress tolerance; transcriptome; lectin receptor-like kinases; expression pattern

1. Introduction

During evolution, plants develop a series of mechanisms to maintain oxidative balance under various stressors [1]. All kinds of receptors located on the cell membrane or in plant cells can specifically recognize bioactive molecules and receive and process external information [2]. The receptor-like kinases (RLKs) are an important and diverse group of cell surface receptor proteins that possess an extracellular domain, a membrane-spanning domain, and an intracellular kinase domain [3]. Based on the variability of extracellular structural domains, *RLKs* can be classified into 17 subgroups [2,4]. Lectin receptor-like kinases belong to the *RLK* gene family, which was named for its lectin/lectin-like ectodomain which can bind carbohydrates [5]. The N-terminal lectin structural domain allows LecRLKs to recognize environmental stimuli, whereas the C-terminal intracellular kinase structural domain phosphorylates downstream proteins to transmit signals. Depending on the type of lectin domain at the N-terminus, LecRLKs are further classified into three subfamilies:

L-type, G-type, and C-type LecRLKs [3]. L-type LecRLKs possess a legume-like lectin domain. G-type LecRLKs have an α-mannose-binding bulb lectin domain, along with an S-locus glycoprotein domain (SLG) and plasminogen apple nematode (PAN) and/or epidermal growth factor (EGF) domains. The SLG domain has been proved to play an important role in the self-incompatibility of plant gametophytes [6]. The PAN motif is associated with protein–protein and protein–carbohydrate interactions, while the EGF domain likely contributes to the formation of disulfide bonds [7]. C-type LecRLKs are marked by a calcium-dependent lectin domain [3,5]. LecRLKs have been identified and discovered in various plants, including 75 members in *Arabidopsis thaliana*, 173 members in rice [4], and 263 members in wheat [8].

LecRLKs play a crucial role in plant growth, stress management, and innate immune responses [9–11]. Crop yield loss due to soil salinization is an escalating threat to global agriculture [12]. Salt stress, either due to soil type or from irrigation water, impacts the growth, development, and grain productivity of crops [13] in arid and semi-arid regions [14]. With increased irrigation, global warming, and rising sea levels, the area of land (including reclaimed land) under salt stress is increasing, while the use of saline land for productive agriculture has become a new trend in the development of agriculture and animal husbandry [13,15].

Many plant processes involve families of genes with identical or related functions, or within a pathway. For example, the glucanase gene family plays a defensive role in response to pathogen attacks [16]; the flavonoid gene family plays a key role in plant growth and development, response to adversity, and interactions with the environment [17]. In some pathways, gene clusters play a role in the specialized metabolism of plants [18]; for the flavonoid pathway, the presence of gene clusters contributes to the concentration of members of different gene families in physical locations [19–21]. The study of gene families, genes in a pathway, and gene clusters not only improves the understanding of plant metabolic pathways and their dynamics, regulation, and evolution, but also provides important molecular resources for crop improvement and the breeding of new varieties.

Plants have developed regulatory mechanisms to cope with salt stress over a long period of time in response to the environment, including regulation of salt stress resistance through maintenance of photosynthesis, ion homeostasis, and synthesis of sugar metabolism, proline, and other compatible solutes that regulate reactive oxygen species (ROS) levels [13,22]. Several reports have presented the involvement of LecRLKs in plant responses against salt stress. A G-type LecRLK known as GsSRK in *Medicago sativa* has the potential to regulate Na^+/K^+ balance during salt stress, possibly through the scavenging of reactive oxygen species (ROS) and the modulation of osmotic homeostasis [23,24]. Under salt stress conditions, *SIT1* (an LecRLK gene) enhances the sensitivity of plants [25]. The overexpression of one *Pisum sativum* LecRLK, PsLecRLK, prevents the accumulation of reactive oxygen species and membrane damage and enhances salt tolerance in plants [9].

Here, we aimed to perform a comprehensive genome-wide identification and characterization of the lectin receptor-like kinase (LecRLK) gene family in *Avena sativa* (oat), with a focus on elucidating their potential roles in salt stress tolerance. Specifically, this study aimed to identify *LecRLK* gene family members and analyze their gene structures, conserved domains, and phylogenetic relationships. Additionally, utilizing transcriptomic data, we investigated the expression profiles of these genes under salt stress conditions. Key candidates were further validated through RT-qPCR analysis and subcellular localization assays, enhancing our understanding of their functional roles in mediating salt stress responses. This work provides new insights into the molecular mechanisms underlying salt stress tolerance in oats and offers valuable resources for future breeding efforts aimed at improving salt tolerance in this crop.

2. Results

2.1. Identification of LecRLK Gene Family Members in Oat

A total of 390 putative AsaLecRLK protein sequences were identified in *A. sativa*, which were further classified into L-type LecRLKs (n = 168), G-type LecRLKs (n = 219), and C-type LecRLKs (n = 3). The physicochemical properties of AsaLecRLK family genes (named L, G, and C then in order on chromosomes) were summarized (Table S1).

Coding sequence size (CDS) length ranged from 915 to 2943 bp, 1338 to 2976 bp, and 1662 to 1680 bp, whereas the protein lengths varied from 304 to 980 aa, 445 to 991 aa, and 553 to 559 aa for L-type, G-type, and C-type proteins, respectively. The molecular weights ranged from 34.10 kDa (AsaLecRLK-L-type-168) to 108.43 kDa (AsaLecRLK-G-type-200), and the isoelectric point (pI) ranged from 4.53 (AsaLecRLK-G-type-28) to 10.20 (AsaLecRLK-L-type-168). In total, 281 LecRLK proteins were acidic (pI < 7); the remaining 106 proteins were alkali (pI > 7). The grand average of hydropathicity (GRAVY) ranged from −0.311 to 0.102, with 344 proteins exhibiting hydrophilicity (GRAVY < 0) and 46 proteins exhibiting hydrophobicity (GRAVY > 0), indicating diverse functions among this protein family. Subcellular localization prediction results were analyzed using the program WoLF PSORT, which showed that 222 proteins were only located in the plasma membrane, and the other 168 were in different cellular compartments, including extracellular, plastid, vacuole, chloroplast, mitochondria, cytoplasm, nucleus, etc. (Table S1), indicating the diverse roles of the LecRLK family in variable environments.

2.2. Chromosomal Distribution and Phylogenetic Analysis of LecRLK Gene Family Members

The chromosomal positions of AsaLecRLKs showed that 390 members of the gene family were distributed (Figure 1) across all 21 chromosomes and the unanchored scaffold. *AsaLecRLK* genes are less abundant in regions with low gene density. The numbers of genes per chromosome ranged from 39 (chromosome 2D) to 9 genes (1C). Gene clusters were observed on each chromosome, perhaps a consequence of segmental duplications.

Evolutionary relationships of the AsaLecRLK members were shown in a phylogenetic tree, with outgroups of three animal Pelles [26] [the animal homologs of Arabidopsis RLKs; a Pelle kinase (DmPelle) in Drosophila, a Pelle-like kinase (CePelle) in Caenorhabditis, and an IRAK in humans], and 390 LecRLKs of *A. sativa* (maximum-likelihood (ML) method; Figure 2). The result showed that the L-type, G-type, and C-type formed monophyletic lineages.

2.3. Conserved Domain, Conserved Motif and Gene Structure Analysis of AsaLecRLKs

Conserved domain analysis of AsaLecRLKs demonstrated that there were significant differences in the structures of *AsaLecRLK* genes within and between the three subgroups (Figures 3 and S1). There were six conserved domains in the G-type AsaLecRLKs, including the protein kinase domain (Pkinase; PF00069), S-locus glycoprotein domain (S_locus_glycop; PF00954), mannose-binding bulb-lectin domain (B_lectin; PF01453), two PAN domains, and a DUF3403 domain. Based on the type and number of conserved domains, the G-type AsaLecRLKs can be classified into seven distinct categories: a total of 33 G-type AsaLecRLKs exhibited two domains including Pkinase and B-lectin, and 20 contained Pkinase with B-lectin and S_locus_glycop. Additionally, 145 genes contain an additional PAN domain, either PAN_1 or PAN_2, beyond the three aforementioned structural domains. A total of four G-type AsaLecRLKs contain the DUF3403 domain at their C-terminus. Fifteen genes are identified to contain a lectin domain, a kinase domain, and a PAN domain. One gene featured two kinase domains. Within these domains, the S-locus_glycoprotein is implicated in the self-incompatibility response in plants; the PAN domain is involved in protein–protein and protein–carbohydrate interactions; and the DUF3403 domain has been previously reported in the literature, although its function remains elusive. In addition, transmembrane domain predictions showed that the majority of AsaLecRLK proteins (96.92%) had transmembrane domains, which is consistent with previous reports in other species. The L-type and C-type have a less complex domain structure: 166 L-type AsaLecRLKs contain a single legume lectin domain (Lectin_legB; PF00139) and Pkinase. Some unique domain architectures were also observed within the L-type AsaLecRLKs: two L-type AsaLecRLKs contain an adh_short domain (PF00106), which is involved in a variety of oxidation–reduction reactions. There was only one calcium-binding lectin (Lectin_C; PF00059) and one Pkinase domain present in three C-type AsaLecRLKs.

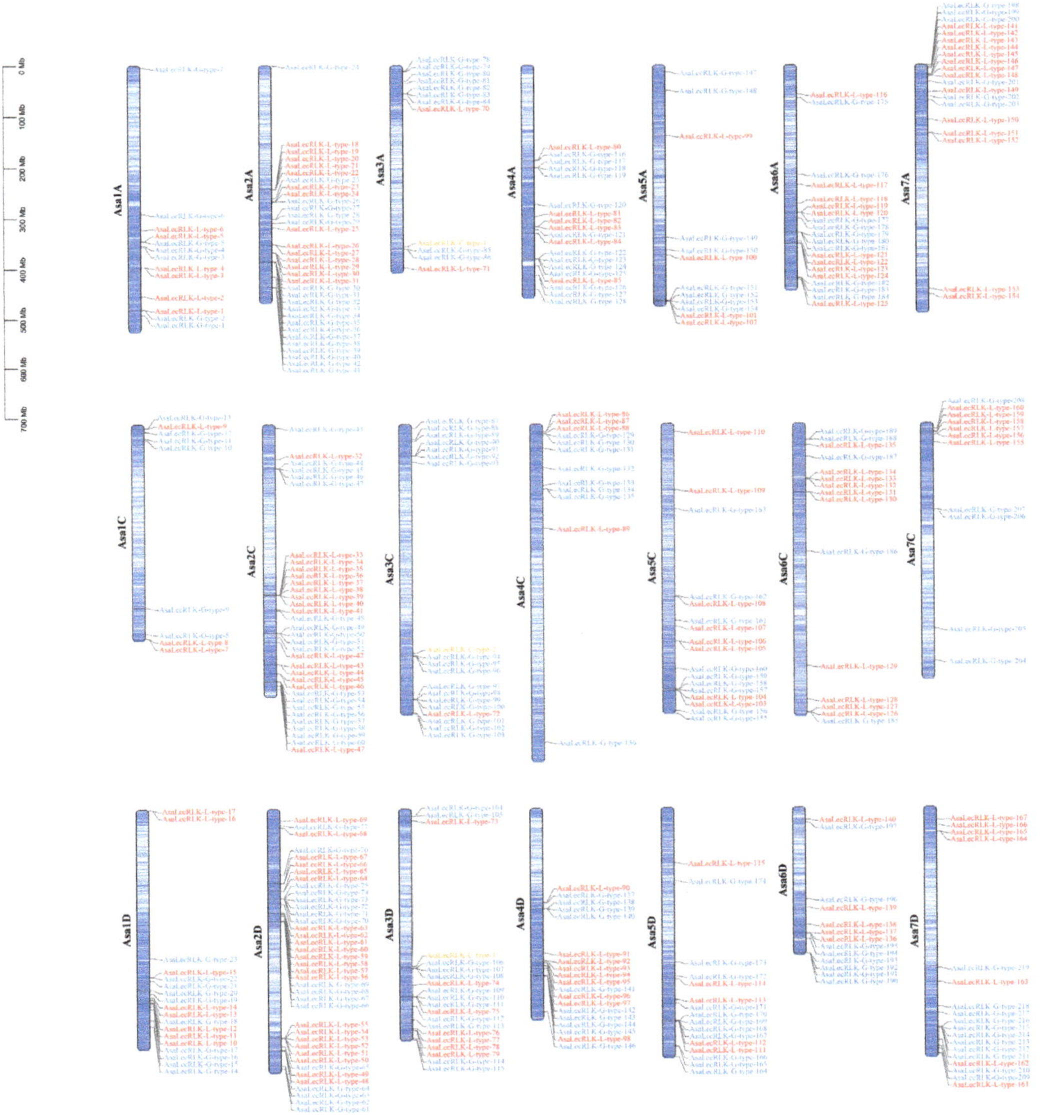

Figure 1. Chromosome location of *LecRLK* gene family of *Avena sativa*; L-type (red), C-type (yellow), and G-type (blue) subfamilies are shown. Chromosome numbers are shown at the left. Center of chromosomes shows the overall gene density. *LecRLK* gene locations are shown on the right.

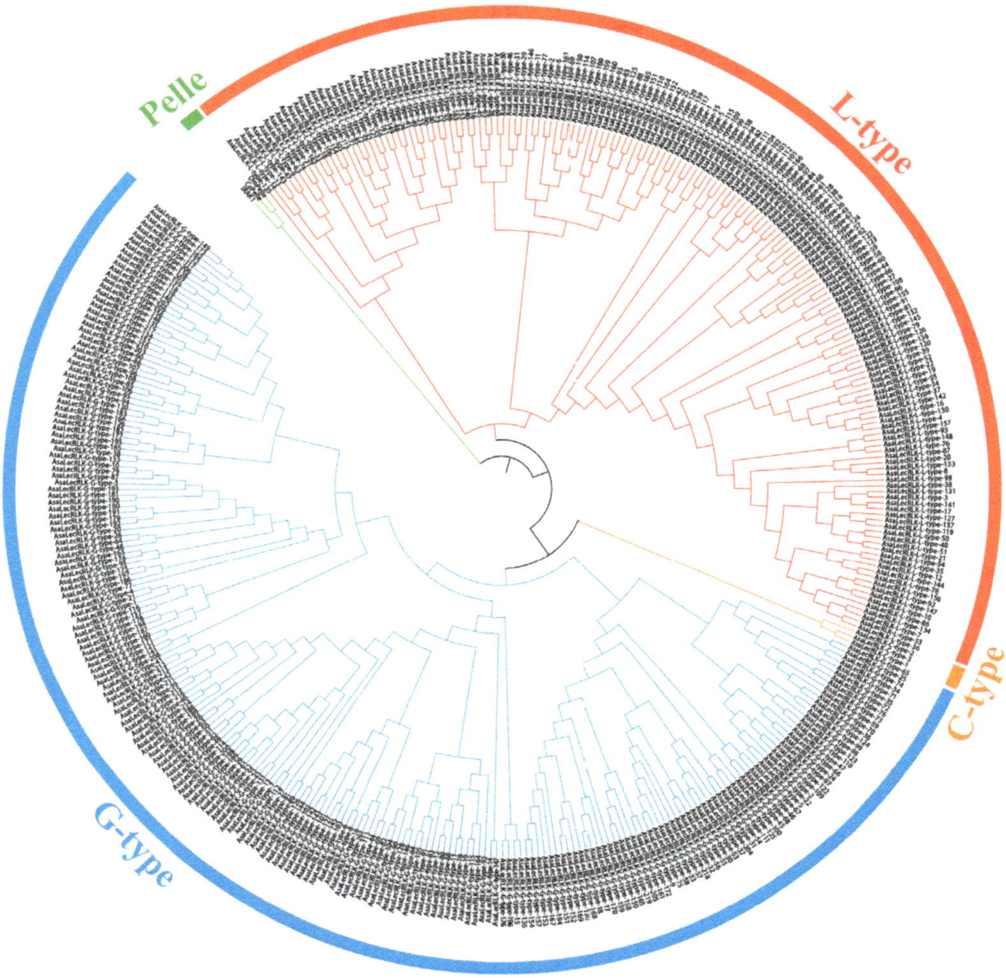

Figure 2. Phylogenetic relationships of LecRLK proteins in *Avena sativa* and three RLK Pelle family proteins in animals. The phylogenetic trees were constructed using the maximum-likelihood method based on predicted protein sequences. L-type (red), C-type (yellow), G-type (blue), Pelle (green).

Ten conserved motifs in the *AsaLecRLK* gene family were identified using MEME online tools [27] and ten motifs were detected in 390 AsaLecRLK proteins from *A. sativa* (Figure S2, Table S2). The results showed that most of these existed in the kinase domain. This confirms that the kinase domain has remained conserved during evolution. All G-type genes of the AsaLecRLK family had B-lectin with the typical motif 10, which is absent in the L-type and C-type genes. AsaLecRLK proteins of the same type had similar domains to the conserved motifs (Table S2) and distributions, indicating functional similarity among members of the same type. The variation in motifs among different types also demonstrates the functional diversity of LecRLKs.

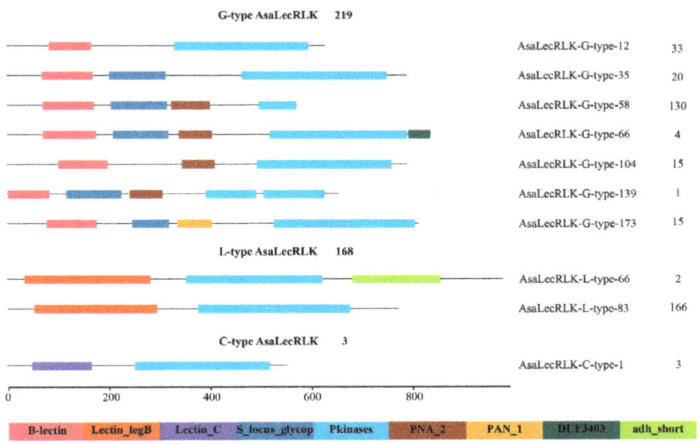

Figure 3. Domain structure prediction of AsaLecRLKs showing the number of genes (right column) with each structure. G-type AsaLecRLKs contain bulb lectin domain, S-locus glycoprotein domain, and PAN domain at the N-terminus and protein kinase domain and DUF3403 domain at the C-terminus; L-type AsaLecRLKs contain the legume lectin domain at the N-terminus and protein kinase domain and adh_short domain at the C-terminus; C-type AsaLecRLK contains the calcium-binding lectin domain at the N-terminus and protein kinase domain at the C-terminus.

The gene structure of the *AsaLecRLK* gene family showed structural diversity. The number of *LecRLK* gene exons of *A. sativa* ranged from 1 to 9 (0 to 8 introns), although most (215 out of 390) genes had a single exon. Intron length varied from 61 bp (*AsaLecRLK-G-type-106*) to 2041 bp (*AsaLecRLK-L-type-27*) (Figure S3), with subfamily branches showing similarities in structure.

2.4. Cis-Acting Element Prediction of AsaLecRLKs

Promoter cis-element analysis provided insights into the tissue-specific expression and stress response patterns of the gene family. Except for the TATA-BOX and CAAT-BOX, a total of 120 cis-acting elements were identified, with a variety of environmental and stress response elements in the *AsaLecRLK* gene family of *A. sativa* (Table S3), using the PlantCARE [28] analysis of the upstream 1500 bp sequence. The analysis associated these elements with wound response, hormone response, light response, promoter and enhancer elements, binding-site elements, and development. Among them, the most abundant element is transcription factor MYB (1086), which has been identified in plant development and responses to stress by combining with MYB cis-elements in promoters of target genes [29]. The second most abundant element is MYC (1054), which is associated with dehydration response and ABA induction [30]. The third most abundant element is ABRE (978), which is associated with hormone response [31]. The fourth most abundant element is STRE (837), which is associated with environmental stress response [32,33]. Additionally, the elements as-1, CGTCA-motif, and TGACG-motif were abundant, all of which are hormone-responsive elements, implying that the expression of *AsaLecRLKs* might be regulated by phytohormones [30,32]. Among the *AsaLecRLK* gene family, AsaLecRLK-L-type-162 has the highest number of cis-acting elements (86).

2.5. Expansion and Selection Pressure Analysis of AsaLecRLKs

Gene family formation, member expansion, and functional diversification are primarily driven by the process of gene duplication. Among 390 *AsaLecRLK* genes, we found that there were 190 possible pairs of segmental duplication (SD) genes and 19 pairs of potential tandem duplication (TD) genes (Figure S4, Table S4). Among these 19 tandem duplication gene pairs, we identified the formation of a gene cluster of tandem duplication on the

Asa7D and Asa4D chromosomes. The results of gene duplication analysis indicated that SD and TD were the main expansion mechanisms of the *AsaLecRLK* gene family. In addition, we calculated the non-synonymous (Ka) and synonymous (Ks) substitution rates, as well as the Ka/Ks ratios. The Ka/Ks ratio varied from 0 to 0.862761 (Table S5), indicating that purification selection plays an important role during geneexpansion.

2.6. Collinearity Analysis of AsaLecRLK Genes

To explore the evolutionary mechanism between the *AsaLecRLK* genes of *Avena sativa* and other species, synteny analyses were performed on four representative plants, including two possible ancestral species of oats (*A. longiglumis* and *A. insularis*), a monocotyledon (*Oryza sativa*), and a dicotyledon (*Arabidopsis thaliana*). Dual synteny analysis revealed that 246 and 356 *AsaLecRLK* syntenic gene pairs were identified in *A. longiglumis* (Table S6) and *A. insularis* (Table S7), respectively, and 116 and 24 syntenic gene pairs each were identified in *Oryza sativa* (Table S8) and *Arabidopsis thaliana* (Table S9). Consistent with known relationships, these results indicated that oat was most distantly related to *Arabidopsis* and most closely related to *Avena insularis*. By comparing the collinear relationship between*A. sativa*, *A. insularis*, and *A. longiglumis*, it was found that collinearity did not only occur on homologous, but also heterologous chromosomes (Figure 4), supporting that oat chromosomes have undergone chromosomal rearrangement and gene duplication events during evolution.

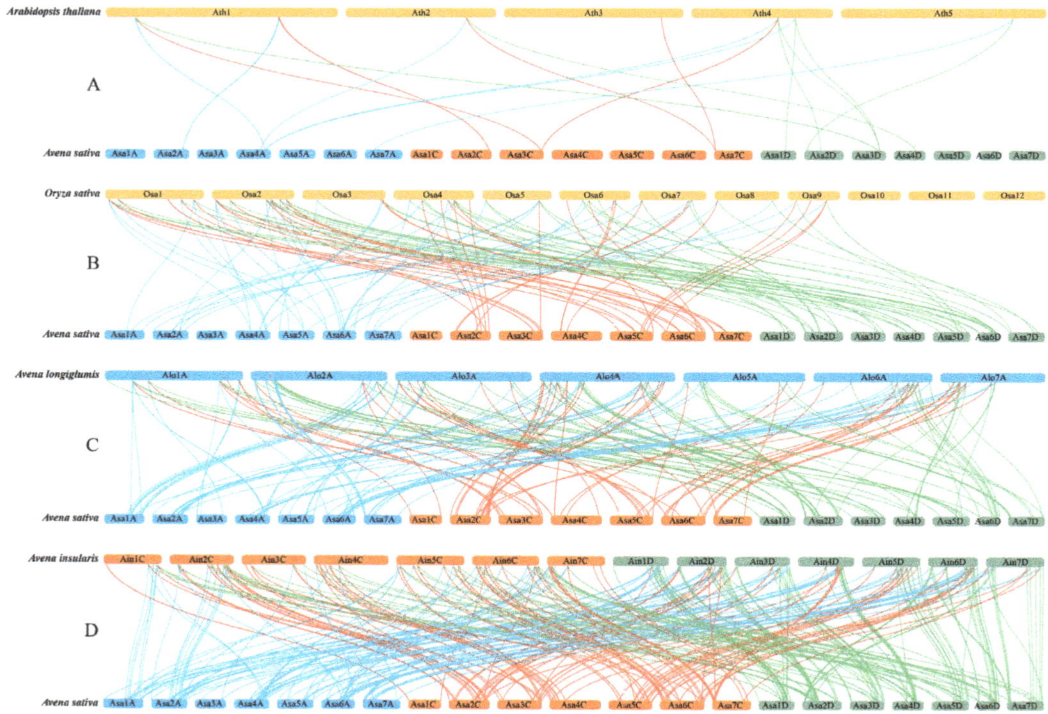

Figure 4. Synteny analyses of *LecRLK* gene family between *Arabidopsis thaliana* and *Avena sativa* (**A**), *Oryza sativa* and *A. sativa* (**B**), *A. longiglumis* and *A. sativa* (**C**), and *A. insularis* and *A. sativa* (**D**). Lines represent collinear gene pairs between genomes of *A. sativa* and other species. Blue line: A genome; red line: C genome; green line: D genome.

2.7. Gene Expression Analysis of AsaLecRLKs' Response to Salt Stress Treatments

To investigate the role of AsaLecRLKs in oat salt stress resistance, we analyzed RNA-sequencing data, focusing on the RPKM-based expression patterns of *AsaLecRLK* genes in the root and leaf to provide insights into their potential functions. We analyzed the expression of the 390 *AsaLecRLK* genes under salt stress, first removing 217 genes with FPKM less than one in both root and leaf before and after salt stress. We generated a heat map of the 173 (G-type: 96; L-type: 74; C-type: 3) expressed *AsaLecRLK* genes using the FPKM values (Figure 5). Among these genes, three of the C-type *AsaLecRLKs* were up-regulated in both the root and leaf under salt stress. In total, 45 G-type genes were down-regulated in the leaf and 50 G-type genes were up-regulated in the leaf; 58 G-types were down-regulated and 37 up-regulated in the root. A total of 34 G-type genes were down-regulated in both the root and leaf, and 27 G-type genes were up-regulated in both the root and leaf. A total of 25 L-type genes were down-regulated in the leaf and 47 L-type genes were up-regulated in the leaf; 30 L-type genes were down-regulated and 42 L-type genes were up-regulated; and 13 L-type genes were down-regulated in both the root and leaf, and 62 L-type genes were up-regulated in both the root and leaf. The results indicate the differential expression of some *AsaLecRLK* genes in the root and leaf, with genes such as *AsaLecRLK-G-type-45/68*, *AsaLecRLK-L-type-70*, and *AsaLecRLK-C-type-2* being expressed more in the root than in the leaf; different *AsaLecRLK* genes exhibit varying expression levels under identical salt stress conditions, and their expression also differs between the root and leaf. A cluster analysis showed that *AsaLecRLK* genes located in the same evolutionary branch showed some similarity. For example, *AsaLecRLK-G-type-45* and *AsaLecRLK-G-type-68* showed similar expression patterns after NaCl treatment. However, there is also disaggregation of genes located in the same evolutionary branch. The differential expression profiles among homologous genes suggest that during the evolution of oats, some genes may have acquired new functions or lost old ones following polyploidization. This phenomenon is referred to as the subfunctionalization of homologous genes by some researchers [34].

To verify the results acquired from RNA-Seq data, we selected 19 *AsaLecRLK* genes from three different subtypes for RT-qPCR analysis. The results showed that the expression patterns of these genes were basically consistent with the RNA-sequencing results after 48 h of salt stress, indicating that these 19 genes responded to salt stress to different degrees. Under salt stress treatment, there were significant differences in the expression of multiple *AsaLecRLK* genes in the root and leaf of *Avena sativa* at different time points (0 h, 6 h, 12 h, 24 h and 48 h). Overall, most genes showed different degrees of up-regulation or down-regulation in both root and leaf tissues, with some genes showing significantly enhanced or attenuated expression at specific time points (such as the significant increase in expression of *AsaLecRLK-C-type-3* in the root at 24 h). Additionally, there were tissue-specific differences in the response of root and leaf tissue to salt stress at different time points, suggesting that these genes may have potential tissue-specific functions in the response to salt stress. For instance, the expression levels of *AsaLecRLK-L-type-7/8* and *AsaLecRLK-G-type-200* genes decreased in root tissue but increased in leaf tissue under stress at the 6 h time point, which may be closely related to their functions and participation in *A. sativa* salt stress at different stages. Furthermore, the timing of peak expression levels differs between root and leaf tissue. For example, the *AsaLecRLK-G-type-110* gene reaches its peak expression in the root 48 h post-stress, whereas in the leaf, the peak is attained 24 h post-stress (Figure 6). The presence of temporal differences in the peak appearance of these genes after salt stress suggests that these genes may play different roles at different times, and it is hypothesized that they interact in response to salt stress. Notably, the expression of all three C-type *AsaLecRLKs* was significantly increased in root tissue after salt stress, suggesting that all three genes may be involved in salt stress.

RNA-sequencing and RT-qPCR analyses revealed the spatiotemporal expression patterns of *AsaLecRLK* genes in oat under salt stress, demonstrating significant tissue specificity and temporal dependency. The differential peak expression levels observed in root and leaf suggest that *AsaLecRLK* genes may contribute to salt stress tolerance through tissue-specific and time-dependent mechanisms, potentially involving synergistic interactions. These findings underscore the potential functional importance of *AsaLecRLK* genes in regulating salt stress responses in oat.

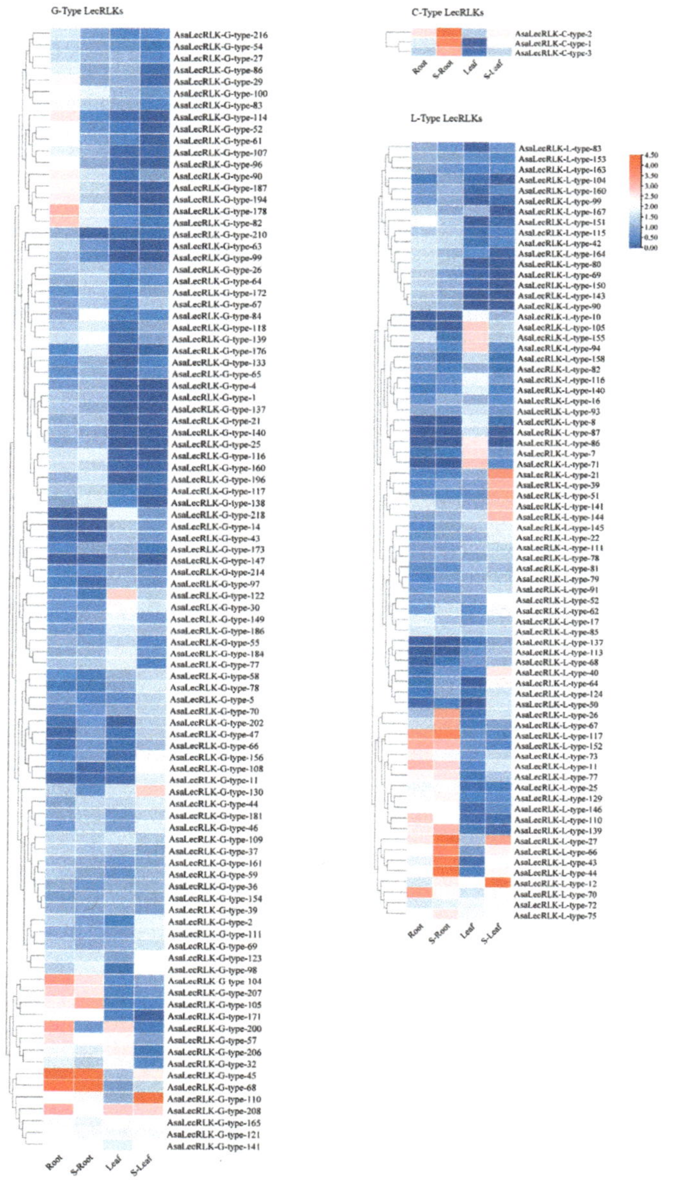

Figure 5. Expression profiles of *AsaLecRLK* genes under different levels of salt stress in root and leaf. RNA-sequencing data on salt stress for *AsaLecRLKs*. The heatmap was generated on the Log2 of (FPKM+1) values using TBtools. Color bar represents normalized FPKM values: red, high expression level; blue, low expression level.

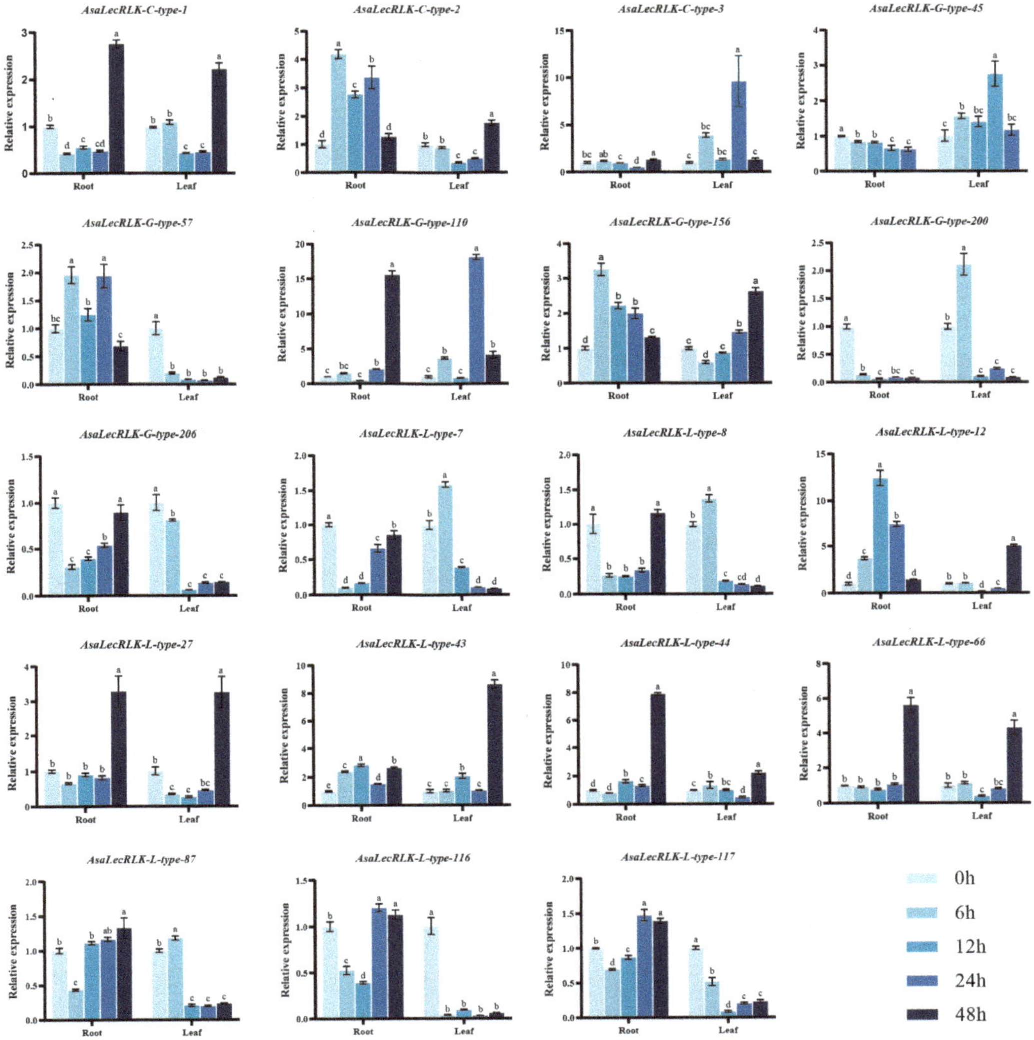

Figure 6. Relative expression level of *AsaLecRLK* genes under salt stress after 0 h, 6 h, 12 h, 24 h and 48 h in root and leaf tissue. Expression level of each gene at 0 h is set as reference. The data represent the mean values of three replicates ± SD. Statistical significance of differences was tested by one-way ANOVA analysis ($p < 0.05$) and is indicated by lowercase letters.

2.8. Subcellular Localization Analysis of AsaLecRLK-G-type-45

In this study, most of the AsaLecRLK proteins were predicted to be localized to the plasma membrane. To validate the subcellular localization predictions and clarify the functional sites of AsaLecRLK proteins, we cloned the coding sequences of AsaLecRLK-L-type-43 (2043 bp long), AsaLecRLK-L-type-44 (2043 bp), and AsaLecRLK-G-type-45 (2550 bp). The subcellular localization result showed that AsaLecRLK-L-type-43, AsaLecRLK-L-type-44, and AsaLecRLK-G-type-45 were localized to the plasma membrane, which was consistent with the predicted results (Figure 7). The 35S::GFP, used as a control, was localized to both nuclear and cell membranes.

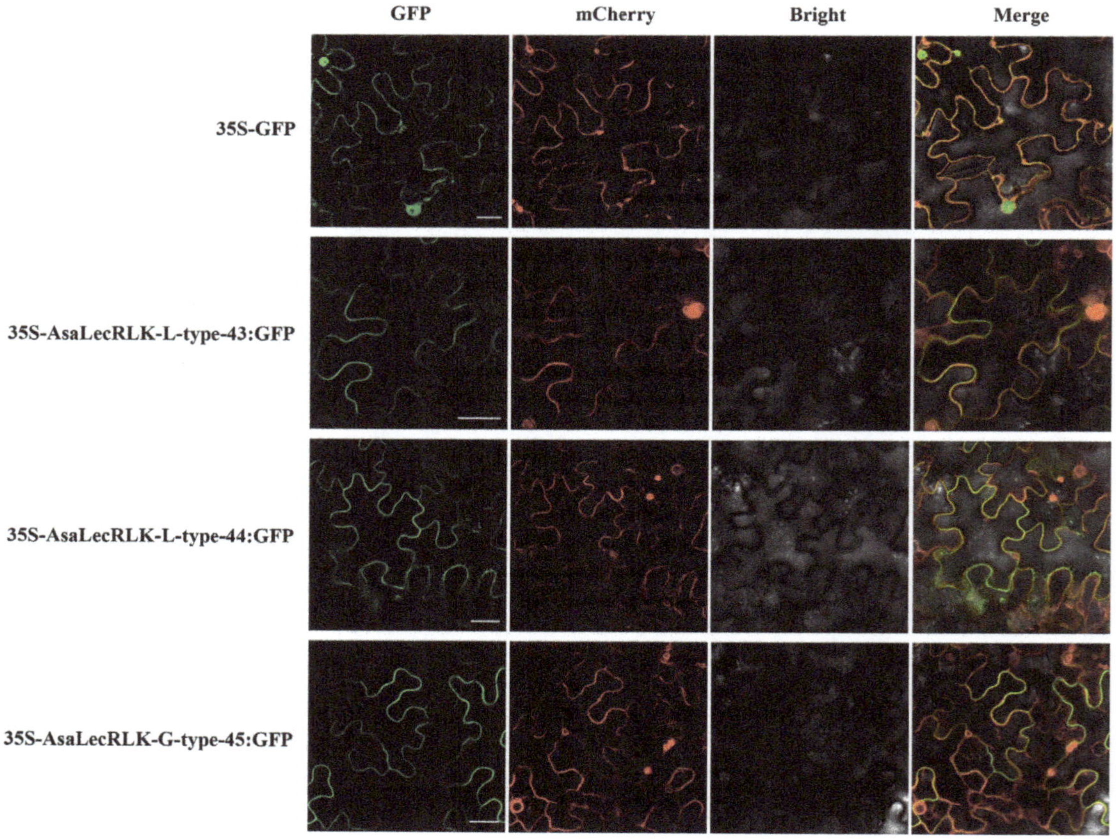

Figure 7. The subcellular localization of *AsaLecRLK-L-type-43*, *AsaLecRLK-L-type-44*, and *AsaLecRLK-G-type-45*. Bars = 25 μm. The figures show confocal images of GFP fluorescence, plasmalemma localization (mCherry), bright field, and composite field.

3. Discussion

As receptor-like kinases at the cell surface, LecRLKs have been demonstrated to play a crucial role in enhancing plant resistance against abiotic (salt, low-temperature, or drought stress and mechanical damage) and biotic (plant disease caused by bacteria, viruses, fungi, and herbivorous insects) stresses as well as in development (seed germination, leaf development, flower organ development, and reproductive development) [35,36]. Here, we identified 390 genes encoding LecRLK proteins in oat, *A. sativa* (Figures 1 and 2, Table S1). The number of *LecRLK* genes in the hexaploid oat was higher than in diploid reference species (e.g., 173 in *Oryza sativa* [4]; 113 in *Hordeum vulgare* [31]; 75 in *Arabidopsis thaliana* [4]), with clear duplication between homoeologous chromosomes (Figure S4 center); and between rice and Arabidopsis (Figure 4A,B).

Gene tandem replication and fragment replication events are considered to be the key mechanisms for increasing gene family diversity. Analysis based on duplication events indicated that segmental duplication was the major factor leading to the amplification of the *AsaLecRLK* gene family. At the same time, 19 tandem duplication pairs were identified in oat, and interestingly, these tandem duplication gene pairs were not strictly clustered in the phylogenetic tree, suggesting that these *AsaLecRLK* genes may have evolved different functions after duplication [37] (Figure 2, Table S4). *Avena sativa* is an allohexaploid, and the diploid species *A. longiglumis* and the tetraploid species *A. insularis* are its potential ancestral relatives [38]. We identified 219 G-type LecRLK members in oat, which exceeds the combined total of 205 in *A. longiglumis* (Figure 4C) and *A. insularis* (Figure 4D). In contrast, the 168 L-type members in oats are fewer than the combined total of 176 in the two ancestral species. The number of C-type members is equal to the combined total, with three in each. This result suggests that *LecRLK* genes in oat may have undergone gene loss (or degeneration) and duplication during the evolutionary hybridization and polyploidization events. This study found that *AsaLecRLK* family members are unevenly distributed across the A (185), C (116), and D (142) subgenomes (Figure 1), not least as a consequence of chromosomal rearrangements between genomes in the hexaploidy [39].

By analyzing the predictions of domain architecture and organization, we observed several interesting features of AsaLecRLKs. Firstly, the majority of AsaLecRLKs possess transmembrane (TM) domains. Additionally, we have identified several lectins with two or three transmembrane domains. It is speculated that AsaLecRLKs function as potential membrane-bound receptors (Figure 7) and may bind to the membrane system in a variety of ways [37]. Interestingly, an adh_short domain was identified at the C-terminus of AsaLecRLK-L-type-66 and AsaLecRLK-L-type-27. The adh_short domain belongs to short-chain dehydrogenase/reductase gene family (Figures 3 and S1), which is one of the largest and oldest NAD(P)(H)-dependent oxidoreductase families [40]. It has been demonstrated that LecRLKs can recognize DNA; LecRK-VI.2 is a potential receptor for extracellular NAD+ (eNAD+) and NAD+ phosphate (eNADP+) [41], suggesting that AsaLecRLK-L-type-66 and AsaLecRLK-L-type-27 may also be potential receptors for NAD+.

The gene structure (Figure S3) and conserved motifs (Figure S2, Table S2) of AsaLecRLKs were analyzed. All G-type AsaLecRLKs have motif 10, indicating that motif 10 plays a crucial regulatory role in G-type AsaLecRLKs. The conserved motif 10 is unique in all G-type RLKs, and is specifically located within the B-lectin domain. This suggests its potential involvement in key functional processes, possibly related to signal transduction or stress response pathways specific to this subfamily. This conserved motif might be a target for further functional analysis to elucidate its role in salt stress tolerance. Intron gain or loss and intron density significantly impact the evolution of large eukaryotic genomes [42]. Gene structural analysis reveals that more than half of *AsaLecRLKs* lack introns. Genes with similar numbers of exons and introns that cluster together on a phylogenetic tree may indicate shared ancestry or conserved gene structures that have been maintained throughout evolution [43]. This conservation could suggest that these genes are subject to similar functional constraints or are involved in similar biological processes. The clustering can also reflect the evolutionary history and potential subfunctionalization or neofunctionalization events that have occurred after gene duplications [44].

Soil salinization represents a significant stressor that adversely affects the majority of plants. Extensive research has established that plants possess the ability to perceive salt stress signals and promptly activate signaling pathways, thereby re-establishing cellular homeostasis through the regulation of growth and metabolic processes [45]. For example, encoding transcriptional factor OsWRKY53 and mitogen-activated protein kinase OsMKK10.2 mediate root Na^+ flux and Na^+ homeostasis [46]. In addition to regulating ion homeostasis, salt stress induces enzymatic and non-enzymatic systems in plants to mitigate reactive oxygen species (ROS) stress [47]. Enzymatic scavengers include superoxide dismutase, ascorbate peroxidase, catalase, guaiacolperoxidase, guaiacol peroxidase, dehydroascorbate reductase, monodehydroascorbate reductase, glutathione peroxidase, and glutathione S-transferase. Salt stress also triggers abscisic acid (ABA) pathways. Salt treatment increases ABA concentrations in plant cells and activates sucrose non-fermenting 1-related protein kinase 2 (SnRK2) kinase activity [15,48].

LecRLKs play a role in tolerance to salt stresses [2]. A rice L-type LecRLK gene, *SIT1*, is activated during salt stress, phosphorylates downstream effectors, triggers ethylene signaling-induced reactive oxygen species accumulation, and enhances plant sensitivity [25]. The overexpression of the rice *OsSIK2* gene leads to enhanced salt stress tolerance in plants and promotes leaf emergence [49]. In 2020, Passricha et al. [50] demonstrated that *OsLecRLKs* play a crucial role in salt stress responses. Overexpression of *OsLecRLKs* activated sucrose non-fermenting-related kinase-1 (SnRK1), which in turn activated the salt overly sensitive 1 (SOS1) channel to transport Na^+ ions out of the cytosol, maintaining ionic balance. In contrast, down-regulation of *OsLecRLKs* led to cytosolic Na^+ accumulation, disrupting ionic homeostasis and normal cell function. In soybean, a G-type lectin receptor-like kinase GsSRLK positively regulates plant tolerance to salt stress [24]. Due to the close homologous relationship between *SIT1* and *AsaLecRLK-L-type-26*, *AsaLecRLK-L-type-43*, *AsaLecRLK-L-type-67*, *AsaLecRLK-L-type-118*, *AsaLecRLK-L-type-128*, and *AsaLecRLK-L-type-138*, it is speculated that these *AsaLecRLKs* may be related to salt stress. In our study, all selected RT-qPCR genes exhibited varying degrees of response to salt stress, with the *AsaLecRLK-G-type-110* gene demonstrating a 15-fold increase in expression in root tissue 48 h post-salt stress and an 18-fold increase in leaf tissue 24 h post-salt stress (Figure 6). The cis-acting elements of the gene promoter regions are involved in the regulation of gene expression. In this study, the promoter sequences of the *AsaLecRLK* genes predicted multiple cis-acting elements related to stress responsiveness (STRE, TC-rich repeats), phytohormone responsiveness (TGAGG-motif, CGTCA-motif, as-1, ABER, TATC-box, TCA-element), drought-inducible MYB binding site elements (MBS, MYB, Myb, MYC, Myc), and plant growth and development (CAT-box) (Figure S3). Among the identified cis-acting elements, those involved in hormone response are predominant, suggesting that there is a close regulatory link between hormone signaling pathways and the activation of AsaLecRLKs in response to environmental stress. LecRLKs have been shown to play an important role in plant responses to biotic and abiotic stresses and innate immunity. The signaling pathways of the *LecRLK* gene family and their potential ligands have lacked systematic study, and so far, only four ligands of LecRLKs have been identified (eATP, eNAD+, eNADP+ and 3-OH-C10:0) [51], and very little is known about the other ligands, so more studies are needed in the future to explore the regulatory mechanisms. The use of single-molecule analyses to analyze the amino acid sequence of proteins (e.g., Nanopore) offers a novel perspective for studying protein structure and function [52] in combination with the improved computational approaches to analysis of protein structures and interactions with ligands (e.g., Deepmind and AlphaFold3 [53]). Detection of kinase activity and post-translational modification sites defines structural and functional diversity in proteins and identifies downstream substrates. With functional information related to sequences, gene editing technologies such as CRISPR-Cas9 can be employed to regulate the expression of LecRLKs, enabling the development of stress-resistant, high-efficiency crop varieties. For example, Wang et al. [54]. demonstrated that knocking out the L-type LecRLK gene *OsCORK1* enhances rice tolerance to copper stress. Polysaccharide microarrays can be used for high-throughput screening of protein–polysaccharide interactions, which is crucial for understanding processes such as cell signaling, immune responses, and pathogen infection. Techniques like yeast two-hybrid and co-immunoprecipitation can be utilized to study the interaction networks of LecRLK proteins, shedding light on their mechanisms of action in cellular signal transduction.

With global environmental change, biotic and abiotic stresses are becoming increasingly critical for plant growth and crop yield, and the regulatory mechanisms mediated by LecRLKs have the potential to enable plants to meet these challenges. For oats, a high-quality human food and animal feed, there are huge opportunities to improve the crop and overcome the significant constraint on oat cultivation caused by progressive increase in soil salinity due to irrigation with brackish water. The utilization of key genes from the *Avena* pangenome provides a promising avenue for improving the oat crop's tolerance to salinity stress, a critical advance for achieving sustainable agricultural practices.

4. Materials and Methods

4.1. Database Search and Retrieval of Lectin Receptor-like Kinase (LecRLK) Protein Sequences in Avena sativa, Avena insularis, and Avena longiglumis

The genomic resources of the hexaploid oat *Avena sativa* cv. 'Sang' and the diploid *A. longiglumis* accession PI 657387 [55] were acquired from the National Center for Biotechnology Information (NCBI) database (https://identifiers.org/ncbi/insdc.gca:GCA_030063025.1, accessed on 11 September 2023). The genomic resources of the tetraploid *A. insularis* cv 'BYU209' were acquired from the GrainGenes database (https://wheat.pw.usda.gov/GG3/content/avena-insularis-download, accessed on 11 September 2023) [56].

The Hidden Markov Model (HMM) profiles of the kinase domain (PF00069), N-terminal domain B_lectin (PF01453), Lectin_legB (PF00139), and Lectin_C (PF00059) were downloaded from the InterPro database (https://www.ebi.ac.uk/interpro/entry/pfam/, accessed on 21 January 2024) [57]. We retrieved genes containing a kinase domain by running the program hmmsearch (HMMER v.3.3.2) [58] with default parameters to search the kinase profile (PF00069) within three genome protein sequences. Then, all the candidate protein sequences were submitted to the National Center for Biotechnology Information by the batch CD-search tool (https://www.ncbi.nlm.nih.gov/Structure/bwrpsb/bwrpsb.cgi, accessed on 23 January 2024) [59] and online software InterPro (https://www.ebi.ac.uk/interpro/entry/pfam/, accessed on 27 January 2024) to verify the conserved domains.

The protein was predicted by TMHMM Server v. 2.0 (http://www.cbs.dtu.dk/services/TMHMM/, accessed on 2 February 2024) [60] based on a deep learning model to determine whether it was a membrane protein. Subcellular localization was predicted using WoLF PSORT (https://wolfpsort.hgc.jp/, accessed on 4 February 2024). The molecular weight, theoretical isoelectric point, and grand average of hydropathicity of each LecRLK protein were obtained using the ExPasy website (http://au.expasy.org, accessed on 7 February 2024). The amino acid length and coding sequence length (bp) of each LecRLK protein were obtained using the software SeqKit v.2.3.0 [61].

4.2. Phylogenetic and Alignment Analysis

Using MEGA-11 (https://www.megasoftware.net/; accessed on 15 February 2024), multiple amino acid sequence alignment (MSA) was performed by the ClustalW [62] algorithm with default settings. Phylogenetic analysis of the LecRLK gene family was constructed with maxlikehood methods by using the FastTree v.2.1 with default settings [63]. The resulting phylogenetic tree was visualized by using iTOL v.4 [64]. Basing on the classification method used for *A. thaliana*, phylogenetic analysis, *AsaLecRLK* genes were further categorized into three subfamilies, C-type, G-type, and L-type, in *A. sativa*.

4.3. Analysis of Motifs, Gene Promotor, Gene Structures, and Conserved Domains

The conserved motifs in the oat LecRLK proteins were determined using the MEME server (https://meme-suite.org/; accessed on 15 February 2024) [27] with a maximum motif number of 10. We used SeqKit v.2.3.0. to extract 1.5 kb sequences upstream of the *LecRLK* genes as promoter regions and submit these sequences to PlantCARE (http://bioinformatics.psb.ugent.be/webtools/PlantCARE/html/; accessed on 23 February 2024) [28] for the analysis of cis-acting regulatory elements. To extract the CDS and UTR locations corresponding to AsaLecRLKs, we used in-house R scripts. The conserved protein domains of AsaLecRLKs were analyzed based on the online software InterPro (https://www.ebi.ac.uk/interpro/entry/pfam/, accessed on 16 February 2024), SMART (https://smart.embl.de/, accessed on 16 February 2024) [65] and NCBI-CDD (https://www.ncbi.nlm.nih.gov/Structure/cdd/wrpsb.cgi, accessed on 16 February 2024) online databases. And the transmembrane domains were predicted using the Prediction of TMHMM (http://www.cbs.dtu.dk/services/TMHMM-2.0/; accessed on 27 February 2024). Physical mapping of conserved and transmembrane structural domains was performed with in-house R scripts.

4.4. Chromosomal Distribution, Gene Duplication, and Ka/Ks Calculation

The chromosomal position information was retrieved from the *Avena sativa* genome sequence files and the corresponding gene structure annotation files. The gene localization of all *AsaLecRLKs* on the chromosome was visualized by TBtools v.2.119 [66].

A gene duplication search for the identified AsaLecRLK members was performed using the MCScanX tool of Tbtools. The rate of synonymous substitutions (Ks) and non-synonymous substitutions (Ka) in the *LecRLK* genes obtained from gene duplication events was calculated using ParaAT 2.0 [67] and KaKs_Calculator v.2.0 [68]. The *AsaLecRLK* gene replications and homologous genetic relationships between species were visualized using the NGenomeSyn v.1.41 [69].

4.5. Plant Material, Growth Conditions, and Treatment

Treatment of hexaploid common oat (*Avena sativa*; Accession QL&PHY 498; Riyue-hacheng, Huangyuan county, Qinghai Province) seeds involved shaking in 1% NaClO solution for 30 min, followed by rinsing twice with distilled water. The seeds were then germinated on moist filter paper in Petri dishes at 25 °C in darkness. Upon reaching a root length of 3–5 cm (approximately 3–4 days), seedlings were transplanted into polyethylene pots filled with distilled water. Seedlings were secured onto PCR plates floating on distilled water in a growth chamber for two days. The growth chamber conditions were set to 25 °C with a 16 h light/8 h dark cycle and 60% humidity. On the third day, half-strength Hoagland nutrient solution was introduced, with solution renewal every 24 h. When plants reached the two-leaf stage, salt stress was initiated by supplementing the half-strength Hoagland nutrient solution with NaCl to a concentration of 200 mM/L for 0, 6, 12, 24, and 48 h. Fresh root and leaf tissues were immediately frozen in liquid nitrogen and stored at –80 °C for RNA extraction, with three biological replicates per treatment group.

4.6. RNA Extraction, cDNA Reverse Transcription, and RT-qPCR Analysis

Leaf and root tissues from *A. sativa* were pulverized into a fine powder using liquid nitrogen for the purpose of RNA extraction, following the procedures outlined in the manual of the FastPure® Cell/Tissue Total RNA Isolation Kit V2 (Vazyme, Nanjing, China). The conversion of RNA into cDNA was facilitated using PrimeScript™ RT Master Mix (Takara, Kusatsu, Japan). Subsequently, the RT-qPCR analysis was conducted using ChamQ Universal SYBR qPCR Master Mix (Vazyme, Nanjing, China). Gene-specific primers, sourced from NCBI, were utilized for this experiment (Table S10). The ADPR gene served as the normalization control. Gene expression levels were determined through the $2^{-\Delta\Delta CT}$ method, with the control group and the root tissue being normalized to 1 in comparison to the NaCl treatments, as well as among various tissues. Three technical replicates were performed for each biological replicate to ensure the accuracy of the results.

4.7. RNA-seq Data Analysis

RNA-seq was performed by the Illumina DNBSEQ platform (carried by BerryGenomics Technology Co., Ltd., Beijing, China). RNA was extracted from the root and leaf of *Avena sativa* using an RNAprep Pure Plant Kit (DP441, TIANGEN, Beijing, China). The library was constructed using the TruSeq RNA v.2 kit (Illumina, San Diego, CA, USA). Transcriptome sequencing was performed by the Illumina platform NovaSeq 6000, and the RNA-seq data can be obtained from the Genome Sequence Archive of the China National Center for Bioinformation, project SRP519844. Low-quality sequences and splice sequences in raw reads were filtered using in-house perl scripts to obtain clean reads. The filtered reads were aligned to the *Avena sativa* cv. 'Sang' genome by Hisat2 [70], and the raw number of aligned reads was calculated using the featureCounts v.2.0.2 [71]. The level of gene expression was calculated by determining the number of expected fragments per kilobase of transcript per million mapped reads (FPKM). Heatmaps were generated using the heatmap package in TBtools v.2.119 with $\log_2$(FPKM + 1) values.

4.8. Subcellular Localization

To investigate the functional sites of the *AsaLecRLK* gene family, we randomly selected three genes, *AsaLecRLK-L-type-43*, *AsaLecRLK-L-type-44*, and *AsaLecRLK-G-type-45*, for a subcellular localization experiment. The coding sequence (CDS) without stop codons was cloned into the p1302 vector, which contained a 35S-driven green fluorescent protein (GFP) promoter. The In-Fusion cloning kit named ClonExpress® Ultra One Step Cloning Kit, which is produced by Vazyme, was used for cloning. SpeI was utilized as the restriction site of the p1302 vector. The Agrobacterium containing 35S-AsaLecRLK-L-type-43:GFP, 35S-AsaLecRLK-L-type-44:GFP and 35S-AsaLecRLK-G-type-45:GFP was transiently coinfiltrated into tobacco leaves with a subcellular localization marker. After 48 h of infiltration, the distribution of fluorescence was visualized under a confocal laser scanning microscope (Leica, Biberach, Germany).

5. Conclusions

We performed a genome-wide analysis of the *LecRLK* gene family in *Avena* ssp. (*A. sativa*, *A. insularis*, and *A. longiglumis*). In this study, we identified a total of 390 lectin genes in *A. sativa* and analyzed their physicochemical properties, gene structure, conserved domain, phylogenetic relationship, selection pressures, and expression patterns under salt stress. RT-qPCR and subcellular location experiments were used to validate the expression profile of the *AsaLecRLK* genes. *AsaLecRLK-G-type-110* and *AsaLecRLK-L-type-12* genes were considered for further functional characterization. We found that these genes play significant roles in salt tolerance. Our research suggests the involvement of *AsaLecRLK* genes in *A. sativa's* growth and response to salt stress, providing a theoretical basis for further biological research of oats.

Supplementary Materials: The following supporting information can be downloaded at: https://www.mdpi.com/article/10.3390/ijms252312754/s1.

Author Contributions: Conceptualization, Q.L. and J.S.H.-H.; methodology, G.X., D.C. and Y.T.; validation, G.X. and Y.T.; investigation, G.X., D.C. and Q.L.; methodology, Q.L., T.S. and J.S.H.-H.; writing—original draft preparation, G.X. and Q.L.; writing—review and editing, Y.T., T.S. and J.S.H.-H.; visualization, G.X.; supervision, Q.L. and J.S.H.-H.; project administration, Q.L., T.S. and J.S.H.-H.; funding acquisition, Q.L. and J.S.H.-H. All authors have read and agreed to the published version of the manuscript.

Funding: This research was funded by the National Natural Science Foundation of China (32070359, 32370402) to Q.L., the Chinese Academy of Sciences (CAS) President's International Fellowship Initiative (2024PVA0028) to J.S.H.H, the Guangdong Flagship Project of Basic and Applied Basic Research (2023B0303050001) to Q.L., and the Innovation Training Programs for Undergraduates, Chinese Academy of Sciences (KCJH-80107-2023-148) to Y.Q.T.

Institutional Review Board Statement: Not applicable.

Informed Consent Statement: Not applicable.

Data Availability Statement: Data are contained within the article and Supplementary Materials.

Acknowledgments: The authors wish to thank Mingzhi Li and Fanrui Meng for guidance on collinearity analysis, and Xunchen Liu, Ming Luo and Jiajia Cai (South China National Botanical Garden, Guangzhou, China) for qPCR and subcellular localization experimental assistance.

Conflicts of Interest: The authors declare no conflicts of interest.

References

1. Gao, W.; Feng, Z.; Bai, Q.; He, J.; Wang, Y. Melatonin-Mediated Regulation of Growth and Antioxidant Capacity in Salt-Tolerant Naked Oat under Salt Stress. *Int. J. Mol. Sci.* **2019**, *20*, 1176. [CrossRef] [PubMed]
2. Zhang, Y.; Fang, Q.; Zheng, J.; Li, Z.; Li, Y.; Feng, Y.; Han, Y.; Li, Y. GmLecRlk, A Lectin Receptor-like Protein Kinase, Contributes to Salt Stress Tolerance by Regulating Salt-Responsive Genes in Soybean. *Int. J. Mol. Sci.* **2022**, *23*, 1030. [CrossRef] [PubMed]
3. Kaur, A.; Sharma, A.; Upadhyay, S.K. Analysis of Lectin Receptor-Like Kinases and Their Functions in Higher Plants. In *Plant Receptor-Like Kinases*; Upadhyay, S.K., Shumayla, Eds.; Academic Press: New York, NY, USA, 2023; pp. 139–154.

4. Vaid, N.; Pandey, P.K.; Tuteja, N. Genome-Wide Analysis of Lectin Receptor-Like Kinase Family from Arabidopsis and Rice. *Plant Mol. Biol.* **2012**, *80*, 365–388. [CrossRef] [PubMed]
5. Sun, Y.; Qiao, Z.; Muchero, W.; Chen, J. Lectin Receptor-Like Kinases: The Sensor and Mediator at the Plant Cell Surface. *Front. Plant Sci.* **2020**, *11*, 596301. [CrossRef] [PubMed]
6. Paetsch, M.; Mayland-Quellhorst, S.; Neuffer, B. Evolution of the Self-Incompatibility System in the Brassicaceae: Identification of S-locus Receptor Kinase (SRK) in Self-Incompatible *Capsella grandiflora*. *Heredity* **2006**, *97*, 283–290. [CrossRef]
7. Sun, Y.; Zhao, X.H.; Gao, Y.H.; Jiao, J.; Sun, Y.; Zhu, D.Z.; Yang, J.J.; Wu, F.L.; Su, H.Y. Genome-Wide Analysis of Lectin Receptor-Like Kinases (LecRLKs) in Sweet Cherry (*Prunus avium*) and Reveals PaLectinL16 Enhances Sweet Cherry Resistance with Salt Stress. *Environ. Exp. Bot.* **2022**, *194*, 104751. [CrossRef]
8. Shumayla; Sharma, S.; Pandey, A.K.; Singh, K.; Upadhyay, S.K. Molecular Characterization and Global Expression Analysis of Lectin Receptor Kinases in Bread Wheat (*Triticum aestivum*). *PLoS ONE* **2016**, *11*, e0153925. [CrossRef]
9. Vaid, N.; Pandey, P.; Srivastava, V.K.; Tuteja, N. Pea Lectin Receptor-Like Kinase Functions in Salinity Adaptation without Yield Penalty, by Alleviating Osmotic and Ionic Stresses and Upregulating Stress-Responsive Genes. *Plant Mol. Biol.* **2015**, *88*, 193–206. [CrossRef]
10. Wang, Y.J.; Subedi, S.; de Vries, H.; Doornenbal, P.; Vels, A.; Hensel, G.; Kumlehn, J.; Johnston, P.A.; Qi, X.; Blilou, I.; et al. Orthologous Receptor Kinases Quantitatively Affect the Host Status of Barley to Leaf Rust Fungi. *Nat. Plants* **2019**, *5*, 1129–1135. [CrossRef]
11. Cheng, X.; Wu, Y.; Guo, J.; Du, B.; Chen, R.; Zhu, L.; He, G. A Rice Lectin Receptor-Like Kinase That Is Involved in Innate Immune Responses Also Contributes to Seed Germination. *Plant J.* **2013**, *76*, 687–698. [CrossRef]
12. Zhang, Y.; Zhou, J.; Ni, X.; Wang, Q.; Jia, Y.; Xu, X.; Wu, H.; Fu, P.; Wen, H.; Guo, Y.; et al. Structural Basis for the Activity Regulation of Salt Overly Sensitive 1 in Arabidopsis salt tolerance. *Nat. Plants* **2023**, *9*, 1915–1923. [CrossRef]
13. Liu, C.; Mao, B.; Yuan, D.; Chu, C.; Duan, M. Salt Tolerance in Rice: Physiological Responses and Molecular Mechanisms. *Crop. J.* **2022**, *10*, 13–25. [CrossRef]
14. Ahmad, R.; Hussain, S.; Anjum, M.A.; Khalid, M.F.; Saqib, M.; Zakir, L.; Hassan, A.; Fahad, S.; Ahmad, S. Oxidative stress and antioxidant defense mechanisms in plants under salt stress. In *Plant Abiotic Stress Tolerance*; Hasanuzzaman, M., Hakeem, K.R., Eds.; Pantanal Editora: Nova Xavantina, Brazil, 2019; pp. 191–205.
15. Zelm, E.; Zhang, Y.; Testerink, C. Salt Tolerance Mechanisms of Plants. *Annu. Rev. Plant Biol.* **2020**, *71*, 403–433. [CrossRef]
16. Balasubramanian, V.; Vashisht, D.; Cletus, J.; Sakthivel, N. Plant β-1,3-Glucanases: Their Biological Functions and Transgenic Expression Against Phytopathogenic Fungi. *Biotechnol. Lett.* **2012**, *34*, 1983–1990. [CrossRef]
17. Kayani, S.I.; Shen, Q.; Rahman, S.; Fu, X.; Li, Y.; Wang, C.; Hassani, D.; Tang, K. Transcriptional Regulation of Flavonoid Biosynthesis in Artemisia Annua by AaYABBY5. *Hortic. Res.* **2021**, *8*, 257. [CrossRef]
18. Watanabe, M.; Tohge, T. Species-Specific 'Specialized' Genomic Region Provides the New Insights into the Functional Genomics Characterizing Metabolic Polymorphisms in Plants. *Curr. Opin. Plant Biol.* **2023**, *75*, 102427. [CrossRef] [PubMed]
19. Smit, S.J.; Lichman, B.R. Plant Biosynthetic Gene Clusters in the Context of Metabolic Evolution. *Nat. Prod. Rep.* **2022**, *39*, 1465–1482. [CrossRef] [PubMed]
20. Cui, D.L.; Xiong, G.; Ye, L.H.; Gornall, R.; Wang, Z.W.; Heslop-Harrison, P.; Liu, Q. Genome-Wide Analysis of Flavonoid Biosynthetic Genes in Musaceae (*Ensete*, *Musella*, and *Musa* Speices) Reveals Amplification of Flavonoid 3′,5′-Hydroxylase. *AoB Plants* **2024**, *16*, plae049. [CrossRef]
21. Liu, W.; Feng, Y.; Yu, S.; Fan, Z.; Li, X.; Li, J.; Yin, H. The Flavonoid Biosynthesis Network in Plants. *Int. J. Mol. Sci.* **2021**, *22*, 12824. [CrossRef]
22. Xu, Z.; Chen, X.; Lu, X.; Zhao, B.; Liu, J. Integrative Analysis of Transcriptome and Metabolome Reveal Mechanism of Tolerance to Salt Stress in Oat (*Avena sativa* L.). *Plant Physiol. Biochem.* **2021**, *160*, 315–328. [CrossRef] [PubMed]
23. Sun, M.; Qian, X.; Chen, C.; Cheng, S.; Jia, B.; Zhu, Y.; Sun, X. Ectopic Expression of GsSRK in *Medicago Sativa* Reveals its Involvement in Plant AArchitecture and Salt Stress Responses. *Front. Plant Sci.* **2018**, *9*, 226. [CrossRef] [PubMed]
24. Sun, X.L.; Yu, Q.Y.; Tang, L.L.; Ji, W.; Bai, X.; Cai, H.; Liu, X.F.; Ding, X.D.; Zhu, Y.M. GsSRK, A G-type Lectin S-receptor-Like Serine/Threonine Protein Kinase, is a Positive Regulator of Plant Tolerance to Salt Stress. *J. Plant Physiol.* **2013**, *170*, 505–515. [CrossRef] [PubMed]
25. Li, C.H.; Wang, G.; Zhao, J.L.; Zhang, L.Q.; Ai, L.F.; Han, Y.F.; Sun, D.Y.; Zhang, S.W.; Sun, Y. The Receptor-Like Kinase SIT1 Mediates Salt Sensitivity by Activating MAPK3/6 and Regulating Ethylene Homeostasis in Rice. *Plant Cell* **2014**, *26*, 2538–2553. [CrossRef] [PubMed]
26. Shiu, S.H.; Bleecker, A.B. Receptor-Like Kinases from Arabidopsis form a Monophyletic Gene Family Related to Animal Receptor Kinases. *Proc. Natl. Acad. Sci. USA* **2001**, *98*, 10763–10768. [CrossRef] [PubMed]
27. Bailey, T.L.; Williams, N.; Misleh, C.; Li, W.W. MEME: Discovering and Analyzing DNA and Protein Sequence Motifs. *Nucleic Acids Res.* **2006**, *34*, W369–W373. [CrossRef]
28. Rombauts, S.; Déhais, P.; Van Montagu, M.; Rouzé, P. PlantCARE, A Plant Cis-Acting Regulatory Element Database. *Nucleic Acids Res.* **1999**, *27*, 295–296. [CrossRef]
29. Wang, X.; Niu, Y.; Zheng, Y. Multiple Functions of MYB Transcription Factors in Abiotic Stress Responses. *Int. J. Mol. Sci.* **2021**, *22*, 6125. [CrossRef]

30. Xue, Y.; Qian, F.; Guan, W.; Ji, G.; Geng, R.; Li, M.; Li, L.; Ullah, N.; Zhang, C.; Cai, G.; et al. Genome-Wide Identification of the ICS Family Genes and its Role in Resistance to Plasmodiophora Brassicae in *Brassica napus* L. *Int. J. Biol. Macromol.* **2024**, *270*, 132206. [CrossRef]
31. Ahmed, F.F.; Dola, F.S.; Islam, M.S.U.; Zohra, F.T.; Akter, N.; Rahman, S.M.; Sarkar, M.A.R. Genome-Wide Comprehensive Identification and In Silico Characterization of Lectin Receptor-Like Kinase Gene Family in Barley (*Hordeum vulgare* L.). *Genet. Res.* **2024**, *2024*, 2924953. [CrossRef]
32. Wang, Q.; Li, B.; Qiu, Z.; Lu, Z.; Hang, Z.; Wu, F.; Chen, X.; Zhu, X. Genome-Wide Identification of MYC Transcription Factors and Their Potential Functions in the Growth and Development Regulation of Tree Peony (*Paeonia suffruticosa*). *Plants* **2024**, *13*, 437. [CrossRef]
33. Basu, D.; Butler, C.; Rollins, M.B.; South, P. Identification and Functional Characterization of Cis-Regulatory Elements of Key Photorespiratory Genes in Response to Short-Term Abiotic Stress Conditions. *Methods Mol. Biol.* **2024**, *2792*, 251–264. [PubMed]
34. Chen, L.; Yang, W.; Liu, S.; Meng, Y.; Zhu, Z.; Liang, R.; Cao, K.; Xie, Y.; Li, X. Genome-Wide Analysis and Identification of Light-Harvesting Chlorophyll a/b Binding (LHC) Gene Family and BSMV-VIGS Silencing TaLHC86 Reduced Salt Tolerance in Wheat. *Int. J. Biol. Macromol.* **2023**, *242*, 124930. [CrossRef]
35. De Coninck, T.; Van Damme, E.J.M. Plant lectins: Handymen at the Cell Surface. *Cell Surf.* **2022**, *8*, 100091. [CrossRef]
36. Zhang, H.; Chen, C.; Li, L.; Tan, X.; Wei, Z.; Li, Y.; Li, J.; Yan, F.; Chen, J.; Sun, Z. A Rice LRR Receptor-Like Protein Associates with Its Adaptor Kinase OsSOBIR1 to Mediate Plant Immunity Against Viral Infection. *Plant Biotechnol. J.* **2021**, *19*, 2319–2332. [CrossRef] [PubMed]
37. Yang, Y.; Labbé, J.; Muchero, W.; Yang, X.; Jawdy, S.S.; Kennedy, M.; Johnson, J.; Sreedasyam, A.; Schmutz, J.; Tuskan, G.A.; et al. Genome-Wide Analysis of Lectin Receptor-Like Kinases in *Populus*. *BMC Genom.* **2016**, *17*, 699. [CrossRef]
38. Kamal, N.; Tsardakas Renhuldt, N.; Bentzer, J.; Gundlach, H.; Haberer, G.; Juhász, A.; Lux, T.; Bose, U.; A Tye-Din, J.; Lang, D.; et al. The Mosaic Oat Genome Gives Insights into A Uniquely Healthy Cereal Crop. *Nature* **2022**, *606*, 113–119. [CrossRef] [PubMed]
39. Liu, Q.; Ye, L.; Li, M.; Wang, Z.W.; Xiong, G.; Ye, Y.S.; Tu, T.Y.; Schwarzacher, T.; Heslop-Harrison, P. Genome-Wide Expansion and Reorganization During Grass Evolution: From 30 Mb Chromosomes in Rice and Brachypodium to 550 Mb in *Avena*. *BMC Plant Biol.* **2023**, *23*, 627. [CrossRef]
40. Moummou, H.; Kallberg, Y.; Tonfack, L.B.; Persson, B.; van der Rest, B. The Plant Short-Chain Dehydrogenase (SDR) SuperFamily: Genome-Wide Inventory and Diversification Patterns. *BMC Plant Biol.* **2012**, *12*, 219. [CrossRef]
41. Wang, C.; Huang, X.; Li, Q.; Zhang, Y.; Li, J.L.; Mou, Z. Extracellular Pyridine Nucleotides Trigger Plant Systemic Immunity Through a Lectin Receptor Kinase/BAK1 Complex. *Nat. Commun.* **2019**, *10*, 4810. [CrossRef]
42. Peng, J.; Liu, S.; Wu, J.; Liu, T.; Liu, B.; Xiong, Y.; Zhao, J.; You, M.; Lei, X.; Ma, X. Genome-Wide Analysis of the Oat (*Avena sativa*) HSP90 Gene Family Reveals Its Identification, Evolution, and Response to Abiotic Stress. *Int. J. Mol. Sci.* **2024**, *25*, 2305. [CrossRef]
43. Keren, H.; Lev-Maor, G.; Ast, G. Alternative Splicing and Evolution: Diversification, Exon Definition and Function. *Nat. Rev. Genet.* **2010**, *11*, 345–355. [CrossRef] [PubMed]
44. Sun, Y.B.; Zhang, X.J.; Zhong, M.C.; Dong, X.; Hu, J.Y. Genome-Wide Identification of WD40 Genes Reveals A Functional Diversification of COP1-Like Genes in Rosaceae. *Plant Mol. Biol.* **2020**, *104*, 81–95. [CrossRef] [PubMed]
45. Zhou, H.; Shi, H.; Yang, Y.; Feng, X.; Chen, X.; Xiao, F.; Lin, H.; Guo, Y. Insights into Plant Salt Stress Signaling and Tolerance. *J. Genet. Genom.* **2024**, *51*, 16–34. [CrossRef] [PubMed]
46. Yu, J.; Zhu, C.; Xuan, W.; An, H.; Tian, Y.; Wang, B.; Chi, W.; Chen, G.; Ge, Y.; Li, J.; et al. Genome-wide Association Studies Identify OsWRKY53 as A Key Regulator of Salt Tolerance in Rice. *Nat. Commun.* **2023**, *14*, 3550. [CrossRef]
47. Yang, Y.; Guo, Y. Elucidating the Molecular Mechanisms Mediating Plant Salt-Stress Responses. *New Phytol.* **2018**, *217*, 523–539. [CrossRef]
48. Krzywińska, E.; Kulik, A.; Bucholc, M.; Fernandez, M.A.; Rodriguez, P.L.; Dobrowolska, G. Protein Phosphatase Type 2C PP2CA Together with ABI1 Inhibits SnRK2.4 Activity and Regulates Plant Responses to Salinity. *Plant Signal Behav.* **2016**, *11*, e1253647. [CrossRef]
49. Chen, L.J.; Wuriyanghan, H.; Zhang, Y.Q.; Duan, K.X.; Chen, H.W.; Li, Q.T.; Lu, X.; He, S.J.; Ma, B.; Zhang, W.K.; et al. An S-Domain Receptor-Like Kinase, OsSIK2, Confers Abiotic Stress Tolerance and Delays Dark-Induced Leaf Senescence in Rice. *Plant Physiol.* **2013**, *163*, 17521765. [CrossRef]
50. Passricha, N.; Saifi, S.K.; Kharb, P.; Tuteja, N. Marker-free Transgenic Rice Plant Overexpressing Pea LecRLK Imparts Salinity Tolerance by Inhibiting Sodium Accumulation. *Plant Mol. Biol.* **2019**, *99*, 265–281. [CrossRef]
51. Liu, L.; Liu, J.; Xu, N. Ligand Recognition and Signal Transduction by Lectin Receptor-Like Kinases in Plant Immunity. *Front. Plant Sci.* **2023**, *14*, 1201805. [CrossRef]
52. Motone, K.; Kontogiorgos-Heintz, D.; Wee, J.; Kurihara, K.; Yang, S.; Roote, G.; Fang, Y.; Cardozo, N.; Nivala, J. Multi-Pass, Single-Molecule Nanopore Reading of Long Protein Strands. *Nature* **2024**, *633*, 662–669. [CrossRef]
53. Abramson, J.; Adler, J.; Dunger, J.; Evans, R.; Green, T.; Pritzel, A.; Ronneberger, O.; Willmore, L.; Ballard, A.; Bambrick, J.; et al. Accurate Structure Prediction of Biomolecular Interactions with AlphaFold 3. *Nature* **2024**, *630*, 493–500. [CrossRef]
54. Wang, K.; Li, S.; Yang, Z.; Chen, C.; Fu, Y.; Du, H.; Sun, H.; Li, J.; Zhao, Q.; Du, C. L-Type Lectin Receptor-Like Kinase OsCORK1 as an Important Negative Regulator Confers Copper Stress Tolerance in Rice. *J. Hazard Mater.* **2023**, *459*, 132214. [CrossRef]
55. Liu, Q.; Xiong, G.; Wang, Z.; Wu, Y.; Tu, T.; Schwarzacher, T.; Heslop-Harrison, P. Chromosome-level Genome Assembly of the Diploid Oat Species *Avena longiglumis*. *Sci. Data* **2024**, *11*, 412. [CrossRef] [PubMed]

56. O'Sullivan, H. GrainGenes. *Methods Mol. Biol.* **2007**, *406*, 301–314. [PubMed]
57. Paysan-Lafosse, T.; Blum, M.; Chuguransky, S.; Grego, T.; Pinto, B.L.; Salazar, G.A.; Bileschi, M.L.; Bork, P.; Bridge, A.; Colwell, L.; et al. InterPro in 2022. *Nucleic Acids Res.* **2022**, *51*, D418–D427. [CrossRef]
58. Eddy, S.R. Accelerated Profile HMM Searches. *PLoS Comput. Biol.* **2011**, *7*, e1002195. [CrossRef] [PubMed]
59. Marchler-Bauer, A.; Bryant, S.H. CD-Search: Protein Domain Annotations on the Fly. *Nucleic Acids Res.* **2004**, *32*, W327–W331. [CrossRef]
60. Hallgren, J.; Tsirigos, K.D.; Pedersen, M.D.; Almagro Armenteros, J.J.; Marcatili, P.; Nielsen, H.; Krogh, A.; Winther, O. DeepTMHMM Predicts Alpha and Beta Transmembrane Proteins Using Deep Neural Networks. *bioRxiv* **2022**, 487609. [CrossRef]
61. Shen, W.; Le, S.; Li, Y.; Hu, F. SeqKit: A Cross-Platform and Ultrafast Toolkit for FASTA/Q File Manipulation. *PLoS ONE* **2016**, *11*, e0163962. [CrossRef]
62. Thompson, J.D.; Higgins, D.G.; Gibson, T.J. CLUSTAL W: Improving the Sensitivity of Progressive Multiple Sequence Alignment through Sequence Weighting, Position-Specific Gap Penalties and Weight Matrix Choice. *Nucleic Acids Res.* **1994**, *22*, 4673–4680. [CrossRef] [PubMed]
63. Price, M.N.; Dehal, P.S.; Arkin, A.P. FastTree: Computing Large Minimum Evolution Trees with Profiles Instead of a Distance Matrix. *Mol. Biol. Evol.* **2009**, *26*, 1641–1650. [CrossRef] [PubMed]
64. Ivica, L.; Peer, B. Interactive Tree of Life (iTOL) v4: Recent Updates and New Developments. *Nucleic Acids Res.* **2019**, *47*, W256–W259.
65. Letunic, I.; Khedkar, S.; Bork, P. SMART: Recent Updates, New Developments and Status in 2020. *Nucleic Acids Res.* **2021**, *49*, D458–D460. [CrossRef] [PubMed]
66. Chen, C.; Wu, Y.; Li, J.; Wang, X.; Zeng, Z.; Xu, J.; Liu, Y.; Feng, J.; Chen, H.; He, Y.; et al. TBtools-II: A "One for All, All for One" Bioinformatics Platform for Biological Big-Data Minin. *Mol. Plant.* **2023**, *16*, 1733–1742. [CrossRef]
67. Zhang, Z.; Xiao, J.; Wu, J.; Zhang, H.; Liu, G.; Wang, X.; Dai, L. ParaAT: A Parallel Tool for Constructing Multiple Protein-Coding DNA Alignments. *Biochem. Biophys. Res. Commun.* **2012**, *419*, 779–781. [CrossRef]
68. Wang, D.; Zhang, Y.; Zhang, Z.; Zhu, J.; Yu, J. KaKs_Calculator 2.0: A Toolkit Incorporating Gamma-Series Methods and Sliding Window Strategies. *Genom. Proteom. Bioin.* **2010**, *8*, 77–80. [CrossRef]
69. He, W.; Yang, J.; Jing, Y.; Xu, L.; Yu, K.; Fang, X. NGenomeSyn: An Easy-to-Use and Flexible Tool for Publication-Ready Visualization of Syntenic Relationships Across Multiple Genomes. *Bioinformatics* **2023**, *39*, btad121. [CrossRef]
70. Kim, D.; Langmead, B.; Salzberg, S.L. HISAT: A Fast Spliced Aligner with Low Memory Requirements. *Nat. Methods* **2015**, *12*, 357–360. [CrossRef]
71. Liao, Y.; Smyth, G.K.; Shi, W. FeatureCounts: An Efficient General Purpose Program for Assigning Sequence Reads to Genomic Features. *Bioinformatics* **2014**, *30*, 923–930. [CrossRef]

International Journal of
Molecular Sciences

MDPI

Article

Comprehensive Genome-Wide Identification and Expression Profiling of *bHLH* Transcription Factors in *Areca catechu* Under Abiotic Stress

Akhtar Ali [1], Noor Muhammad Khan [1], Yiqi Jiang [1], Guangzhen Zhou [2] and Yinglang Wan [2,*]

[1] Hainan Key Laboratory for Sustainable Utilization of Tropical Bioresources, School of Tropical Agriculture and Forestry, Hainan University, Haikou 570228, China; anaakhtar631@gmail.com (A.A.); noormaghmoom@gmail.com (N.M.K.); jyq217317@163.com (Y.J.)

[2] The Ministry of Education Key Laboratory for Ecology of Tropical Islands, Key Laboratory of Tropical Animal and Plant Ecology of Hainan Province, College of Life Sciences, Hainan Normal University, Haikou 571158, China; gzzhou@hainanu.edu.cn

* Correspondence: ylwan@hainanu.edu.cn

Abstract: The basic helix-loop-helix (*bHLH*) transcription factor (TF) family, the second-largest among eukaryotes, is known for its evolutionary and functional diversity across plant species. However, *bHLH* genes have not yet been characterized in *Areca catechu*. In this study, we identified 76 *AcbHLH* genes, which exhibit a variety of physicochemical properties. Phylogenetic analysis revealed evolutionary relationships between *Arabidopsis thaliana bHLH* genes (*AtbHLH*) and their counterparts in *A. catechu* (*AcbHLH*). These analyses also highlighted conserved amino acid motifs (S, R, K, P, L, A, G, and D), conserved domains, and evolutionary changes, such as insertions, deletions, and exon gains or losses. Promoter analysis of *AcbHLH* genes revealed 76 cis-elements related to growth, phytohormones, light, and stress. Gene duplication analysis revealed four tandem duplications and twenty-three segmental duplications, while *AcbHLH63* in the Areca genome exhibited significant synteny with *bHLH* genes from *A. thaliana, Vitis vinifera, Solanum lycopersicum, Brachypodium distachyon, Oryza sativa*, and *Zea mays*. Furthermore, relative expression analysis showed that under drought stress (DS), *AcbHLH22, AcbHLH39, AcbHLH45*, and *AcbHLH58* showed distinct upregulation in leaves at specific time points, while all nine *AcbHLH* genes were upregulated in roots. Under salt stress (SS), *AcbHLH22, AcbHLH39, AcbHLH45*, and *AcbHLH58* were upregulated in leaves, and *AcbHLH22, AcbHLH34*, and *AcbHLH39* exhibited differential expression in roots at various time points. This study provides valuable insights into the bHLH superfamily in *A. catechu*, offering a solid foundation for further investigation into its role in responding to abiotic stresses.

Keywords: *Areca catechu*; *bHLH* gene family; genome-wide analysis; expression pattern; abiotic stresses

1. Introduction

Transcription factors (TFs) are pivotal in regulating plant growth and development by controlling gene expression [1,2]. They influence various cellular processes by interacting with other proteins involved in transcription [3]. The bHLH is the second-most abundant group of TFs family found in plants, animals, and microorganisms within this group of proteins [4]. The bHLH transcription factors domain consists of 50–60 amino acids in length and is divided into two functional regions, the N-terminal basic domain and the C-terminal (HLH) domain region [5]. The N-terminal basic region consists of 10–15 bp amino acids and is responsible for binding to cis-elements E-box (CANNTC), while the C-terminal region consists of 40–50 bp amino acids and is responsible for the formation of homo- and heterodimer protein complexes [6]. Moreover, MYC-like bHLH proteins possess an MYB-interacting region and extra N-terminal MYC domain, enabling attachment to bHLH and R2R3-MYB domain proteins, respectively [7].

Many physiological and biochemical processes in plants have been linked to bHLH TFs, as shown by the extensive study on these proteins [8]. Key developmental processes such as seed coat differentiation [9], stomata differentiation [10,11], trichome/root hair creation [12], fruit development [13,14], and carpel edge development are regulated by transcription. Certain bHLH TFs also have a role in premature seedling photomorphogenesis [15,16]. Additionally, a growing body of research indicates the involvement of bHLH proteins in plant actions connected to the heterogeneity of abiotic stressors, for instance, salinity, drought, low temperature, and mechanical injury [14,17]. In addition, bHLH proteins have been extensively studied in plants for their regulatory functions in forming secondary metabolites, including alkaloids [18], flavonoids [19], phenolic acids [20], and terpenoids [21]. The advancement in sequencing technologies facilitated the recognition and characterization of numerous families (bHLH) in plant genomes. Genome-wide analyses of the bHLH family have been identified in various plants, including *Arabidopsis* [22] *O. sativa* [23] *S. lycopersicum* [24], *Z. mays* [25], *T. aestivum* [26], *P. alba* [27], *S. tuberosum* [28], and other plants based on genome sequences [25].

A. catechu, is a vital evergreen vascular monocot tree belonging to the Areceae family [29]. *A. catechu* possesses significant medicinal and economic importance and is extensively cultivated in China, India, Thailand, Indonesia, Malaysia, and Cambodia [30,31]. In China, the areca nut is utilized in traditional Chinese medicine [32]. Due to its high arecoline, the areca nut has become a significant cash crop in Southeast Asia and Africa as the fourth most addictive substance in the world after alcohol, caffeine, and nicotine. Additionally, the study of bHLH protein *A. catechu,* a rare monocot tree, not only carries substantial economic value, but also holds theoretical importance in deciphering abiotic stress responses within this plant species [33]. However, a comprehensive study of *AcbHLH* gene expression patterns within the *Areca* genome remains an understudied area within the scientific literature. This research categorized 76 *AcbHLHs* genes from the *A. catechu* genome. The comprehensive analyses were conducted on their phylogenic relationships, sequence characteristics, gene structures, promoter sequences, and collinearity. Furthermore, we investigated the specific expression of these genes across various parts in response to drought and salt stresses. This work provides a base for future research on molecular mechanisms of *AcbHLH* genes in response to abiotic stresses in *A. catechu.*

2. Results

2.1. Identification of bHLH Genes in A. catechu

The present study identified and characterized 76 *AcbHLH* genes in the *A. catechu* genome. The gene names were assigned from *AcbHLH01* to *AcbHLH76* based on phylogenic connection with already known *A. thaliana bHLH* genes (162) (Table S1A). Additionally, characterizing these genes provided information on their molecular weight, isoelectric points, protein length, domain composition, and subcellular localization (Table S1B). Among 76 *AcbHLH* genes, the amino acid lengths ranged from 149 amino acids (*AcbHLH17*) up to 685 amino acids (*AcbHLH59*), while the moderate range of amino acids was 363.08 kDa. The molecular mass of the proteins ranged from 16.28 kDa (*AcbHLH17*) to 76.32 kDa (*AcbHLH59*), whereas their molecular mass ranged from 16.28 kDa (*AcbHLH17*) to 76.32 kDa (*AcbHLH59*). The calculated isoelectric point (pI) of *AcbHLH25* started from 4.99 to 10.30 (*AcbHLH57*), with a mean of 6.96 pI of *AcbHLH* genes. The subcellular localization prediction result revealed that 70 *AcbHLHs* are present in the nucleus, cytoplasm 01, chloroplast 04, and mitochondria 01 (Table S1B).

2.2. Phylogenetic and Multiple Sequence Alignment

The phylogenetic relationships among 76 *AcbHLH* and 162 *AtbHLH* genes were inferred using the Neighbor Joining (NJ) method with 1000 bootstrap replicates. The analysis was based on amino acid sequence data (Figure 1 and Table S1B). *bHLH* genes were classified into 24 groups based on topological structures. According to Pires and Gabriela's presented classification method and topological outline of the tree, the phylogenetic tree of 238 *bHLH* genes was classified into 24 groups (1–24) [34]. The phylogenic tree shows a closed relationship between the *AcbHLHs* and *AtbHLHs* proteins with a bootstrap support of 80 or higher. The study reveals that in Group 19, three pairs of genes showed close evolutionary ties: *AtbHLH47* with *AcbHLH31*, *AtbHLH11* with *AcbHLH61*, and *AtbHLH105* with *AcbHLH69*. Group 8 displayed a close relationship between *AtbHLH21* and *AcbHLH77*, while Group 9 linked *AtbHLH13* with *AcbHLH83*. In Group 4, *AtbHLH98* showed a close relationship with *AcbHLH25*. Group 13 connected *AtbHLH71* with *AcbHLH64*, and Group 2 included *AtbHLH162* with *AcbHLH35*. Group 3 revealed a relationship between *AtbHLH155* and *AcbHLH42*, and Group 14 linked *AtbHLH60* with *AcbHLH47*. In Group 22, *AtbHLH78* was closely related to *AcbHLH5*, while Group 15 included two pairs: *AtbHLH77* with *AcbHLH75* and *AtbHLH62* with *AcbHLH49*. Group 20 also presented two connections, with *AtbHLH24* linked to *AcbHLH12* and *AtbHLH8* to *AcbHLH50*. Additionally, Group 17 showed an evolutionary relationship between *AtbHLH87* and *AcbHLH3*, while Group 16 connected *AtbHLH66* with *AcbHLH44*. These findings underscore the evolutionary connections between *bHLH* genes in *Arabidopsis* and their counterparts, suggesting functional and ancestral relationships across these groups.

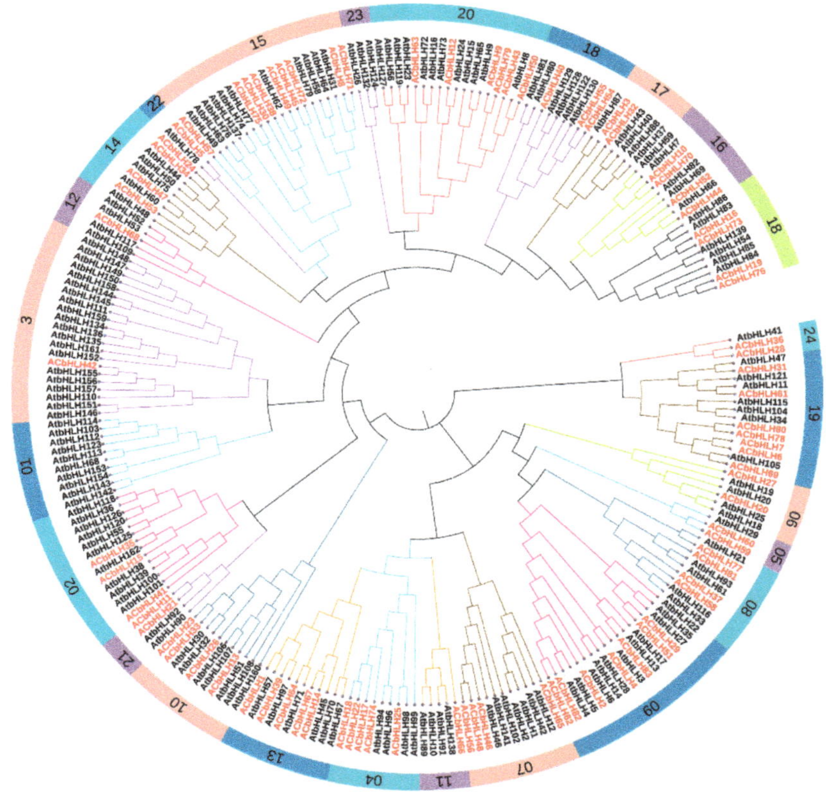

Figure 1. A phylogenic tree illustrates the relationship among *bHLH* domains of *A. catechu* and *A. thaliana*. The black color presents (*AtbHLH*) *A. thaliana*, and red represents *AcbHLH* of *A. catechu* *bHLH* protein.

As illustrated in Figure 2, our sequence alignment analysis of *AcbHLH* genes in *A. catechu* revealed consistent patterns of amino acid motifs (S, R, K, P, L, A, G, E, Y, I, V, H,

M, Q, N, and D). As previously observed, these motifs are shared by bHLH proteins from various plant species.

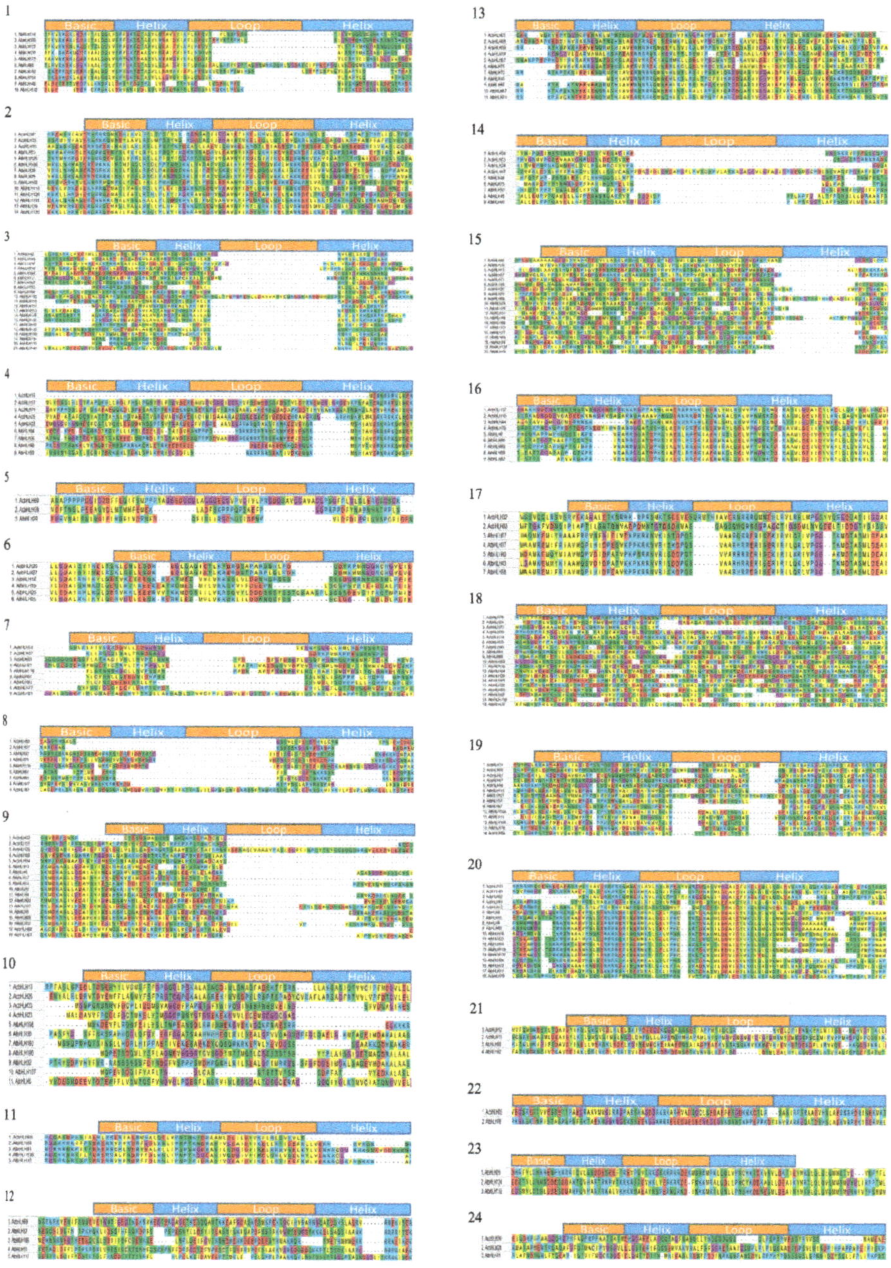

Figure 2. Multiple sequence alignments of *A. catechu* and 24 subgroups *A. thaliana*. *A. catechu* is devoid of subgroup 24. The location and boundaries of the bHLH domain are indicated at the top of each subgroup.

2.3. Analysis of Gene Structure and Conserved Domains of AcbHLH Genes

To reveal the *AcbHLH* genes' phylogeny and intron–exon structures, we compared 76 *AcbHLH* genes, ranging from 1 to 13, showing differences in exon–intron structures (Figure 3A,B and Table S1B). Of the 76 *AcbHLH* genes studied, 2 genes (2.63%) possess a single exon, while the rest exhibit two or more exons in their structures. Some intron-free genes were identified within Subgroups 2 and 6. Among the remaining 69 genes, three intron patterns were most prevalent. *AcbHLH40* exhibited the highest number of introns (13), while genes in Group 3 contained either 0 introns or a single intron. Group 3 exhibited the most significant variant in the number of exons, ranging from a single intron in *AcbHLH22* to 18 introns in *AcbHLH40*. As revealed by comparative analysis, Subgroup 3, among the analyzed *AcbHLH* genes, exhibited the most diverse range of intron counts.

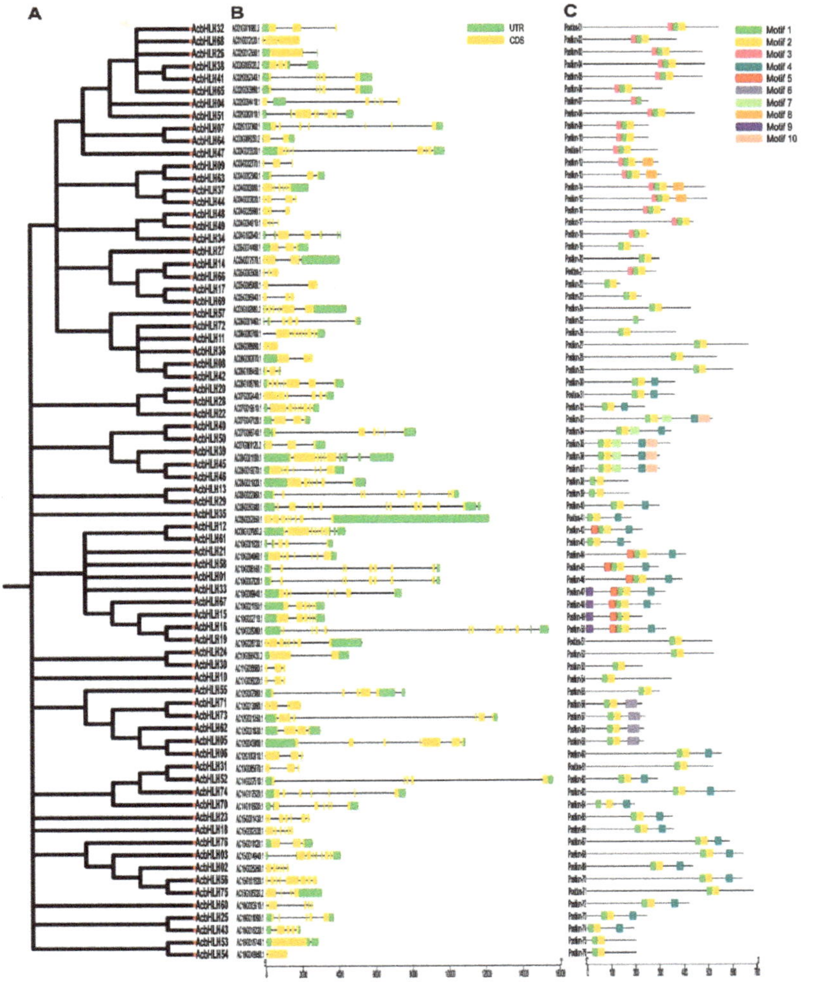

Figure 3. Comparative analysis of *A. catechu AcbHLH* gene phylogeny, structure, and motifs. (**A**) Phylogeny was inferred based on the NJ method with 1000 bootstrap replicates. (**B**) Introns and exons are visually represented as yellow and black lines. (**C**) Amino acid motifs (1–10) are indicated by colored, relative protein lengths represented with black lines.

Distribution analysis of motifs among *AcbHLH* gene, as shown in (Figure 3C), revealed that motifs 1, 2, and 3 were prevalent and abundant in all groups, while motifs 6, 8, 9, and 10 exhibited the least across the *AcbHLH* proteins. Distinct motif compositions characterized *AcbHLH* subgroups. For instance, Motifs 1, 2, and 3 were commonly found in Subgroup 1 and 2 proteins. Subgroup 4 proteins, on the other hand, were characterized by the presence of motifs 1, 2, 5, and 9. Some motifs were notably prevalent within particular subgroups, such as motifs 8, 10, and 6 in subgroups 2, 3, and 7, respectively. Motifs 1 and 2 mainly were observed across multiple subgroups. Additional analysis identified distinct patterns in the arrangement and relative positions of the motifs. For instance, motif 4 exclusively appears at the end of subgroups 3, 5, and 6, while a specific arrangement has been observed in motifs 1, 2, and 10.

2.4. Characterization of Cis-Acting Elements Within AcbHLH Promoter Regions

To investigate the functional activities and regulatory regions of *AcbHLH* genes, we analyzed the 2000 bp upstream promoter regions for potential cis-elements using Plant-CARE (Figure 4). Our analysis identified significant stress-related cis-elements within these regions, suggesting a critical role for *AcbHLH* genes in stress response pathways. We categorized the cis-acting elements into four functional groups as depicted in Figure 4B: (I) growth-related, (II) phytohormone-responsive, (III) light-responsive, and (IV) stress-responsive elements. The growth-related group includes motifs such as AAGAA-motif, as-1, and CAT-box. The phytohormone-responsive group contains elements associated with various hormones, such as ABRE, ABRE3a, ABRE4, AuxRR-core, and GARE-motif. Light-responsive elements include AE-box, ARE, Box 4, GA-motif, Gap-box, GATA-motif, G-box, GT1-motif, I-box, Lamp-element, MRE, Sp1, TCCC-motif, and TCT-motif. The stress-responsive category includes DRE, ERE, LTR, MBS, Myb, Myb-binding sites, Myc, STRE, W box, and WRE3.

This analysis highlights the potential involvement of *AcbHLH* genes in regulating responses to environmental stimuli. Notably, cis-elements associated with abscisic acid (ABA) and salicylic acid (SA) were identified within the promoter regions of *AcbHLH* genes linked to abiotic stress responses, such as extreme temperatures, wounding, and drought. These hormone-responsive elements further suggest involvement in pathways regulated by SA, gibberellin, methyl jasmonate (MeJA), auxin, and ethylene (Figure 4A).

2.5. Chromosomal Distribution and Gene Duplication of AcbHLH Genes

The latest *A. catechu* genome database enabled the creation of a precise physical map that offers the specific positions of *AcbHLH* genes within the genome (Figure 5A and Table S2). The allocation of 76 *AcbHLH* genes across Chr1 to 16 displayed uneven distribution patterns. The name of each *AcbHLH* gene was assigned based on its sequential physical location along *A. Catechu* Chr1-10, from top to bottom. Chr10 possesses the greatest number of *AcbHLH* genes (10 genes, ~13.16%), followed by Chr 2, 8, and 16, each containing 7, respectively (~9.21%), while Chr13 contained the least (1, ~3.31%) and Chr9 contain zero genes. Chr4, Chr5, Chr6, and Chr12 each contained 6 (~7.9%) *AcbHLH* genes. Chr7 and Chr15 each contained 5 (~6.6%) *AcbHLH* genes. Chr3 and Chr14 contained 3 (~3.94%) and chr11 (2~2.6%) *AcbHLH* genes. In addition, we discussed the chr of *AcbHLH* gene-duplication instances. The chromosomal region spanning 30 Mb and 60 Mb exhibiting two homogenous genomic segments is clarified as a tandem duplication event [35]. On Chr8 and 11, we observed 04 tandem duplication segments involving two genes *AcbHLH* (Figure 5A). *AcbHLH38* and *AcbHLH41* on Chr 8, and *AcbHLH53* and *AcbHLH54* on chromosome 11, each had tandem repeat events, while Chr10 had two tandem duplication events involving four *AcbHLH* genes (*AcbHLH45*, *AcbHLH46*, *AcbHLH48* and *AcbHLH49*). The gene involved in the tandem repeat occurrences originated exclusively within identical subfamilies. For instance, genes *AcbHLH38* and *AcHLH41* had tandem repeat sequences and clustered together in subfamilies 15 and 2, *AcbHLH53* and *AcbHLH54* were clustered together in subfamily 14 and 18, while *AcbHLH45*, *AcbHLH46*, *AcbHLH48*, and *AcbHLH49* were clustered together in subfamily 15, 7, 7, and 15, respectively (Figure 5B and Table S2).

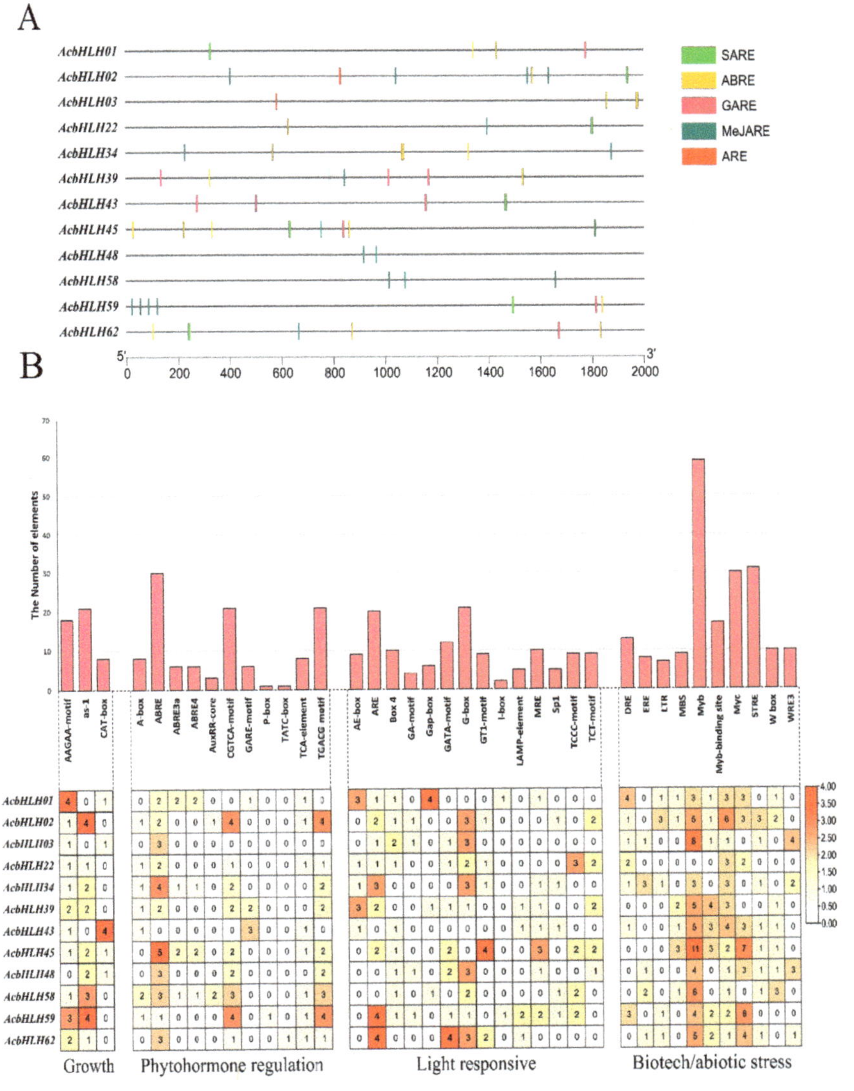

Figure 4. (**A**) Cis-regulatory elements in the 2000 bp upstream region of *AcbHLH* gene promoters. (**B**) Distribution of cis—regulatory elements of *AcbHLH* gene members. Colored rectangles visually depict the various cis-acting elements.

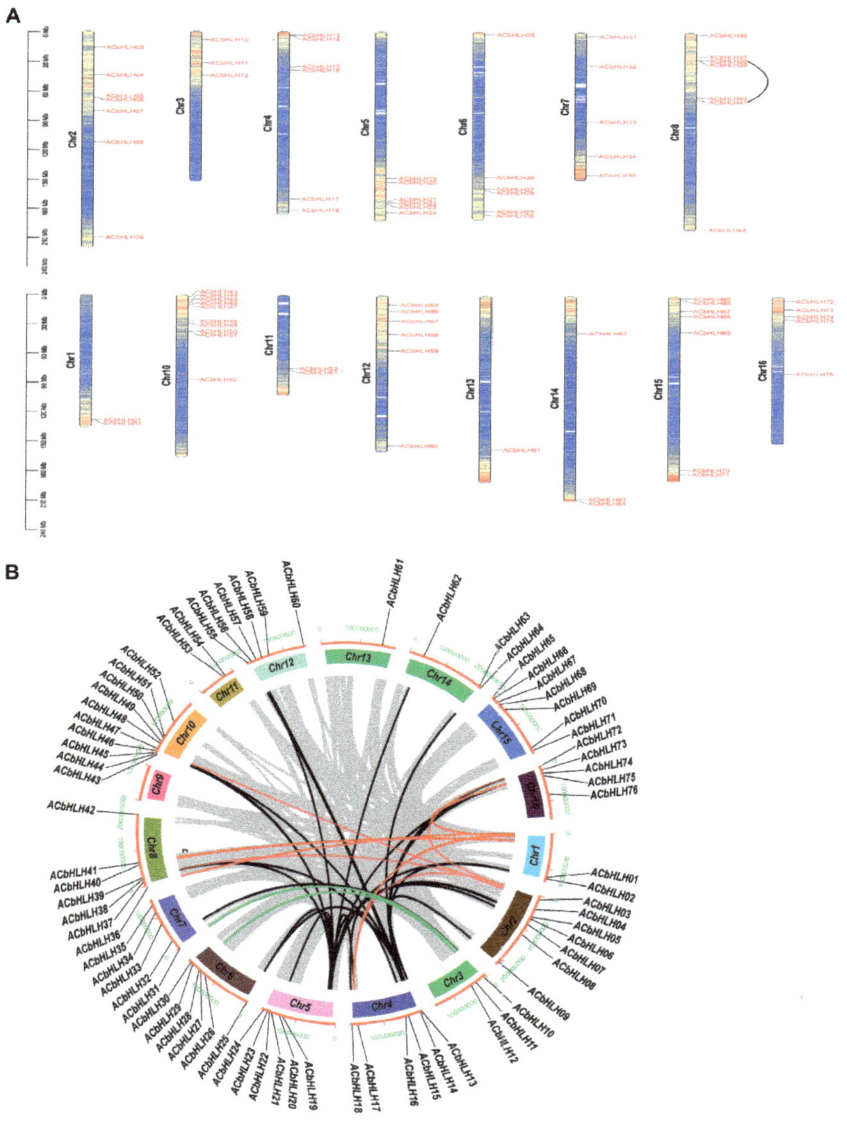

Figure 5. (**A**) Location of 76 *AcbHLH* genes across 16 *A. catechu* chromosomes. The left-hand scale indicates chromosomal length. (**B**) The schematic diagram represents the distribution of *A. catechu* chromosomes and interchromosomal interaction. Distinct colored lines within the diagram represent gene pairs. Red lines indicate *AcbHLH* gene pairs. *A. catechu* are labeled outside the chromosome circles, while chromosome numbers are indicated within.

Additionally, a total of 23 segmental duplications were identified within the *AcbHLH* gene family. As illustrated in Figure 6, the *AcbHLH* gene family contains 10 (13.15%) paralogs, indicating a common evolutionary origin of these *AcbHLH* members. The distribution of *AcbHLH* genes across the 16 linkage groups (LGs) of *A. catechu* was notably uneven (Figure 4). Some linkage groups, specifically LG10, d followed by LG8, exhibited a higher count of *AcbHLH* genes than others. LG2 boasted the highest count of *AcbHLH* genes, at fourteen, while LG5 contained the fewest count of one *AcbHLH*. Further analysis of *bHLH* families observed that many were significantly linked within their respective subfamilies, except *AcbHLH61/01* and *AcbHLH062/42*. Within all identified *AcbHLH* genes, group 15 exhibited the highest count of linked genes, encompassing 9/76 genes. Additionally, group 15 has nine genes, while groups 22, 12, 11, and 3 have only one, and groups 23 and 1 have no genes (Table S2). The present study implies that genes *AcbHLH* indicate gene duplication events and have been a significant factor in the *AcbHLH* genes in *A. catechu*, resulting in the emergence of novel functions and the expansion of the *AcbHLH* gene family.

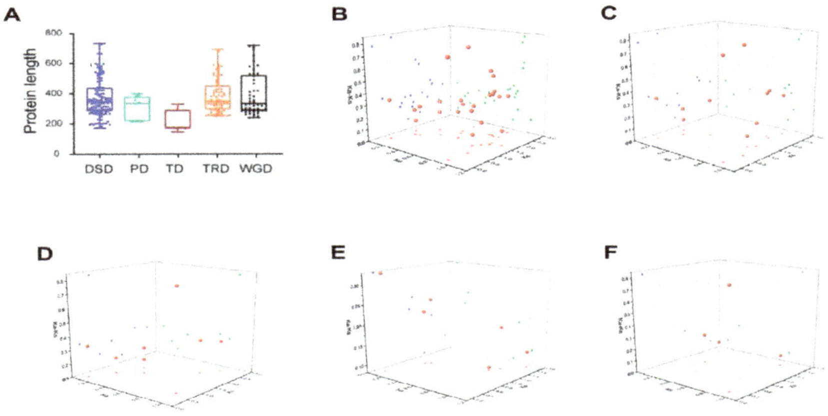

Figure 6. Gene *AcbHLH* duplication events. (**A**) Length distribution of *AcbHLHs* across five duplication events. (**B–F**) distribution of Ka, Ks, and Ka/Ks value in duplicated genes across five duplication events. (**B–F**) DSD, WGD, TRD, TD and PD event, respectively. Further details on the duplicated genes across the five duplication events are provided in Table S9.

2.6. Analysis of Gene Duplication Events of AcbHLH Genes

To investigate the possible driving force for the diversifications of the *AcbHLH* gene family, the Dup-206 Gene finder was used to examine gene-duplication segments, inclusive of DSD, WGD, TRD, TD and PD. The result revealed that there was high variation in the number of duplicated genes and the distribution of their protein. Furthermore, the largest number of duplicated genes, at 23, is in DSD, while the lowest (4) were observed in PD. Additionally, 11 duplicated genes were identified in WGD (Figure 6 and Table S3A–C). The value of the (Ks) and (Ka) substitution rate, and the Ka/Ks ratio of the duplicated genes over five replication events was calculated, to provide insight into the selection pressure on *AcbHLH* genes. The Ka/Ks value was less than 1, indicating that clear selection of these genes has taken place, as shown in Figure 6B–F.

2.7. Synteny Analysis of AcbHLH Genes

To elucidate the phylogenic processes underlying the *A. catechu bHLH* gene family, we developed seven comparative synteny maps for *A. catechu* association, encompassing three dicots, (*A. thaliana, V. vinifera,* and *S. lycopersicum*) and four monocots (*B. distachyon, O. sativa, Z. mays,* and *C. nucifera*) (Figure 7 and Table S4A–F). Through collinear analysis, 63 genes of *AcbHLH* were identified, with the gene set of *A. thaliana* (41), *V. vinifera* (66), *S. lycopersicum* (59), *B. distachyon* (193), *O. sativa* (150), and *Z. mays* (170). The homologous logarithmic values for six species were shown as *A. thaliana* (44), *V. vinifera* (66), *S. lycopersicum* (60), *B. distachyon* (194), *O. sativa* (207), *Z. mays* (273) and *C. nucifera* (160). A notable finding is that several *AcbHLH* were connected with at least three synthetic gene pairs, emphasizing the link between *A. catechu* and *V. vinifera*, *AcbHLH43*, and *AcbHLH64*. *AcbHLH* gene, along with *A. thaliana* (54.0%), *B. distachyon* (53.73%), and *O. sativa* (55.56%). With regard to two or more synthetic gene pairs, the presence of these genes within the homologous gene pairs, comprising over 50% of the total, underscores their importance in evolutionary pathways.

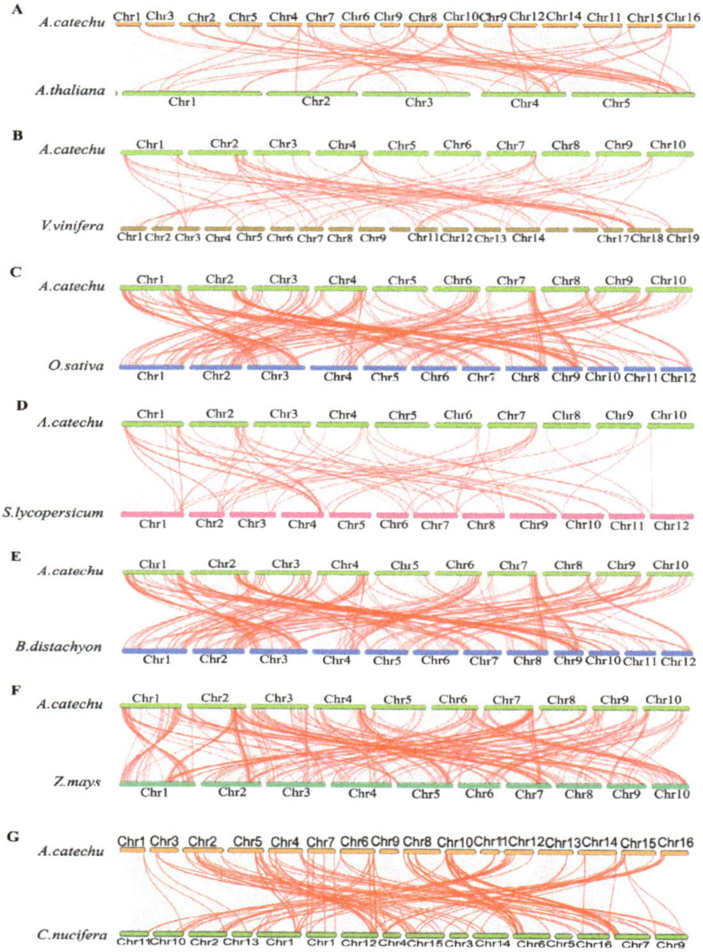

Figure 7. (**A**–**G**) Comparative synteny analysis of *AcbHLH* genes between *A. catechu* and six representative plant species (*A. thaliana*, *V. vinifera*, *S. lycopersicum*, *B. distachyon*, *O. sativa* sub sp. *indica*, *Z. mays* and *C. nucifera*). Gray lines show conserved syntenic blocks between *A. catechu* and other plant genomes. Red lines indicate that *AcbHLH* pairs of genes are syntenic across species.

As expected, some *AcbHLH* genes, especially *AcbHLH73*, *AcbHLH71*, *AcbHLH67*, and *AcbHLH29*, formed entirely homologous genes, with three dicotyledons. The fact that all these genes, such as *AcbHLH73*, *AcbHLH71*, *AcbHLH67*, and *AcbHLH29* exhibited the presence of homologous gene pairs with three typical dicotyledons indicated a possible evolutionary lineage leading to the development of dicotyledons. A specific *AcbHLH* gene homology with at least one gene in the five specified species was observed; for instance, *AcbHLH37*, *AcbHLH30*, *AcbHLH21*, *AcbHLH73*, *AcbHLH17*, *AcbHLH42*, *AcbHLH3*, and *AcbHLH41*. This observation suggested that these might be crucial primordial gene which have either been lost or which showed highly significant differentiation during the *A. catechu* long-term evolutionary history.

2.8. The Expression Patterns of AcbHLH Genes Differ Across Tissues

A. catechu plants were treated to drought (25% PEG6000) and salt (5% NaCl) stress, and transcriptome data were used to analyze the expression levels of *AcbHLH* genes in various tissues (Figure 8A and Table S5A,C). Under salt stress, *AcbHLH22* was the most highly expressed in all tissues, followed by *AcbHLH07*, *AcbHLH31*, and *AcbHLH44* in the pericarp, endosperm, and male and female flowers. *AcbHLH70*, *AcbHLH74*, *AcbHLH59*, *AcbHLH66*, *AcbHLH40*, *AcbHLH39*, and *AcbHLH13* were found to be highly expressed in flowers, whereas *AcbHLH74*, *AcbHLH59*, *AcbHLH52*, and *AcbHLH37* were prominent in the endosperm. *AcbHLH02*, *AcbHLH13*, *AcbHLH26*, and *AcbHLH37* were highly expressed in the pericarp, whereas *AcbHLH57*, *AcbHLH40*, *AcbHLH39*, *AcbHLH13*, and *AcbHLH08* were substantially expressed in the leaves. *AcbHLH02*, *AcbHLH34*, *AcbHLH37*, *AcbHLH44*, and *AcbHLH56* were highly expressed in roots.

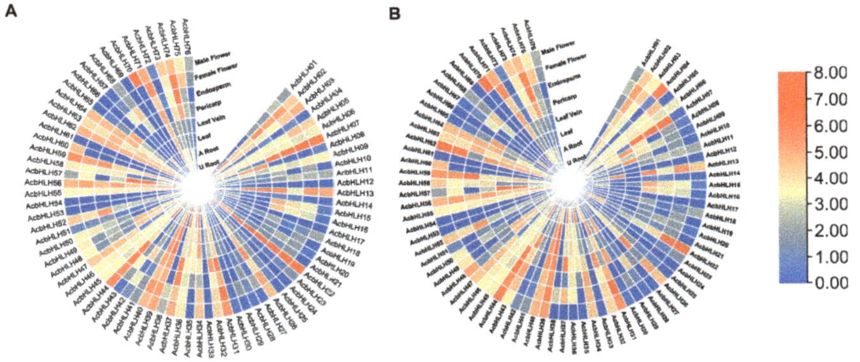

Figure 8. Heatmap of *AcbHLH* gene expression in *A. catechu* during salt stress (SS) (**A**) and drought stress (DS) (**B**). Red indicates larger log2FPKM values, whereas blue indicates lower values.

AcbHLH22, *AcbHLH07*, *AcbHLH44*, *AcbHLH59*, and *AcbHLH62* were the most expressed genes in male and female flowers during drought stress, followed by *AcbHLH04* and *AcbHLH70*. *AcbHLH07*, *AcbHLH44*, *AcbHLH59*, and *AcbHLH74* were abundant in the endosperm. *AcbHLH07*, *AcbHLH44*, *AcbHLH13*, *AcbHLH26*, *AcbHLH37*, and *AcbHLH63* all demonstrated high expression levels in the pericarp. *AcbHLH39* and *AcbHLH57* were highly expressed in the leaf veins, followed by *AcbHLH40* (Figure 8B and Table S5B,D).

2.9. Phenotypic and Physiological Changes

To analyze the phenotypic changes of plants under control (CK), salt stress (SS), and drought stress (DS), observations were made on Day 1 and Day 28. The results showed that all plants exhibited normal growth on Day 1 of exposure to abiotic stress. However, by Day 28, plants in the CK group displayed robust growth, while plants subjected to SS and DS exhibited visible signs of stress, including yellowing and leaf dryness (Figure 9A).

The plant height in the CK group was observed to be greater at 28 days compared to Day 1 and higher than in the SS and DS groups at 28 days (Figure 9B). However, the fresh weight of plants in the CK, SS, and DS groups was higher on Day 1 than at 28 days, with the CK plants showing the greatest fresh weight overall, compared to SS and DS. Additionally, the dry weight of CK plants at 28 days was significantly higher than that of CK, SS, and DS plants measured on both Day 1 and at 28 days.

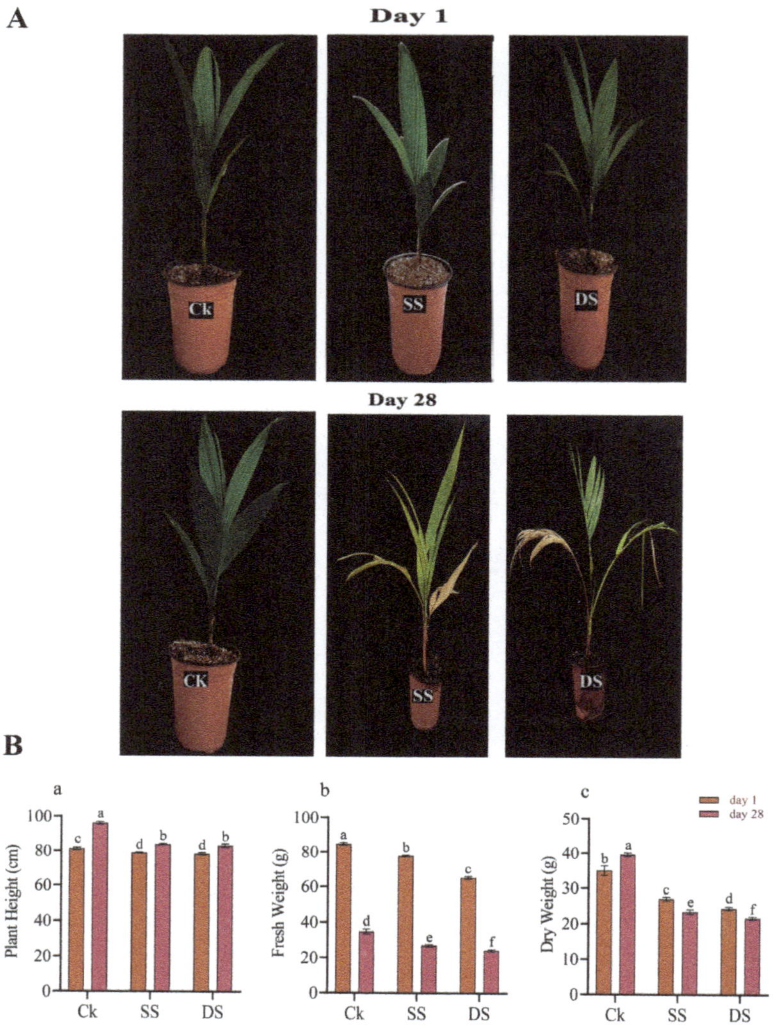

Figure 9. (**A**) The presentation of phenotypic changes observed on Day 1 and Day 28 of abiotic stress. (**B**) The physiological changes observed on Day 1 and Day 23 under abiotic stress (salt and drought): (a) plant height, (b) plant fresh weight, and (c) plant dry weight for CK, SS, and DS groups. Different lowercase letters represent the significant statistical level $p < 0.05$.

2.10. Effects of Different Treatments on AcbHLH Expression

To investigate the role of *AcbHLH* genes in response to abiotic stress, we evaluated the expression levels of nine selected *AcbHLH* genes (*AcbHLH22*, *AcbHLH34*, *AcbHLH39*, *AcbHLH43*, *AcbHLH45*, *AcbHLH48*, *AcbHLH58*, *AcbHLH59*, and *AcbHLH62*) under SS and DS in both roots and leaves, using quantitative reverse-transcription PCR (qRT-PCR) (Figure 10A–D). The results revealed a variety of gene expression responses to both stress treatments, with distinct activation or inhibition patterns observed at different time points. Under DS, the expression of *AcbHLH22* was significantly upregulated in leaves at 28 days, while *AcbHLH39* showed a peak at 14 days, and AcbHLH45 was upregulated at 7, 21, and 28 days. Additionally, *AcbHLH58* showed significant upregulation in leaves at 21 days. In the roots, the expression of all nine genes was significantly upregulated at various time points under DS.

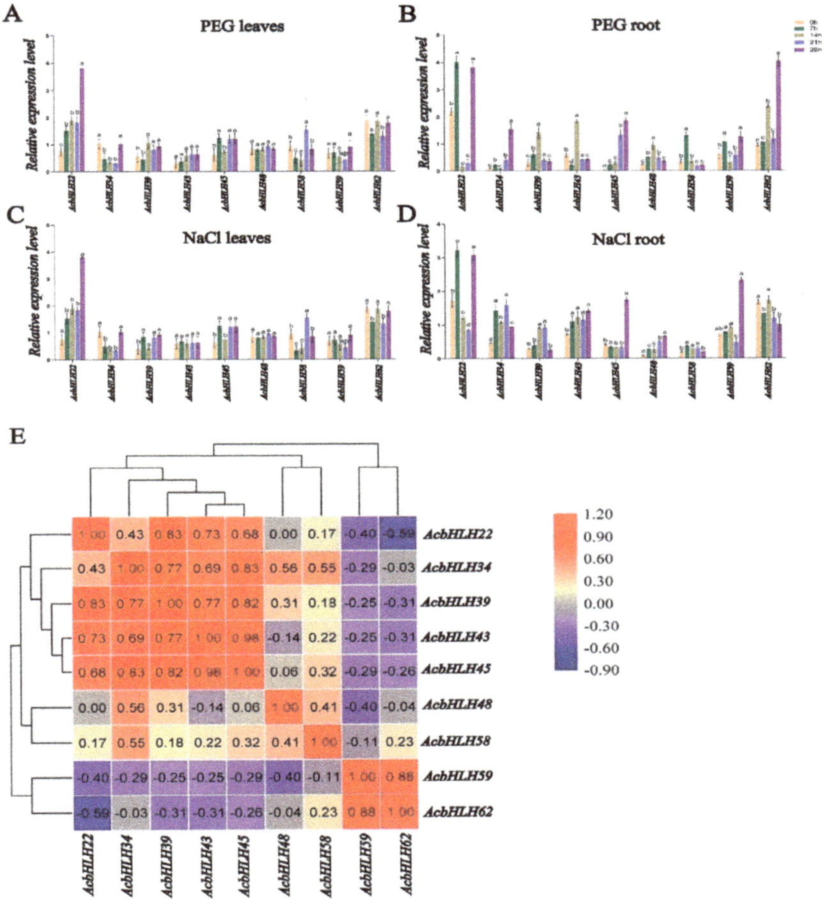

Figure 10. (**A–D**) The effects of abiotic stress (NaCl and PEG) detected by qRT-PCR on the expression of nine *AcbHLH* genes in roots and leaves of young *A. catechu* seedlings (0, 7, 14, 21, and 28 h) ($p < 0.05$). (**E**) Heatmap of nine *AcbHLH* genes. Different letters indicate statistically significant group differences ($p < 0.05$, LSD).

Under SS, specific genes exhibited differential expression in both leaves and roots. *AcbHLH22* was significantly upregulated in leaves at 21 and 28 days, *AcbHLH39* at 28 days, *AcbHLH45* at 7, 21, and 28 days, and *AcbHLH58* at 21 days. In roots, *AcbHLH22* was significantly upregulated at 7 and 28 days, *AcbHLH34* at 7 and 28 days, and *AcbHLH39* at 14 and 21 days. Other genes, such as *AcbHLH43*, *AcbHLH45*, *AcbHLH28*, and *AcbHLH59*, were significantly upregulated at 28 days, while *AcbHLH58* showed significant upregulation at 7 days. These findings demonstrate the temporal and tissue-specific regulation of *AcbHLH* genes under salt- and drought-stress conditions, shedding light on their potential roles in stress adaptation.

3. Discussion

Transcription factors control several biological processes by controlling the expression of certain genes, such as growth, development, and stress responses in plants [36]. The latest investigations show that the *A. catechu* plant has 31,571 protein-coding genes on 16 chromosomes [37]. Among these genes, a significant portion are classified as transcription factors [31]. Only a small number of the transcription factor families, including the DOF [38], WRKY, and MYB families (paper under review), have been systematically investigated in *A. catechu*. The bHLH TFs family, the second-largest among eukaryotes, is recognized for its evolutionary and functional diversity among plant species. However, the *bHLH* genes have not yet been characterized in *A. catechu*. In this study, we identified 76 *AcbHLH* genes that display a range of physicochemical characteristics, including differences in coding sequence length, chromosome number, molecular weight, protein length, GRAVY, isoelectric point (pI), number of exons/introns, subcellular localization, instability and aliphatic index (Table S1B). We used GRAVY values to indicate the average hydropathy of each peptide or protein. Our hydrophobicity research found that the *bHLH* gene family has negative GRAVY values, implying that they are hydrophilic, which aligns with those previously investigated [39].

Based on phylogenetic analysis, we categorized 76 *AcbHLH* genes and 162 *AtbHLH* proteins into 24 sub-families, emphasizing their evolutionary links and functional variety (Figure 1 and Table S1A). Our findings highlight evolutionary connections between *A. thaliana bHLH* genes (*AtbHLH*) and their counterparts in *A. catechu* (*AcbHLH*), with specific groupings showing close gene pair interactions, such as *AtbHLH47-AcbHLH31*, *AtbHLH21-AcbHLH77*, *AtbHLH13-AcbHLH83*, and *AtbHLH78-AcbHLH5*, among others. These findings, consistent with previous research into tomato, maize, and barley [40,41], indicated that *QbHLH* transcription factors have preserved evolutionary and functional properties across plant species, highlighting their extensive role in plant adaptation and development. Sequence alignment results of *AcbHLH* genes in *A. catechu* revealed consistent patterns of amino acid motifs (S, R, K, P, L, A, G, E, Y, I, V, H, M, Q, N, and D), as shown in Figure 2. As previously observed, these motifs were shared by *bHLH* proteins from various plant species, such as *A. thaliana* [42,43], *F. tataricum* [44], and *S. lycopersicum* [45].

The phylogeny and gene structure analysis of *AcbHLH* genes, as shown in Figure 3A,B and Table S1A, show that genes from the same subfamily have similar intron–exon structures. This indicated that changes in exon–intron distribution influence the evolution and functional diversity of *AcbHLH* proteins. Notably, the study discovered that the intron–exon number of *AcbHLH* genes varies from 1 to 13 (Figure 3B), indicating evolutionary changes such as insertions, deletions, and exon gains/losses. Furthermore, the relationship between intron number and gene expression in plants suggests that a more compact gene structure may allow for rapid gene expression in response to environmental stimuli [46], providing insight into the evolutionary adaptations of these genes in various biological processes. The conserved motif of AcbHLH proteins contained bHLH domains 1 and 2, as shown in Figure 3C. Additionally, the composition of other motifs was distinctive, and they were conserved across subgroups. For instance, motif 4 is exclusively present at the end of subgroups 3, 5, and 6. Other plants have been observed to exhibit similar phenomena [26,47].

Cis-acting elements, which are substances that bind to trans-acting factors, play a crucial role in regulating the activity of target genes [48]. These elements played an important role in molecular controlling genes, especially in response to stress expression [49]. Promoter analysis of the 2000 bp upstream regions of *AcbHLH* genes revealed 76 cis-elements classified into four main categories, as shown in Figure 4B: (I) growth-related (AAGAA, as-1), (II) phytohormone-responsive (ABRE, AuxRR-core), (III) light-responsive (G-box, TCT-motif), and (IV) stress-responsive (DRE, Myb, Myc). These elements are linked to responses to abiotic stresses like drought, temperature extremes, and wounding, as well as hormones like ABA and SA, implying that *AcbHLH* genes play an important role in regulating stress responses via hormone signaling (Figure 4A); this is as described by previous research, in which the *bHLH* gene, specifically *AtbHLH122* in *A. thaliana*, plays a vital role in ABA signaling pathways, increasing drought tolerance by regulating ABA metabolism and catabolism.

Research has revealed that gene duplication events are essential in several gene families' rapid growth and expansion [50]. In the genome of *A. catechu*, four tandem duplications and twenty-three segmental duplication events were found in the *AcbHLH* family (Figure 5 and Table S2). These results provide strong evidence that segmental and tandem duplication had a role in the growth and divergence of the *AcbHLH* gene. Previous research showed segmental duplication has a more substantial role in genome evolution than tandem duplication [51]. We used Dup-206 Gene Finder to analyze gene duplication events in the *AcbHLH* gene family, identifying 23 duplicated genes in DSD, 11 in WGD, and 4 in PD (Figure 6 and Table S3D). Ka/Ks ratios for duplicated genes were all less than 1, indicating strong purifying selection (Figure 6B–F), aligning with previous study [52].

Gene functions within a gene family are frequently conserved across plant species, although this is not always possible. As a result, it is important to precisely determine the orthologs across plant species using synteny analysis. The result showed that *AcbHLH63* in the Areca genome exhibited significant synteny with *A. thaliana*, *V. vinifera*, *S. lycopersicum*, *B. distachyon*, *O. sativa*, and *Z. mays* (Figure 7 and Table S4D). These results indicate that these genes may have played a crucial role in the evolutionary history of *A. catechu*, potentially undergoing loss or significant differentiation.

Furthermore, to investigate the role of *AcbHLH* genes in response to abiotic stress, we evaluated the expression levels of nine selected *AcbHLH* genes (*AcbHLH22*, *AcbHLH34*, *AcbHLH39*, *AcbHLH43*, *AcbHLH45*, *AcbHLH48*, *AcbHLH58*, *AcbHLH59* and *AcbHLH62*) under NaCl and PEG stresses in both roots and leaves, using quantitative reverse-transcription PCR (qRT-PCR) (Figure 10A–D). Under PEG stress, *AcbHLH22*, *AcbHLH39*, *AcbHLH45*, and *AcbHLH58* exhibited distinct upregulation in leaves at specific time points, with *AcbHLH22* being most strongly expressed at 28 days, *AcbHLH39* peaking at 14 days, and *AcbHLH45* showing sustained expression at 7, 21, and 28 days. These results are consistent with previous findings in other species, where *bHLH* genes demonstrated differential expression patterns under abiotic stress [53]. For example, in *C. quinoa*, eleven *bHLH* genes were significantly upregulated in response to cold stress, while eight were downregulated, suggesting a complex regulation of *bHLH* genes under stress [7,25]. Similarly, in *S. bicolor*, a range of *bHLH* genes exhibited both up- and downregulation in response to various abiotic stresses, supporting the idea of *bHLH* genes as key regulators in stress responses [9,54]. Our findings shows that *AcbHLH* genes, particularly *AcbHLH22* and *AcbHLH39*, are strongly upregulated under DS, suggesting their role in drought tolerance mechanisms.

In the roots, the expression of all nine *AcbHLH* genes was upregulated under PEG stress at various time points, which aligns with the general response of *bHLH* genes to abiotic stresses in other plants [55]. These results imply that *AcbHLH* genes are involved in leaf stress responses and play a crucial role in root adaptation to osmotic stress [10,56]. Under SS, our study found that *AcbHLH22*, *AcbHLH39*, *AcbHLH45*, and *AcbHLH58* exhibited upregulation in leaves, with *AcbHLH22* showing significant expression at both 21 and 28 days. In roots, *AcbHLH22* was upregulated at 7 and 28 days, while *AcbHLH34* was upregulated at 7 and 28 days, and *AcbHLH39* showed peak expression at 14 and 21 days. These results support the fact that *bHLH* genes regulate tissue-specific stress responses, as seen in other species like *Z. mays*, where *bHLH* genes such as *ZmbHLH180* and *ZmbHLH23* exhibited synergistic expression patterns under stress [11,57].

Interestingly, *AcbHLH22* was significantly upregulated under both salt and drought stress in our study, which contrasts with findings in *A. thaliana*, where *bHLH* genes like *AtbHLH22* did not show similar upregulation under abiotic stress [12,58]. This discrepancy suggests that *AcbHLH22* might have a unique role in *A. catechu* that warrants further investigation, particularly in relation to its involvement in stress-induced stomatal regulation. Previous studies have linked bHLH transcription factors to regulating leaf stomatal development, which is crucial for maintaining water balance under stress [13,59]. This opens avenues for exploring how *AcbHLH22* may influence water use efficiency and osmotic regulation in response to environmental stresses. This suggests a potential synergistic interaction between the two genes, likely coordinating their physiological roles, particularly under abiotic stress and during vegetative development [14,60]. This study provides the comprehensive analysis of the bHLH transcription factor family in *A. catechu*, offering valuable insights into their evolutionary diversity and potential roles in abiotic stress responses. While the findings lay a strong foundation, addressing the outlined limitations and exploring the suggested future directions will further enhance our understanding of *AcbHLH* gene functions, enabling their application in crop improvement strategies for stress resilience.

4. Materials and Methods

4.1. Identification of AcbHLH Genes in A. catechu

The complete genome sequence of *A. catechu* L. was obtained from the National Center for Biotechnology Information (NCBI) using the following (ID: JAHSVC000000000; BioSample: SAMN19591864; Accession: PRJNA735650). The bHLH protein domain (Pfam ID: PF00010) was retrieved via the Pfam database. The conserved bHLH protein domain (Pfam ID: PF00010) was retrieved from the Pfam database. A two-step approach was employed to identify *AcbHLH* within *A. catechu*. The first approach, the protein sequences of *Arabidopsis* bHLH used as query sequences downloaded from TAIR (http://www.arabidopsis.org) (accessed on 15 March 2024), apply BLASTp with a score value of at least 100 and an E-value of less than or equal to 1×10^{-10}. In the second approach, the *bHLH* candidates were initially identified within the *A. catechu* genome through a hidden Markov model (HMM) search, employing an e-value lower than 10^{-5}, based on a previously used method [15]. Finally, unidentified conserved sequence motifs were manually removed from the dataset. To validate the presence of bHLH domains, HMMER R3.0 (http://plants.ensembl.org/hmmer/index.html) (accessed on 20 March 2024) [16] with default parameters and a 0.01 cutoff was used for protein homology, and confirmation was achieved using the SMART domain database (http://smart.embl-heidelberg.de/) (accessed on 24 March 2024).

4.2. Analysis of Physicochemical Properties of AcbHLH

The ExPasy ProtParam server (http://web.expasy.org/protparam/) (accessed on 30 March 2024) was utilized to characterize the fundamental characteristics of *AcbHLH* tri-helix proteins, sequence length, isoelectric point (pI), and molecular mass within the *AcbHLH* gene family. A protein subcellular localization prediction tool (PSORT) was used to determine the likely cellular location of the predicted proteins (https://www.genscript.com/psort.html) (accessed on 8 April 2024) [17].

4.3. The bHLH Gene Structure, Conserved Motif, and Cis-Acting Elements

The characterized *AcbHLH* proteins were aligned manually using ClustalW, and GeneDoc software version 2.7 was used to improve the alignment quality manually. An in-depth analysis of the intron–exon structure of *AcbHLH* genes in *A. catechu* was achieved by the Gene Structure Display Server (GSDS) (http://GSDS.cbi.pku.edu.cn/) (accessed on 14 April 2024) by leveraging the GFF3 annotation data. The online tool MEME was used to identify motifs of AcbHLH proteins (http://meme-suite.org/tools/meme) (accessed on 20 April 2024). Promoter sequences of *AcbHLH* genes were retrieved from the *A. catechu* genome (http://bioinformatics.psb.ugent.be/webtools/plantcare/html/) (accessed on 27 April 2024). Upstream regulatory sequences of these genes were identified as the 2000 base pairs preceding the ATG codon. The distribution of cis-elements within the promoter regions was analyzed using the PlantCARE database (http://bioinformatics.psb.ugent.be/webtools/plantcare/html/) (accessed on 8 May 2024).

4.4. Chromosomal Distribution and Gene Duplication

Chromosomal localization of *AcbHLH* was determined by mapping their physical location to the *A. catechu* genomic database and Circular Genome Data Visualization tool (Circos). The Multiple Collinearity Scan toolkit (MCScanX) was employed to analyze gene duplication events, using its parameters [18]. A dual synteny plotter was used to analyze gene homogeneity among *A. catechu* and *A. thaliana* (https://github.com/CJ-Chen/TBtools) (accessed on 12 May 2024). Nonsynonymous and synonymous substitution rates were calculated for each duplicated *bHLH* gene using the Ka/Ks Calculator 2.0 [19].

4.5. Evolutionary Relationships and Classification of AcbHLH Gene Family

All the recognized genes, *AcbHLH*, were classified based on *AtbHLH* categorization. The phylogenic tree was created utilizing the Neighbor Joining (NJ) method in MEGA-X software 11.0.10; using 1000 bootstrap replicates and the NJ method, the phylogenetic tree was created and visualized using iTOL [19]. For phylogenetic analysis, the full-length amino acid sequences of *A. thaliana bHLH* proteins were utilized, (https://itol.embl.de/) (accessed on 15 May 2024).

4.6. RNA-Seq Data Analysis

The expression profile study of *AcbHLH* genes was conducted using Illumina Hisequence 4000 RNA-sequence data submitted to the NCBI database (accession number: PRJNA767949). Genes exhibiting a log2FC fold change greater than 1, and a false discovery rate (FDR) of 0.05 with 3 biological replicates per treatment and differential expression were identified using a significant threshold of p-value < 0.05. The expression pattern of *AcbHLH* genes was plotted using TBtools. Gene expression studies were conducted using the BMK Cloud platform (https://www.biocloud.net/) (accessed on 20 May 2024).

4.7. Plant Materials, Growth Conditions, and Abiotic Stress in A. catechu

The research utilized seedlings of *A.catechu* (Reyan No. 1) from the Coconut Research Institute of the Chinese Academy of Tropical Agricultural Sciences, located in Wenchang, Hainan province, China. Seedlings were grown in pots (12 cm × 12 cm) at 14/10 h day/night at 28/25 °C. The plants were treated, including drought (DS) (25% PEG6000) and salt (SS) (5% NaCl) [7]. Leaf and root samples were collected at 1, 7, 14, 21, and 28 days for each treatment and preserved at −80 °C for s subsequent analysis.

4.8. Phentypic and Physiological Analysis

Phenotypic changes were observed on Day 1 and Day 28 in the CK, SS, and DS groups. Plant height was measured on both days, using a measuring tape. Fresh weight was determined using a weighing balance for samples collected on Day 1 and Day 28. For dry weight analysis, plant samples from the control and treatment groups were dried in an oven (DGG-9053A, Shanghai Enxin Instrument Co., Ltd., Shanghai, China) at 65 °C for 72 h and then weighed using a balance [61].

4.9. Total RNA Extraction, cDNA and qRT-PCR

Total RNA extraction from plant samples was performed using a specialized extraction kit (TIANGEN, Beijing, China). RNA concentration and purity were checked using NanoDrop 2000 (KAIAO, Beijing, China). First-strand cDNA synthesis was performed using the TIANScript RT Kit (TIANGEN, Beijing, China). Quantitative real-time PCR (qRT-PCR) was used to analyze gene expression using Vazyme Master Mix (Vazyme, Nanjing, China). PCR conditions were set according to the company manual. Primers were designed via Primer 6.0 software, and *Actin* was used as an internal control for gene expression analysis (Table S6). The $2^{-\Delta\Delta CT}$ method was utilized to analyze qRT-PCR data and calculate relative gene expression levels.

4.10. Statistical Analysis

Analysis of variance (ANOVA) was analyzed by Statistics 8.1 software, followed by the Tukey LSD test, and compared with the least significant difference (LSD) ($p \geq 0.05$). The histogram was drawn using Origin 8.0 software (OriginLab, Northampton, MA, USA).

5. Conclusions

This research represents the first comprehensive and systematic evaluation of the *AcbHLH* gene family within the *A. catechu* genome, encompassing a wide range of analytical approaches and verifications. A total of 76 *AcbHLH* gene–proteins were identified and categorized into 24 distinct subgroups based on their protein domain profiles and gene structure, which carry the categorization accuracy based on phylogenic analysis and are irregularly distributed on 16 chromosomes (Chr00). The distribution of *AcbHLH* genes across the chromosomes was not evenly distributed. Some *AcbHLH* genes are involved in gene replication events, and fragment repeat contributes more favorably than tandem duplication. A significant homologous phenomenon revealed the homology between one *AcbHLH* gene and multiple *bHLH* genes in *A. Catechu*. The *AcbHLH* genes from *A. catechu* are the most closely related genes among the six representative plant species, as determined by sequence comparison. Finally, qRT-PCR analysis revealed that nine *AcbHLH* genes were differentially expressed in response to abiotic stress conditions. The expression mechanism of these genes was investigated throughout the developmental stages, and *AcbHLH22* and *AcbHLH144* demonstrated a notable impact on abiotic stress resistance. It is speculated that *AcbHLH22* and *AcbHLH62* are essential components of the *A. catechu* life cycle, contributing to its viability and development. In conclusion, this research offers valuable insights into the biological functions and development of *AcbHLH* in *A. catechu* and will facilitate future research on the functions of *AcbHLHs*.

Supplementary Materials: The following supporting information can be downloaded at: https://www.mdpi.com/article/10.3390/ijms252312936/s1.

Author Contributions: Conceptualization, A.A. and Y.W.; methodology, A.A., N.M.K. and Y.J.; software, A.A., Y.J. and G.Z.; validation, A.A. and N.M.K.; formal analysis, A.A., N.M.K. and Y.J.; investigation, A.A., G.Z. and Y.J.; resources, A.A. and N.M.K.; data curation, A.A., Y.J. and G.Z.; writing—original draft preparation, A.A. and Y.W.; writing—review and editing, A.A., Y.W., N.M.K. and G.Z.; visualization, A.A. and Y.W.; supervision, Y.W.; project administration, Y.W.; funding acquisition, Y.W. All authors have read and agreed to the published version of the manuscript.

Funding: This work was financially supported by the National Natural Science Foundation of China (No. 31960064).

Institutional Review Board Statement: Not applicable.

Informed Consent Statement: Not applicable.

Data Availability Statement: The data generated and analyzed in this study are available on the request.

Conflicts of Interest: The authors declare no conflicts of interest.

References

1. Xu, Y.; Zhang, H.; Zhong, Y.; Jiang, N.; Zhong, X.; Zhang, Q.; Chai, S.; Li, H.; Zhang, Z. Comparative Genomics Analysis of bHLH Genes in Cucurbits Identifies a Novel Gene Regulating Cucurbitacin Biosynthesis. *Hortic. Res.* **2022**, *9*, uhac038. [CrossRef] [PubMed]
2. Li, J.; Li, X.; Han, P.; Liu, H.; Gong, J.; Zhou, W.; Shi, B.; Liu, A.; Xu, L. Genome-Wide Investigation of bHLH Genes and Expression Analysis under Different Biotic and Abiotic Stresses in *Helianthus annuus* L. *Int. J. Biol. Macromol.* **2021**, *189*, 72–83. [CrossRef] [PubMed]
3. Lai, X.; Vega-Léon, R.; Hugouvieux, V.; Blanc-Mathieu, R.; van Der Wal, F.; Lucas, J.; Silva, C.S.; Jourdain, A.; Muino, J.M.; Nanao, M.H. The Intervening Domain Is Required for DNA-Binding and Functional Identity of Plant MADS Transcription Factors. *Nat. Commun.* **2021**, *12*, 4760. [CrossRef] [PubMed]

4. Pires, N.; Dolan, L. Origin and Diversification of Basic-Helix-Loop-Helix Proteins in Plants. *Mol. Biol. Evol.* **2010**, *27*, 862–874. [CrossRef]
5. Zhang, Z.; Chen, J.; Liang, C.; Liu, F.; Hou, X.; Zou, X. Genome-Wide Identification and Characterization of the bHLH Transcription Factor Family in Pepper (*Capsicum annuum* L.). *Front. Genet.* **2020**, *11*, 570156. [CrossRef]
6. Grove, C.A. A Multiparameter Network Reveals Extensive Divergence Between C. Elegans bHLH Transcription Factors: A Dissertation. Ph.D. Thesis, University of Massachusetts (USA), Boston, MA, USA, 11 September 2009.
7. Qin, Y.; Li, J.; Chen, J.; Yao, S.; Li, L.; Huang, R.; Tan, Y.; Ming, R.; Huang, D. Genome-Wide Characterization of the bHLH Gene Family in *Gynostemma pentaphyllum* Reveals Its Potential Role in the Regulation of Gypenoside Biosynthesis. *BMC Plant Biol.* **2024**, *24*, 205. [CrossRef]
8. Guo, J.; Sun, B.; He, H.; Zhang, Y.; Tian, H.; Wang, B. Current Understanding of bHLH Transcription Factors in Plant Abiotic Stress Tolerance. *Int. J. Mol. Sci.* **2021**, *22*, 4921. [CrossRef]
9. Gonzalez, A.; Mendenhall, J.; Huo, Y.; Lloyd, A. TTG1 Complex MYBs, MYB5 and TT2, Control Outer Seed Coat Differentiation. *Dev. Biol.* **2009**, *325*, 412–421. [CrossRef]
10. Chang, G.; Ma, J.; Wang, S.; Tang, M.; Zhang, B.; Ma, Y.; Li, L.; Sun, G.; Dong, S.; Liu, Y. Liverwort bHLH Transcription Factors and the Origin of Stomata in Plants. *Curr. Biol.* **2023**, *33*, 2806–2813.e6. [CrossRef]
11. Qiu, Y.; Tao, R.; Feng, Y.; Xiao, Z.; Zhang, D.; Peng, Y.; Wen, X.; Wang, Y.; Guo, H. EIN3 and RSL4 Interfere with an MYB–bHLH–WD40 Complex to Mediate Ethylene-Induced Ectopic Root Hair Formation in *Arabidopsis*. *Proc. Natl. Acad. Sci.* **2021**, *118*, e2110004118. [CrossRef]
12. Groszmann, M.; Paicu, T.; Alvarez, J.P.; Swain, S.M.; Smyth, D.R. SPATULA and ALCATRAZ, Are Partially Redundant, Functionally Diverging bHLH Genes Required for *Arabidopsis gynoecium* and Fruit Development. *Plant J.* **2011**, *68*, 816–829. [CrossRef]
13. Leivar, P.; Monte, E.; Oka, Y.; Liu, T.; Carle, C.; Castillon, A.; Huq, E.; Quail, P.H. Multiple Phytochrome-Interacting bHLH Transcription Factors Repress Premature Seedling Photomorphogenesis in Darkness. *Curr. Biol.* **2008**, *18*, 1815–1823. [CrossRef]
14. Liang, J.; Fang, Y.; An, C.; Yao, Y.; Wang, X.; Zhang, W.; Liu, R.; Wang, L.; Aslam, M.; Cheng, Y. Genome-Wide Identification and Expression Analysis of the bHLH Gene Family in Passion Fruit (*Passiflora edulis*) and Its Response to Abiotic Stress. *Int. J. Biol. Macromol.* **2023**, *225*, 389–403. [CrossRef] [PubMed]
15. Su, L.; Zhang, Y.; Yu, S.; Geng, L.; Lin, S.; Ouyang, L.; Jiang, X. RcbHLH59-RcPRs Module Enhances Salinity Stress Tolerance by Balancing Na+/K+ through Callose Deposition in Rose (*Rosa chinensis*). *Hortic. Res.* **2023**, *10*, uhac291. [CrossRef]
16. Xue, L.; Wei, Z.; Zhai, H.; Xing, S.; Wang, Y.; He, S.; Gao, S.; Zhao, N.; Zhang, H.; Liu, Q. The IbPYL8–IbbHLH66–IbbHLH118 Complex Mediates the Abscisic Acid—dependent Drought Response in Sweet Potato. *New Phytol.* **2022**, *236*, 2151–2171. [CrossRef] [PubMed]
17. Zhang, L.; Xiang, Z.; Li, J.; Wang, S.; Chen, Y.; Liu, Y.; Mao, D.; Luan, S.; Chen, L. *bHLH57* Confers Chilling Tolerance and Grain Yield Improvement in Rice. *Plant. Cell Environ.* **2023**, *46*, 1402–1418. [CrossRef]
18. Singh, S.K.; Patra, B.; Paul, P.; Liu, Y.; Pattanaik, S.; Yuan, L. *bHLH IRIDOID SYNTHESIS 3* Is a Member of a bHLH Gene Cluster Regulating Terpenoid Indole Alkaloid Biosynthesis in *Catharanthus roseus*. *Plant Direct* **2021**, *5*, e00305. [CrossRef] [PubMed]
19. Li, X.; Cao, L.; Jiao, B.; Yang, H.; Ma, C.; Liang, Y. The bHLH Transcription Factor AcB2 Regulates Anthocyanin Biosynthesis in Onion (*Allium cepa* L.). *Hortic. Res.* **2022**, *9*, uhac128. [CrossRef] [PubMed]
20. Liu, S.; Wang, Y.; Shi, M.; Maoz, I.; Gao, X.; Sun, M.; Yuan, T.; Li, K.; Zhou, W.; Guo, X. SmbHLH60 and SmMYC2 Antagonistically Regulate Phenolic Acids and Anthocyanins Biosynthesis in *Salvia miltiorrhiza*. *J. Adv. Res.* **2022**, *42*, 205–219. [CrossRef]
21. Mohammad; Hurrah, I.M.; Kumar, A.; Abbas, N. Synergistic Interaction of Two bHLH Transcription Factors Positively Regulates Artemisinin Biosynthetic Pathway in *Artemisia annua* L. *Physiol. Plant.* **2023**, *175*, e13849. [CrossRef]
22. Hao, Y.; Zong, X.; Ren, P.; Qian, Y.; Fu, A. Basic Helix-Loop-Helix (bHLH) Transcription Factors Regulate a Wide Range of Functions in Arabidopsis. *Int. J. Mol. Sci.* **2021**, *22*, 7152. [CrossRef] [PubMed]
23. Li, X.; Duan, X.; Jiang, H.; Sun, Y.; Tang, Y.; Yuan, Z.; Guo, J.; Liang, W.; Chen, L.; Yin, J. Genome-Wide Analysis of Basic/Helix-Loop-Helix Transcription Factor Family in Rice and Arabidopsis. *Plant Physiol.* **2006**, *141*, 1167–1184. [CrossRef]
24. Wang, J.; Hu, Z.; Zhao, T.; Yang, Y.; Chen, T.; Yang, M.; Yu, W.; Zhang, B. Genome-Wide Analysis of bHLH Transcription Factor and Involvement in the Infection by Yellow Leaf Curl Virus in Tomato (*Solanum lycopersicum*). *BMC Genom.* **2015**, *16*, 1–14. [CrossRef]
25. Zhang, T.; Lv, W.; Zhang, H.; Ma, L.; Li, P.; Ge, L.; Li, G. Genome-Wide Analysis of the Basic Helix-Loop-Helix (bHLH) Transcription Factor Family in Maize. *BMC Plant Biol.* **2018**, *18*, 1–14. [CrossRef]
26. Wang, L.; Xiang, L.; Hong, J.; Xie, Z.; Li, B. Genome-Wide Analysis of bHLH Transcription Factor Family Reveals Their Involvement in Biotic and Abiotic Stress Responses in Wheat (*Triticum aestivum* L.). *3 Biotech* **2019**, *9*, 1–12. [CrossRef]
27. Zhao, K.; Li, S.; Yao, W.; Zhou, B.; Li, R.; Jiang, T. Characterization of the Basic Helix–Loop–Helix Gene Family and Its Tissue-Differential Expression in Response to Salt Stress in Poplar. *PeerJ* **2018**, *6*, e4502. [CrossRef] [PubMed]
28. Wang, R.; Zhao, P.; Kong, N.; Lu, R.; Pei, Y.; Huang, C.; Ma, H.; Chen, Q. Genome-Wide Identification and Characterization of the Potato bHLH Transcription Factor Family. *Genes* **2018**, *9*, 54. [CrossRef]
29. Wang, Y.; Zhou, G.; Luo, H.; Li, X.; Zhang, K.; Wan, Y. Genome-Wide Identification of PIN and PILS Gene Families in *Areca catechu* and the Potential Role of AcPIN6 in Lateral Brace Root Formation. *Plants* **2022**, *12*, 33. [CrossRef]

30. Salehi, B.; Konovalov, D.A.; Fru, P.; Kapewangolo, P.; Peron, G.; Ksenija, M.S.; Cardoso, S.M.; Pereira, O.R.; Nigam, M.; Nicola, S. *Areca catechu*—From Farm to Food and Biomedical Applications. *Phyther. Res.* **2020**, *34*, 2140–2158. [CrossRef] [PubMed]
31. Zhou, G.; Yin, H.; Chen, F.; Wang, Y.; Gao, Q.; Yang, F.; He, C.; Zhang, L.; Wan, Y. The Genome of *Areca catechu* Provides Insights into Sex Determination of Monoecious Plants. *New Phytol.* **2022**, *236*, 2327–2343. [CrossRef]
32. Wang, Z.; Guo, Z.; Luo, Y.; Ma, L.; Hu, X.; Chen, F.; Li, D. A Review of the Traditional Uses, Pharmacology, and Toxicology of Areca Nut. *Phytomedicine* **2024**, 156005. [CrossRef] [PubMed]
33. Xue, G.; Fan, Y.; Zheng, C.; Yang, H.; Feng, L.; Chen, X.; Yang, Y.; Yao, X.; Weng, W.; Kong, L. bHLH Transcription Factor Family Identification, Phylogeny, and Its Response to Abiotic Stress in *Chenopodium quinoa*. *Front. Plant Sci.* **2023**, *14*, 1171518. [CrossRef] [PubMed]
34. Li, K.; Duan, L.; Zhang, Y.; Shi, M.; Chen, S.; Yang, M.; Ding, Y.; Peng, Y.; Dong, Y.; Yang, H. Genome-Wide Identification and Expression Profile Analysis of Trihelix Transcription Factor Family Genes in Response to Abiotic Stress in Sorghum [*Sorghum bicolor* (L.) Moench]. *BMC Genom.* **2021**, *22*, 738. [CrossRef] [PubMed]
35. Chen, F.; Hu, Y.; Vannozzi, A.; Wu, K.; Cai, H.; Qin, Y.; Mullis, A.; Lin, Z.; Zhang, L. The WRKY Transcription Factor Family in Model Plants and Crops. *Crit. Rev. Plant Sci.* **2017**, *36*, 311–335. [CrossRef]
36. Zhan, H.; Liu, H.; Ai, W.; Han, X.; Wang, Y.; Lu, X. Genome-Wide Identification and Expression Analysis of the bHLH Transcription Factor Family and Its Response to Abiotic Stress in *Mongolian oak* (*Quercus mongolica*). *Curr. Issues Mol. Biol.* **2023**, *45*, 1127–1148. [CrossRef] [PubMed]
37. Yang, Y.; Huang, L.; Xu, C.; Qi, L.; Wu, Z.; Li, J.; Chen, H.; Wu, Y.; Fu, T.; Zhu, H. Chromosome—Scale Genome Assembly of Areca Palm (*Areca catechu*). *Mol. Ecol. Resour.* **2021**, *21*, 2504–2519. [CrossRef]
38. Li, J.; Jia, X.; Yang, Y.; Chen, Y.; Wang, L.; Liu, L.; Li, M. Genome-Wide Identification of the DOF Gene Family Involved in Fruitlet Abscission in *Areca catechu* L. *Int. J. Mol. Sci.* **2022**, *23*, 11768. [CrossRef]
39. Amrhein, S.; Oelmeier, S.A.; Dismer, F.; Hubbuch, J. Molecular Dynamics Simulations Approach for the Characterization of Peptides with Respect to Hydrophobicity. *J. Phys. Chem. B* **2014**, *118*, 1707–1714. [CrossRef]
40. Ke, Q.; Tao, W.; Li, T.; Pan, W.; Chen, X.; Wu, X.; Nie, X.; Cui, L. Genome-Wide Identification, Evolution and Expression Analysis of Basic Helix-Loop-Helix (bHLH) Gene Family in Barley (*Hordeum vulgare* L.). *Curr. Genom.* **2020**, *21*, 624–644. [CrossRef]
41. Sun, H.; Fan, H.-J.; Ling, H.-Q. Genome-Wide Identification and Characterization of the *bHLH* Gene Family in Tomato. *BMC Genom.* **2015**, *16*, 9. [CrossRef]
42. Toledo-Ortiz, G.; Huq, E.; Quail, P.H. The Arabidopsis Basic/Helix-Loop-Helix Transcription Factor Family. *Plant Cell* **2003**, *15*, 1749–1770. [CrossRef] [PubMed]
43. Fan, Y.; Lai, D.; Yang, H.; Xue, G.; He, A.; Chen, L.; Feng, L.; Ruan, J.; Xiang, D.; Yan, J. Genome-Wide Identification and Expression Analysis of the *bHLH* Transcription Factor Family and Its Response to Abiotic Stress in Foxtail Millet (*Setaria italica* L.). *BMC Genom.* **2021**, *22*, 778. [CrossRef] [PubMed]
44. Sun, W.; Jin, X.; Ma, Z.; Chen, H.; Liu, M. Basic Helix–Loop–Helix (bHLH) Gene Family in Tartary Buckwheat (*Fagopyrum Tataricum*): Genome-Wide Identification, Phylogeny, Evolutionary Expansion and Expression Analyses. *Int. J. Biol. Macromol.* **2020**, *155*, 1478–1490. [CrossRef]
45. Wray, G.A.; Hahn, M.W.; Abouheif, E.; Balhoff, J.P.; Pizer, M.; Rockman, M.V.; Romano, L.A. The Evolution of Transcriptional Regulation in Eukaryotes. *Mol. Biol. Evol.* **2003**, *20*, 1377–1419. [CrossRef] [PubMed]
46. Kavas, M.; Baloğlu, M.C.; Atabay, E.S.; Ziplar, U.T.; Daşgan, H.Y.; Ünver, T. Genome-Wide Characterization and Expression Analysis of Common Bean bHLH Transcription Factors in Response to Excess Salt Concentration. *Mol. Genet. Genom.* **2016**, *291*, 129–143. [CrossRef]
47. Liu, M.; Ma, Z.; Sun, W.; Huang, L.; Wu, Q.; Tang, Z.; Bu, T.; Li, C.; Chen, H. Genome-Wide Analysis of the NAC Transcription Factor Family in Tartary Buckwheat (*Fagopyrum tataricum*). *BMC Genom.* **2019**, *20*, 113. [CrossRef]
48. Nakashima, K.; Ito, Y.; Yamaguchi-Shinozaki, K. Transcriptional Regulatory Networks in Response to Abiotic Stresses in Arabidopsis and Grasses. *Plant Physiol.* **2009**, *149*, 88–95. [CrossRef]
49. Rivera, A.M.; Swanson, W.J. The Importance of Gene Duplication and Domain Repeat Expansion for the Function and Evolution of Fertilization Proteins. *Front. Cell Dev. Biol.* **2022**, *10*, 827454. [CrossRef]
50. Song, M.; Wang, H.; Wang, Z.; Huang, H.; Chen, S.; Ma, H. Genome-Wide Characterization and Analysis of bHLH Transcription Factors Related to Anthocyanin Biosynthesis in Fig (*Ficus carica* L.). *Front. Plant Sci.* **2021**, *12*, 730692. [CrossRef]
51. Starr, T.K.; Jameson, S.C.; Hogquist, K.A. Positive and Negative Selection of T Cells. *Annu. Rev. Immunol.* **2003**, *21*, 139–176. [CrossRef]
52. Qian, Y.; Zhang, T.; Yu, Y.; Gou, L.; Yang, J.; Xu, J.; Pi, E. Regulatory Mechanisms of bHLH Transcription Factors in Plant Adaptive Responses to Various Abiotic Stresses. *Front. Plant Sci.* **2021**, *12*, 677611. [CrossRef]
53. Ye, D.; Liu, J.; Tian, X.; Wen, X.; Zhang, Y.; Zhang, X.; Sun, G.; Xia, H.; Liang, D. Genome-Wide Identification of bHLH Gene Family and Screening of Candidate Gene in Response to Salt Stress in Kiwifruit. *Environ. Exp. Bot.* **2024**, *222*, 105774. [CrossRef]
54. Rajappa, S.; Krishnamurthy, P.; Kumar, P.P. Regulation of *AtKUP2* Expression by bHLH and WRKY Transcription Factors Helps to Confer Increased Salt Tolerance to *Arabidopsis thaliana* Plants. *Front. Plant Sci.* **2020**, *11*, 1311. [CrossRef]
55. Castilhos, G.; Lazzarotto, F.; Spagnolo-Fonini, L.; Bodanese-Zanettini, M.H.; Margis-Pinheiro, M. Possible Roles of Basic Helix-Loop-Helix Transcription Factors in Adaptation to Drought. *Plant Sci.* **2014**, *223*, 1–7. [CrossRef] [PubMed]

56. Khan, N.; Bano, A.; Ali, S.; Babar, M.A. Crosstalk amongst Phytohormones from Planta and PGPR under Biotic and Abiotic Stresses. *Plant Growth Regul.* **2020**, *90*, 189–203. [CrossRef]
57. Letunic, I.; Bork, P. 20 Years of the SMART Protein Domain Annotation Resource. *Nucleic Acids Res.* **2018**, *46*, D493–D496. [CrossRef]
58. Xie, T.; Chen, C.; Li, C.; Liu, J.; Liu, C.; He, Y. Genome-Wide Investigation of WRKY Gene Family in Pineapple: Evolution and Expression Profiles during Development and Stress. *BMC Genom.* **2018**, *19*, 490. [CrossRef]
59. Wang, Y.; Tang, H.; DeBarry, J.D.; Tan, X.; Li, J.; Wang, X.; Lee, T.; Jin, H.; Marler, B.; Guo, H. MCScanX: A Toolkit for Detection and Evolutionary Analysis of Gene Synteny and Collinearity. *Nucleic Acids Res.* **2012**, *40*, e49. [CrossRef]
60. Sudhakar Reddy, P.; Srinivas Reddy, D.; Sivasakthi, K.; Bhatnagar-Mathur, P.; Vadez, V.; Sharma, K.K. Evaluation of Sorghum [*Sorghum bicolor* (L.)] Reference Genes in Various Tissues and under Abiotic Stress Conditions for Quantitative Real-Time PCR Data Normalization. *Front. Plant Sci.* **2016**, *7*, 529. [CrossRef]
61. Farooq, M.A.; Zeeshan Ul Haq, M.; Zhang, L.; Wu, S.; Mushtaq, N.; Tahir, H.; Wang, Z. Transcriptomic Insights into Salt Stress Response in Two Pepper Species: The Role of MAPK and Plant Hormone Signaling Pathways. *Int. J. Mol. Sci.* **2024**, *25*, 9355. [CrossRef]

International Journal of
Molecular Sciences

MDPI

Article

Integrated Transcriptomic and Metabolomic Analysis of *G. hirsutum* and *G. barbadense* Responses to *Verticillium* Wilt Infection

Baoguang Xing [1,2,†], Pengtao Li [2,†], Yanfang Li [1,2], Bingkai Cui [1,2], Zhihao Sun [2], Yu Chen [2], Shaoliang Zhang [1,2], Qiankun Liu [2], Aiming Zhang [1,2], Liuan Hao [2], Xue Du [2], Xiaoyan Liu [2], Bei Wu [2], Renhai Peng [2,*] and Shoulin Hu [1,*]

[1] College of Agricultural, Tarim University, Alar 843300, China; 19937820150@163.com (B.X.); sybks10lyf25@163.com (Y.L.); cui20210521@163.com (B.C.); zsl001224@icloud.com (S.Z.); m904780569y@163.com (A.Z.)

[2] School of Biotechnology and Food Engineering, Anyang Institute of Technology, Anyang 455000, China; lipengtao1056@126.com (P.L.); zhpaxl@163.com (Z.S.); cyu990324@163.com (Y.C.); liuthundering@163.com (Q.L.); hla19971016@163.com (L.H.); duxue101199@163.com (X.D.); ysjgd548@163.com (X.L.); 18734321011@163.com (B.W.)

* Correspondence: aydxprh@163.com (R.P.); hushoulinghu@163.com (S.H.)

† These authors contributed equally to this work.

Abstract: *Verticillium* wilt (VW) caused by *Verticillium dahliae* (Vd) is a devastating fungal cotton disease characterized by high pathogenicity, widespread distribution, and frequent variation. It leads to significant losses in both the yield and quality of cotton. Identifying key non-synonymous single nucleotide polymorphism (SNP) markers and crucial genes associated with VW resistance in *Gossypium hirsutum* and *Gossypium barbadense*, and subsequently breeding new disease-resistant varieties, are essential for VW management. Here, we sequenced the transcriptome and metabolome of roots of TM-1 (*G. hirsutum*) and Hai7124 (*G. barbadense*) after 0, 1, and 2 days of V991 inoculation. Transcriptome analysis identified a total of 72,752 genes, with 5814 differentially expressed genes (DEGs) determined through multiple group comparisons. KEGG enrichment analysis revealed that the key pathways enriched by DEGs obtained from both longitudinal and transverse comparisons contained the glutathione metabolism pathway. Metabolome analysis identified 995 metabolites, and 22 differentially accumulated metabolites (DAMs), which were correlated to pathways including glutathione metabolism, degradation of valine, leucine, and isoleucine, and biosynthesis of terpenoids, alkaloids, pyridine, and piperidine. The conjoint analysis of transcriptomic and metabolomic sequencing revealed DAMs and DEGs associated with the glutathione metabolism pathway, and the key candidate gene *GH_D11G2329* (glutathione S-transferase, *GSTF8*) potentially associated with cotton response to VW infection was selected. These findings establish a basis for investigating the mechanisms underlying the cotton plant's resistance to VW.

Keywords: *Verticillium* wilt; transcriptome; metabolome; glutathione metabolism pathway

1. Introduction

Cotton is a major source of renewable natural fibers for the textile industry and of vegetable oil for the food/feed industry [1,2]. Of the more than 50 *Gossypium* species identified thus far, four were domesticated for fiber production around the world, namely *G. herbaceum* (A1), *G. arboreum* (A2), *G. hirsutum* (AD1), *G. barbadense* (AD2) [3]. However, cotton growth and development were greatly affected by a range of biotic and abiotic stresses, such as saline-alkali soil, high temperatures, drought, pests, and pathogens, resulting in considerable yield losses and fiber quality reduction [4–6]. *Verticillium* wilt (VW), a soil-borne vascular disease that severely affects plants [7], is principally attributed to the pathogen *Verticillium dahliae* (Vd), which has a broad range of hosts, including more than 700 plant species in the Malvaceae, Fabaceae, Solanaceae, and Rosaceae families [8]. Once colonized in cotton plants, Vd mycelia or spores quickly spread through the xylem vessels via transpiration, obstructing the transport of water and nutrients in the vascular tissues. It also secreted toxic substances, eventually leading to symptoms of vascular bundle browning, leaf yellowing, wilting, defoliation, and even plant death [9,10]. Given the presence of Vd hyphae within plant vascular tissues and the dormancy of spores in the form of microsclerotia, conventional soil amendments, solar irradiation exposure, or fungicide treatments have been proved ineffective in VW disease control [4,11,12]. According to statistical data from the Agricultural Technology Extension Service Center of the Ministry of Agriculture and Rural Affairs of China in 2021, the losses caused by VW accounted for 32.94% of total cotton losses caused by various diseases in China [13]. In recent years, climate change, continuous monoculture, and frequent introduction of new cotton varieties and hybrids have exacerbated the occurrence of VW worldwide [14].

Despite being derived from the full hybridization between the same diploid ancestors, *G. hirsutum* and *G. barbadense* exhibit significant differences in plant traits such as fiber quality, yield, and environmental adaptability [15]. *G. hirsutum*, approximately accounting for 95% of global cotton production [16], is the most extensively cultivated species due to its advantages of high yield and strong adaptation, while it lacks resistant germplasm and superior fiber quality genes. *G. barbadense*, on the contrary, is renowned for its superior fiber length, strength, and ultra fineness, and VW resistance, contributing less than 5% to global cotton production [17]. Given these characteristics, it remains of great significance to introduce resistance genes from *G. barbadense* into *G. hirsutum* to significantly enhance its resistance to VW [18]. Therefore, investigating the distinctions between *G. hirsutum* and *G. barbadense* at the transcriptional and metabolic levels could help us understand the resistance mechanism of cotton to VW, identify key genes for VW resistance, and provide genetic materials for molecular breeding of VW resistance.

Transcriptomics is a discipline that studies the transcriptional levels in the cells or tissues of target organisms at specific developmental stages, locations, or physiological conditions, to dissect the molecular mechanisms behind complex biological pathways and trait regulatory networks [19,20]. Currently, transcriptome sequencing technology (RNA-seq) has been widely applied in plant disease resistance mechanisms to understand the interactions between plants and pathogens, and to identify relevant metabolic pathways and key genes [21]. Some transcriptomic results have identified metabolic pathways and differentially expressed genes (DEGs) associated with plant disease resistance; among these, hypersensitive response regulation and potassium ion transmembrane transport were highly activated in rice plant defense against blast disease [22], while chitinase, cytochrome *P450*, and *GST* were highly up-regulated in peanut defense against rust disease [23]. Similarity, we also noticed that the *GST* gene cluster was involved in cotton-resistant response against VW [8]. Metabolomics can detect changes in metabolites in cells or tissues, which helps determine the relationship between metabolites and plant biochemistry. Metabolome sequencing is also a key link connecting gene function and phenotypic variations, making it an important discipline after genomics, transcriptomics, and proteomics [24,25]. When pathogens invaded organs or tissues, plant metabolites were subjected to a series of changes, which might generate signaling molecules to activate the immune system to suppress or kill pathogens. In the metabolomic studies of *Solanaceae* bacterial wilt, 63 differentially accumulated metabolites (DAMs) were found to be closely related to the biosynthetic pathways of plant hormones, phenylpropanoids, and flavonoids [26]. In the metabolites of *Zanthoxylum bungeanum* Maxim rust disease, most of the metabolites were proved to belong to lipid and lipid-like molecules, phenylpropanoids, and polyketones, organic heterocyclic compounds [27]. In the interaction between susceptible/resistant watermelon varieties and *Fusarium oxysporum*, significant changes were observed in related metabolites, such as jasmonic acid-isoleucine, methyl jasmonate, melatonin, and lysine, implying their important roles in pathogen defense [28]. To understand the complex and variable biological processes, the conjoint analyses of RNA-seq and metabolomics could better reveal the molecular mechanisms of regulatory processes at different expression levels. Referring to the previous studies, it was observed through joint analysis that silver nanoparticles (AgNPs) activate plant hormone signaling and glutathione metabolism pathways, protecting rice seedlings from blast fungus infection [29]; it was also observed that HmF6′H1, an acetyl-coenzyme A 6′-hydroxylase, can promote the accumulation of simple coumarins in *Heracleum moellendorffii* Hance to resist infection by powdery mildew [30], and that DEGs and DAMs related to glutathione metabolism pathways can make great contributions to the response of rice to salt stress [31].

In this study, VW-susceptible TM-1 and VW-resistant Hai7124 were selected to perform transcriptome and metabonomic analyses. Root samples were separately collected at 0, 1, and 2 days after V991 inoculation for sequencing analysis. Multiple pair comparisons among the samples generated a number of DEGs and DAMs, which were then submitted to enrichment analysis in Kyoto Encyclopedia of Genes and Genomes (KEGG) pathways. Combining identified key candidate genes and metabolites, this study lays the foundation for the functional validation of candidate genes in subsequent research.

2. Results

2.1. Sequencing of the Transcriptome and Analysis of Quality Control

To systematically elucidate the critical genes and signaling pathways in plant defense mechanisms against VW infection, 18 RNA-seq libraries were constructed from root samples of the representative varieties TM-1 and Hai7124. The samples were taken at 0, 1, and 2 days after inoculation (DAI) of V991. After removing low-quality reads, a total of 802.88 million clean reads were obtained, with an average clean read (M) of 38.12–49.15 per sample. The Q30 values of all the samples exceeded 91.95%, and the GC content ranged from 43.43% to 45.61%. Mapping the clean reads of each sample to the reference genome showed alignment efficiencies ranging from 94.04% to 97.27% (Table S1).

Based on the annotated reference genome, a total of 72,752 expressed genes were identified in our RNA-Seq dataset. The expression levels of these genes were quantified using the number of fragments per kilobase of transcript per million mapped reads (FPKM). To confirm the accuracy of the RNA-seq data, principal component analysis (PCA) across the 18 samples was performed, and two main components, namely PC1 and PC2, were obtained, which contributed 34.71% and 22.85% of the total variation, respectively. The three biological replicates at different time points showed a high similarity between different samples (Figure 1A). Pearson's correlation coefficient (PCC) analysis on all the RNA-seq samples showed a gene expression correlation of over 90% for each sample (Figure 1B). These results suggest that the transcriptome data are of high quality which meets the standards for the subsequent DEG identification and their functional enrichment analysis.

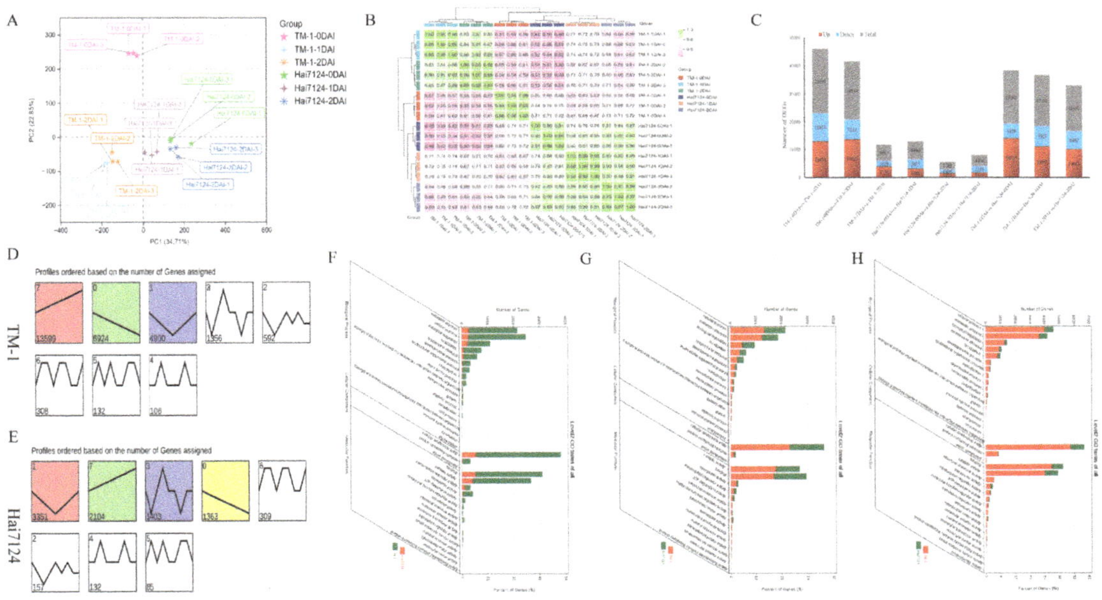

Figure 1. Analysis of Expression Patterns and Identification of Genes with Differential Expression (**A**) PCA analysis of the 18 samples. (**B**) Correlation clustering heatmap of the 18 samples. (**C**) DEGs in the 9 pairwise comparisons. (**D**) Trend analysis results of DEGs in TM-1. (**E**) Trend analysis results of DEGs in Hai7124. (**F**) GO enrichment results of DEGs in expression profile 0 for TM-1 and Hai7124. (**G**) GO enrichment results of DEGs in expression profile 1 for TM-1 and Hai7124. (**H**) GO enrichment results of DEGs in expression profile 7 for TM-1 and Hai7124. Totally, 1692 DEGs were identified in the comparison between different infection stages of TM-1 at 0, 1, and 2 days after V991 inoculation (Figure 2A). KEGG enrichment analysis of these DEGs showed that the enriched pathways included ribosome, terpenoid biosynthesis, plant hormone signal transduction, circadian rhythm-plant and glutathione metabolism (Figure 2B). Similarly, 293 DEGs were identified in the comparison between different infection stages of Hai7124 at 0, 1, and 2 days after V991 inoculation (Figure 2C). KEGG enrichment analysis of these DEGs showed that the enriched pathways included biosynthesis of secondary metabolites, terpenoid biosynthesis, isoflavonoid biosynthesis, carotenoid biosynthesis, plant–pathogen interactions, and glutathione metabolism (Figure 2D). The comparison between the two varieties at the same time point after V991 inoculation resulted in 5814 DEGs (Figure 2E), and the enriched KEGG pathways included biosynthesis of secondary metabolites, glutathione metabolism, sesquiterpenoid and triterpenoid/terpenoid compound biosynthesis, and isoflavonoid biosynthesis (Figure 2F). The metabolism of glutathione was enriched in both comparisons of DEGs between time points and between varieties at same time point, suggesting its potentially important roles in cotton resistance against Vd invasion.

2.2. DEG Identification and Expression Pattern Analysis

Nine pairwise comparisons were performed on all the samples, resulting in 39,976 DEGs (Figure 1C). By comparing the different infection stages within the susceptible TM-1 and within the resistant Hai7124, there were more DEGs identified in the former (TM-1-0DAI vs. TM-1-1DAI, TM-1-0DAI vs. TM-1-2DAI, and TM-1-1DAI vs. TM-1-2DAI) than in the latter (Hai7124-0DAI vs. Hai7124-1DAI, Hai7124-0DAI vs. Hai7124-2DAI, and Hai7124-1DAI vs. Hai7124-2DAI). A similar phenomenon was also observed during early infection (0DAI to 1DAI). The difference between TM-1 and Hai7124 was also manifested in the period when the minimum number of DEGs appears, with the former showing the smallest DEGs in the comparison of TM-1-1DAI vs. TM-1-2DAI, while the latter showed

Hai7124-0DAI vs. Hai7124-2DAI. For the pairwise comparisons of different varieties at the same infection stage (DAI), it was observed that the pair TM-1-0DAI vs. Hai7124-0DAI had the highest number of DEGs, while the pair TM-1-2DAI vs. Hai7124-2DAI had the lowest number of DEGs. In addition, more up-regulated DEGs were identified in the resistant variety than in susceptible varieties.

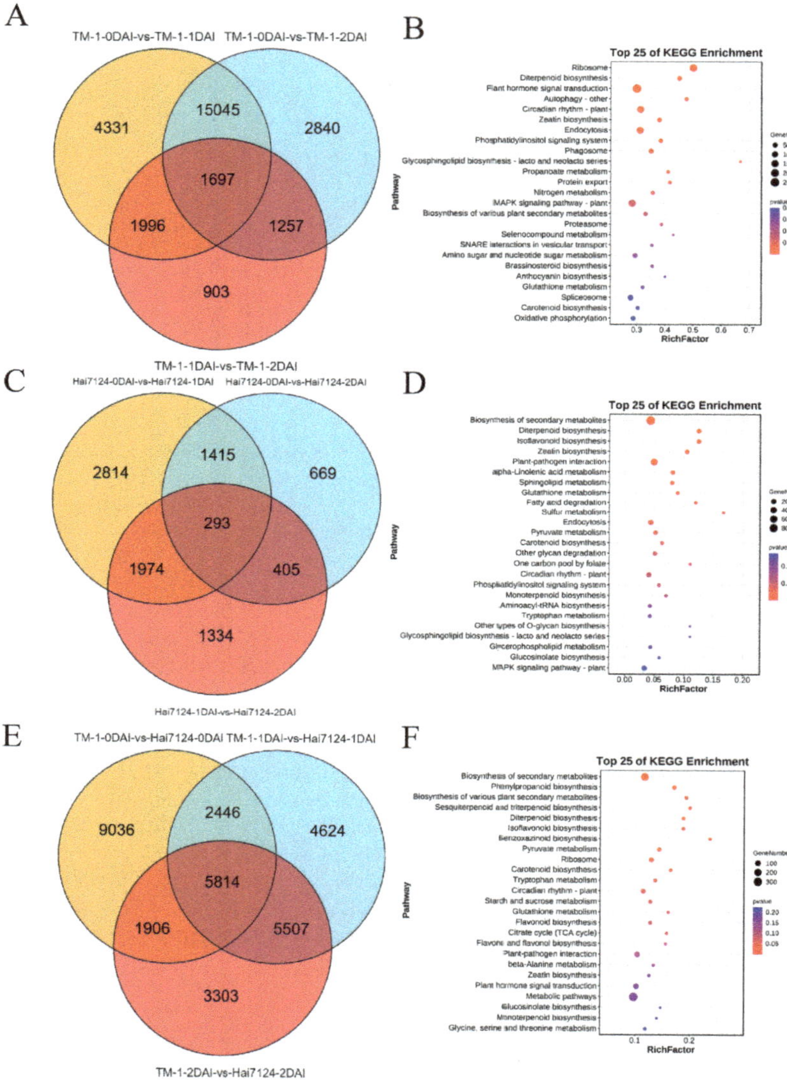

Figure 2. Analysis of Functional Enrichment for Genes with Differential Expression (**A**) Venn diagram of 1697 DEGs across three different infection stages in TM-1. (**B**) KEGG enrichment analysis results of in common DEGs in TM-1. (**C**) Venn diagram of 293 DEGs across three different infection stages in Hai7124. (**D**) KEGG enrichment analysis results of in common DEGs in Hai7124. (**E**) Venn diagram of 5814 DEGs at the same infection stage across different materials. (**F**) Enrichment analysis results in common DEGs at the same infection stage across different material.

To investigate the spatiotemporal expression patterns of genes, 28,069 DEGs in TM-1 and 8904 genes in Hai7124 were separately submitted to short time-series expression miner (STEM) analysis (Figure 1D,E). The findings indicated that a substantial majority of the DEGs in TM-1, 90.57% of the total 28,069 DEGs, were significantly clustered into 3 distinct expression profiles. Among these profiles, in profile 7, which clustered 13,599 DEGs, 48.44%, showed a continuous up-regulation trend; in profile 0, which clustered 6924 DEGs, 24.67%, showed an overall down-regulation trend; and in profile 1, which clustered 4900 DEGs, 17.46%, showed an initial down-regulation followed by an up-regulation trend. On the other hand, 8221 DEGs (92.33% of the total) in Hai7124 were clustered in 4 expression profiles. Among these profiles, profile 1, which clustered3,351 DEGs, 37.63%, showed an initial down-regulation followed by sustained up-regulation trend; profile 7, which clustered 2104 DEGs, 23.63%, showed a continuous up-regulation trend; profile 0, which clustered 1363 DEGs, 15.31%, demonstrated a sustained down-regulation trend; and profile 3, which clustered 1403 DEGs, 15.76%, showed a periodic pattern of initial down-regulation, followed by up-regulation, and then another down-regulation.

2.3. Functional Enrichment Analysis of DEGs

The functional enrichment analyses were performed on the DEGs in the significantly clustered expression profiles (0, 1, and 7) in TM-1 and Hai7124 using the GO and KEGG databases. In profile 0, the DEGs were enriched in metabolic processes, cellular processes, cellular anatomical entities, and catalytic activities in the first three GO terms of biological processes (BP), cellular components (CC), and molecular functions (MF) (Figure 1F). As profile 0 exhibited a sustained down-regulation trend, which was more prevalent in susceptible TM-1 than in resistant Hai7124, KEGG enrichment analysis showed that the common pathways in both TM-1 and Hai7124 included hormone signal transduction, circadian rhythm-plant, MAPK signaling pathway-plant, and phosphatidylinositol signaling system (Figure S1A).

In profile 1, the DEGs were enriched in the GO terms of processes related to cellular processes, metabolic processes, cellular anatomical entities, catalytic activities, and binding (Figure 1G). KEGG enrichment analysis revealed that the common pathways in both genotypes included flavonoid biosynthesis, circadian rhythm-plant, biosynthesis of various plant secondary metabolites, and plant hormone signal transduction (Figure S1B). In particular, the significant enrichment of profile 1 DEGs in the biosynthesis of secondary metabolites and plant hormone signal transduction suggested their crucial role in resisting pathogen invasion.

In profile 7, the DEGs were enriched in the GO terms of cellular processes, metabolic processes, cellular anatomical entities, binding, and catalytic activities (Figure 1H). KEGG enrichment analysis revealed that the key pathways in TM-1 were related to ribosome biogenesis, carotenoid biosynthesis, base excision repair, nitrogen metabolism, and endocytosis, which are associated with growth and development. In contrast, the key pathways in Hai7124 were associated with plant defense, such as ABC transporters, phenylpropanoid biosynthesis, flavonoid biosynthesis, and plant–pathogen interactions (Figure S1C).

2.4. Metabolomic Comparative Analysis of Different Strains

To investigate the metabolic differences between TM-1 and Hai7124 during cotton VW infection, untargeted metabolomic sequencing was performed on the 18 samples, and quality control analysis was performed on the sequencing data. PCA analysis showed that PC1 scored 20.81% and PC2 scored 13.81%, indicating significant metabolic phenotype differences between the two samples, and high similarity among the three biological replicates (Figure S2A). The correlation coefficients exceeding 0.8 in the correlation analysis of the

QC samples suggested significant consistency (Figure S2B). A total of 995 metabolites were identified in the 18 samples, which were categorized into 11 major classes: alkaloids and derivatives (7, 0.7%); benzenoids (38, 3.82%); lignans, neolignans, and related compounds (1, 0.1%); lipids and lipid-like molecules (215, 21.51%); nucleosides, nucleotides, and analogues (7, 0.7%); organic acids and derivatives (84, 8.44%); organic nitrogen compounds (13, 1.31%); organic oxygen compounds (30, 3.02%;, organoheterocyclic compounds (60, 6.03%); others (514, 51.66%); and phenylpropanoids and polyketides (27, 2.71%) (Figure 3A).

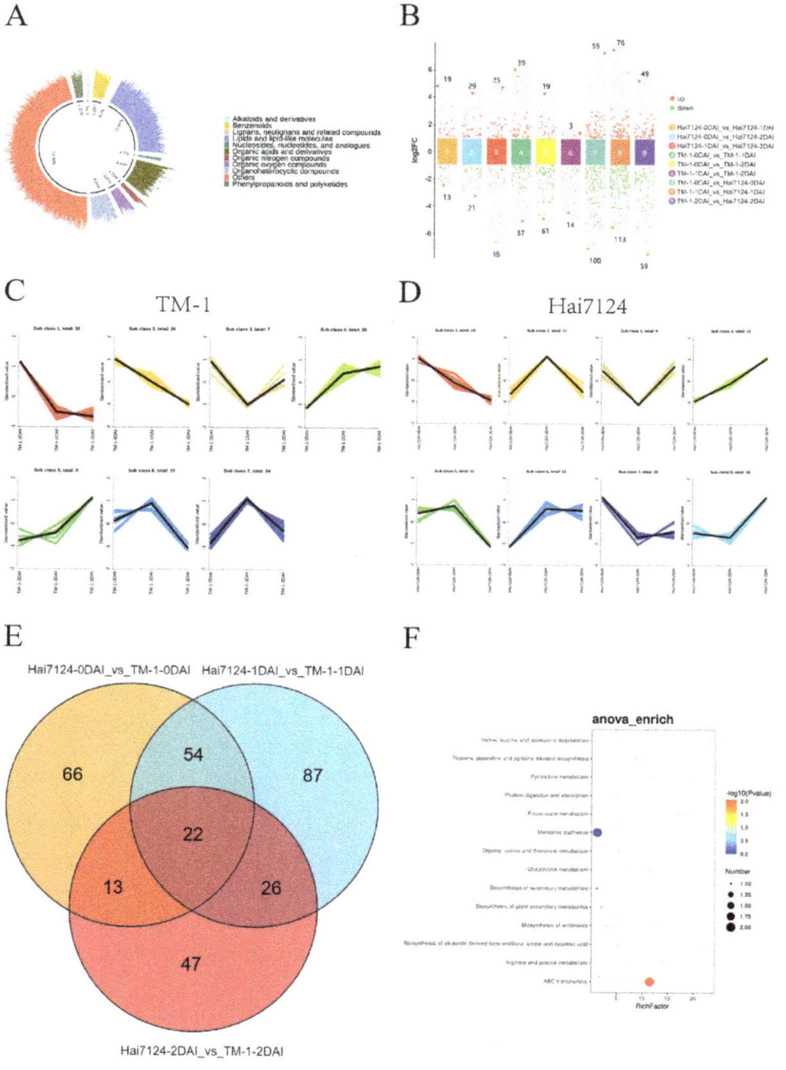

Figure 3. Comparative Metabolomics Analysis of Different Strains (**A**) Secondary classification of 995 metabolites. (**B**) Scatter plots of differential analysis across 9 groups. (**C**) Clustering analysis results of DAMs in TM-1. (**D**) Clustering analysis results of DAMs in Hai7124. (**E**) Venn diagram of 22 common DAMs at the same infection stage across different materials. (**F**) Enrichment analysis results of in common DAMs at the same infection stage across different materials.

Based on the general threshold criteria of Variable Importance in Projection (VIP) equal to or greater than 1, Fold Change Screening Value equals to 2, and *p*-value less than 0.05, a total of 402 DAMs were identified through nine groups of pairwise comparisons. In the comparisons between different time points after V991 inoculation in the same variety, it was found that the largest numbers of DAMs were identified in the groups of TM-1-0DAI vs. TM-1-2DAI (80 DAMs) and Hai7124-1DAI vs. Hai7124-2DAI (50 DAMs), while the least numbers of DAMs were observed in TM-1-1DAI vs. TM-1-2DAI and Hai7124-0DAI vs. Hai7124-1DAI (Figure 3B). Significantly, different varieties showed DAMs at the same infecting time points; the maximum and minimum DAMs were identified in the groups of TM-1-1DAI vs. Hai7124-1DAI (189 ones) and TM-1-2DAI vs. Hai7124-2DAI (108 ones), respectively. After removing duplicates, a total of 137 and 97 DAMs were separately obtained in TM-1 and Hai7124 via different infection time point comparisons. Subsequently, via K-means clustering analysis, the DAMs of TM-1 and Hai7124, were classified into 7 and 8 clusters, respectively (Figure 3C,D). In TM-1, subclusters 4 and 5 demonstrated a consistent increasing expression trend with time advances after V991 infection. Conversely, subclusters 1 and 2 showed a decreasing expression trend with time advances after infection. In Hai7124, subcluster 4 displayed a consistent increasing expression trend with time advances after infection. Subcluster 1 showed a decreasing expression trend with time advances. In both TM-1 and Hai7124, carboxylic acid derivatives and sphingolipids showed a decreasing expression trend with time advances after V991 infection V991, while glycerophospholipids and fatty acyls exhibited a continuous increasing expression trend (Table S2). It could be speculated that glycerophospholipids and fatty acyls play important roles in early resistant responses against Vd invasion. Comparison of DAMs between TM-1 and Hai7124 at the identical time point identified 22 common DAMs (Figure 3E). KEGG enrichment analysis indicated that the majority of common DAMs were significant enriched in pathways of valine, leucine, and isoleucine degradation, glutathione metabolism, metabolic pathways, and propanoate metabolism (Figure 3F).

2.5. Conjoint Analysis of DEGs and DAMs

To dissect the macroscopic developmental processes of biological systems, a joint analysis of transcriptomic and metabolomic data of the two cultivars was performed in this study. The results revealed that the pathways that DAMs and DEGs co-enriched included glutathione metabolism, valine, leucine, and isoleucine degradation, propionate metabolism, ABC transport, secondary metabolite biosynthesis, and metabolic pathways (Figure S3). Further calculation of Pearson correlation coefficients was performed for the 5814 DEGs and 22 DAMs, generating nine quadrant plots (Figure 4A–C) and a correlation clustering heatmap (Figure 4D). The expression patterns of DEGs and DAMs in the first and ninth quadrants were opposite, implying that the changes in these DAMs might be negatively regulated by DEGs. In the third and seventh quadrants, the expression patterns of DEGs and DAMs showed consistency, implying that the variations of these DAMs might be positively regulated by the corresponding DEGs. In addition, during the processes of V991 infection in TM-1 and Hai7124 (0, 1, and 2 days), the metabolite putrescine was positively regulated by *GH_D04G0634*, *GH_D11G2329*, and *GH_D13G1106*. The correlation clustering heatmap results showed that the red part represents a positive correlation between DEGs and DAMs, while the green part represents a negative correlation. Notably, putrescine was significantly positively correlated with *GH_D11G2329*.

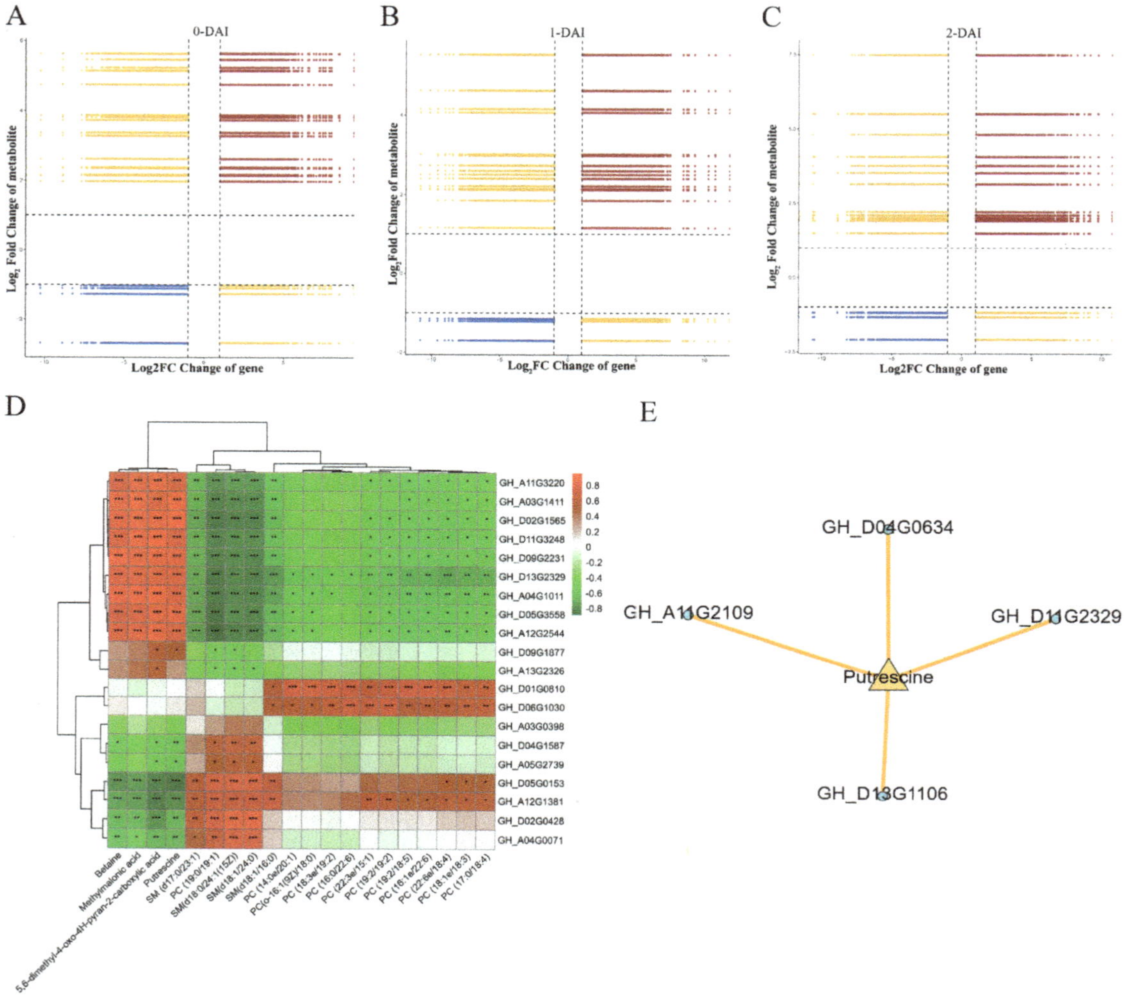

Figure 4. Conjoint analysis of DEGs and DAMs (**A**) Nine-quadrant plot of DEGs and DAMs in response to V991 infection in cotton at day 0 (In the third quadrant, the expression patterns of DEGs and DAMs are consistent and up-regulated at the same time. In the seventh quadrant, the expression patterns of DEGs and DAMs are consistent and down-regulated simultaneously. In the first and ninth quadrants, the expression patterns of DEGs and DAMs are opposite, one up-regulated and the other down-regulated. (**B**,**C**) comments are the same). (**B**) Nine-quadrant plot of DEGs and DAMs in response to V991 infection in cotton at day 1. (**C**) Nine-quadrant plot of DEGs and DAMs in response to V991 infection in cotton at day 2. (**D**) Correlation clustering results of DEGs and DAMs in response to V991 infection in cotton(Use *p*-values to indicate significant differences. * means the significant level is less than 0.05, ** means the significant level is less than 0.01, and *** means the significant level is less than 0.001). (**E**) Network diagram of annotated DEGs involved in glutathione metabolism.

Based on the joint analysis of DEGs and DAMs, it could be speculated that the glutathione metabolism pathway is vital for resisting Vd invasion. The correlation analysis revealed a significant association between putrescine and corresponding DEGs, such as the gene-encoding glutathione S-transferase (GST). Network analysis demonstrated that DEGs and DAMs in the glutathione metabolism signaling pathway formed hub genes in the network, supporting the aforementioned correlation analysis (Figure 4E and Table S3). Additionally, a schematic diagram of the glutathione metabolism pathway was constructed (Figure 5). In total, 9 DEGs were screened in TM-1 and Hai7124 at 0, 1, and 2 DAI through transcriptomic analysis, which belonged to GST; glutathione dehydrogenase/transferase (DHAR); peroxiredoxin 6 (PRDX6); isocitrate dehydrogenase 1 (IDH1); and γ-glutamylcyclotransferase (GGCT). Meanwhile, 5 DAMs involved in this pathway were identified by metabolomic analysis, namely putrescine, spermine, glycine, L-cysteine, and 5-oxoproline.

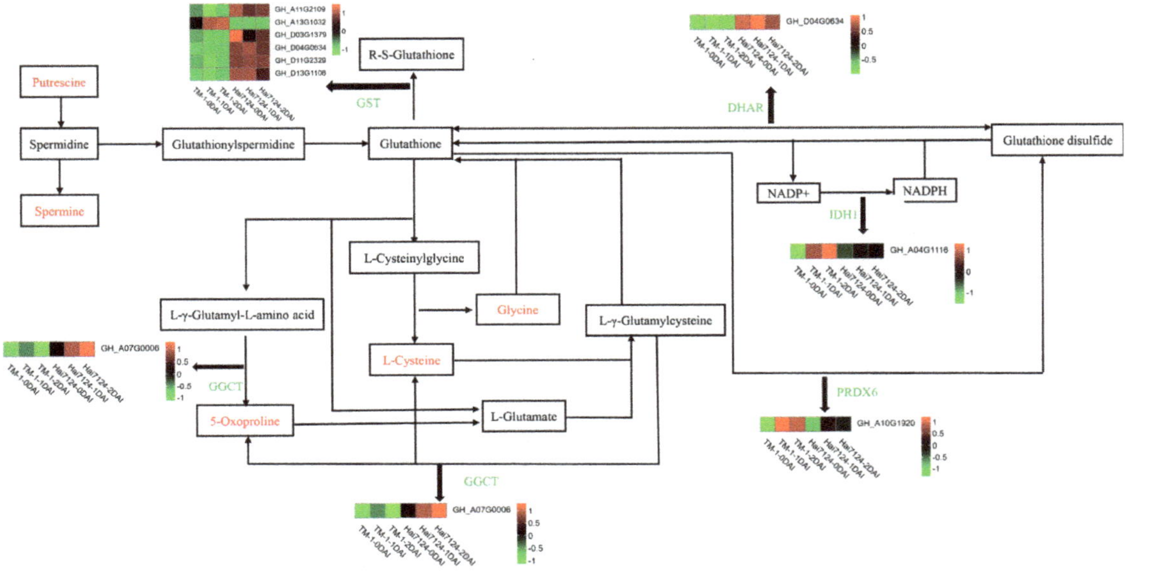

Figure 5. Signaling Pathway of Glutathione Metabolism and Heatmap of Selected Annotated DEGs and DAMs.

In the findings from the joint analysis of transcriptomics and metabolomics, the DEG *GH_D11G2329* (*GHGSTF8*) participating in the glutathione metabolism pathway showed an up-regulated expression pattern in TM-1 during V991 infection, indicating a potential role in VW resistance.

2.6. Verification of Gene Expression qRT-PCR

To validate the accuracy of the transcriptomic analysis results, 20 genes were randomly selected and analyzed using quantitative real-time polymerase chain reaction (qRT-PCR) experiments (Figure 6). Detailed information on the primer design is listed in Table S4. The housekeeping gene *GhUBQ7* (DQ116441) served as a reference control for the analysis of relative quantification. The results confirmed that the expression trends observed in qRT-PCR were consistent with the transcriptomic sequencing results, indicating the reliability of the transcriptomic sequencing.

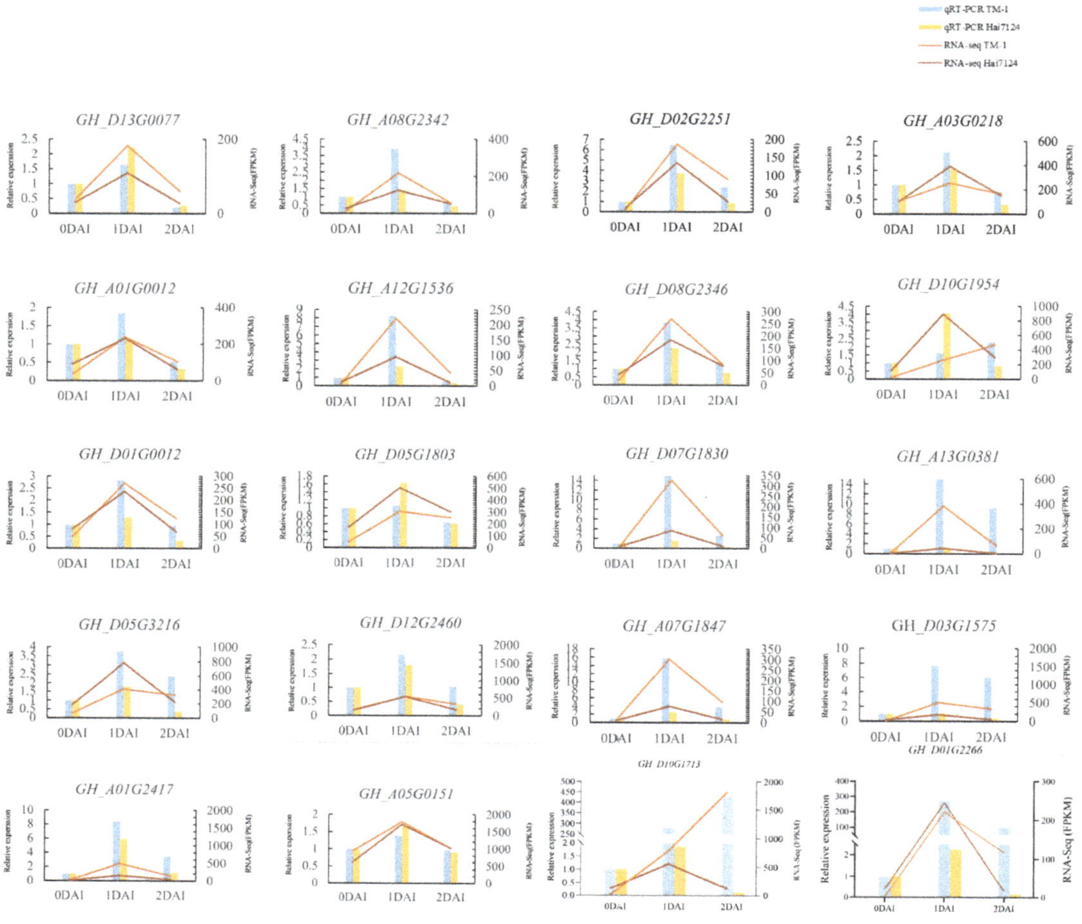

Figure 6. Results from qRT-PCR Randomly Chosen Genes.

3. Discussion

Cotton, as an important economic crop, makes great contributions to global economic development. Currently, more than 50 species have been identified within the cotton genus, among which *G. hirsutum* and *G. barbadense* are widely cultivated allotetraploid species [32]. *G. hirsutum* has the traits of high yields yet average fiber quality, while *G. barbadense* exhibits the opposite characteristics [33]. However, the VW disease could cause annual yield losses of approximately 10–35%, since the majority of *G. hirsutum* varieties are susceptible to VW, ultimately posing a significant economic threat worldwide [34]. Conventional control measures such as crop rotation or chemical fumigation have been proved ineffective against VW, due to fact that the pathogen survives in the environment in the form of hyphae, spores, and microsclerotia. It is stimulated by cotton root exudates and begins to initiate infection at the root tips. During the infection process, the hyphae penetrate the root epidermal cells, move through the xylem, and colonize the vascular tissues, eventually reaching the petiole base. This can lead to browning of the cotton vascular bundles, leaf yellowing, wilting, defoliation, and even plant death [6,35]. Moreover, the pathogens exist in the soil in the form of dormant spores, enabling them to infect multiple generations of crops [36]. Therefore, it is necessary for breeders to focus on developing resistant varieties to meet production needs. However, the condition we faced is that *G. hirsutum* varieties lack resistance genes and quantitative trait loci (QTL) associated with VW [2]. To address this issue, introducing genes of VW resistance from *G. barbadense* into *G. hirsutum* provides a useful strategy to significantly enhance cotton defense [18]. Meanwhile, key SNP loci related to VW resistance in *G. hirsutum* and *G. barbadense* could be used to identify resistance genes and breed resistant varieties. For instance, critical SNP substitutions in *GhWAKL14* (A/C) resulted in amino acid changes (S/R), enhancing resistance of *G. hirsutum* but rendering *G. barbadense* susceptible to *Fusarium wilt* [37]. Similarly, critical SNP substitutions in *GhGLR4.8* (C/A) led to amino acid alterations (L/I), enhancing *Fusarium wilt* resistance in *G. hirsutum* [38]. Similar methods have also been applied to the defense mechanisms of economically important crops against pathogens, such as onions [39], wheat [40], soybeans [41], and citrus [42].

3.1. RNA-Seq of Cotton in Response to VW

Currently, RNA-seq has been widely applied to study the interaction mechanisms between plants and pathogens, such as in maize [43], wheat [44], and rice [45]. In this study, RNA-seq analysis was performed on the roots of TM-1 and Hai7124 at 0, 1, and 2 DPAs after VW infection. Pairwise analysis between different samples identified a total of 39,976 DEGs. Trend analysis of temporal expression patterns of TM-1 and Hai7124 DEGs revealed three common expression patterns. Expression profile 0 showed a continuous downregulation trend, which was more prevalent in the susceptible TM-1 than in the resistant Hai7124. Expression profile 7 exhibited a continuous upregulation trend, with TM-1 showing more key pathways related to growth and development, including ribosome replication, base excision repair, nitrogen metabolism, and endocytosis. In contrast, Hai7124 showed more active pathways related to plant defense, including secondary metabolite biosynthesis, phenylpropane biosynthesis, flavonoid biosynthesis, and plant–pathogen interactions. The findings indicate that Hai7124 may possess the key genes conferring resistance to VW. Research indicates that during VW infection, cotton notably increases the accumulation of secondary metabolites [46]. The phenylpropanoid metabolic pathway produces numerous natural compounds, such as flavonoids and lignin, which are extensively involved in various physiological activities in plants [47]. In cotton, the silencing of the *GhCOMT* gene results in reduced lignin content and decreased resistance to Vd [48]. Conversely, overexpression of *GhLac1* leads to increased lignin content and enhanced resistance to Vd [49]. Flavonoid biosynthesis is crucial for muskmelon defense against downy mildew [50]. KEGG-enriched pathways via longitudinal and transverse comparisons of TM-1 and Hai7124 DEGs included the glutathione metabolism pathway. Studies indicated that glutathione metabolism is crucial for antioxidant stress and pathogenicity within *Fusarium graminearum* [51], cadmium tolerance in *Arabidopsis* [52], and early resistance to *Bacterial wilt* in tobacco [53]. This suggested that our RNA-seq results highly align with these findings, indicating them to be crucial for combating VW infection.

3.2. Metabolomic Analysis of Cotton in Response to VW

Currently, metabolomics has been primarily applied to analyze the types, quantities, intrinsic factors, and variations in plant metabolites under different environmental conditions, making it a valuable tool for studying plant–pathogen interactions [54]. In this study, we identified altogether 402 DAMs through pairwise comparisons. K-means clustering analysis revealed that the expression levels of glycerophospholipids and fatty acyls in TM-1 and Hai7124 increased continuously after the Vd infection. Previous studies have indicated that the accumulation of glycerophospholipids is vital for barley under salt stress [55]; compounds composed of flavonoids, amino acids, and glycerophospholipids in Tibetan hulless barley respond to osmotic stress (cold, high salt, and drought) [56]; metabolites such as fatty acyls are significantly induced in salt-tolerant wild rice compared to salt-sensitive rice [57]. Therefore, it is speculated that glycerophospholipids and fatty acyls play key roles in early resistance to VW invasion. In the longitudinal comparison analysis of TM-1 and Hai7124 infected with V991, a total of 22 DAMs were identified. KEGG analysis of these 22 DAMs indicated a significant enrichment in the glutathione metabolism pathway. Currently, research on cotton VW has mainly focused on gene expression levels, with fewer studies on the metabolite level. The findings of this study complement this aspect and establish a basis for cotton resistance to VW. Metabolomic analysis has been widely applied in other plant disease resistance studies. For example, under *Phytophthora infestans* stress, resistant potato varieties accumulate salicylic acid, phenylpropanoids, terpenoids, and arachidonic acid, which is vital for thickening the cell wall [58]. In multiomics studies, metabolites such as linolenic acid, vitamin E, choline, and riboflavin contribute to enhancing resistance to rice blast disease [59]. In *Fusarium wilt* disease research in wheat, the accumulation of proline (Pro) and alanine (Ala) enhances wheat resistance, whereas a decrease in cysteine (Cys) exacerbates sensitivity [60]. After infection with tomato yellow leaf curl virus, significant changes in the abundance of polyamines, phenols, and indoles were detected in resistant and susceptible tomato varieties. Salicylic acid accumulated significantly in resistant tomatoes, suggesting its important role in tomato disease resistance [61].

3.3. Conjoint Analysis of Metabolome and Transcriptome in Cotton Response to VW

When pathogens invade plants, they trigger the expression of genes associated with plant defense mechanisms, thereby synthesizing disease-resistant proteins or other compounds that resist pathogen invasion. Therefore, the combined analysis of transcriptome and metabolome is utilized to screen target genes and to elucidate target metabolic pathways [62]. For example, the enzymes encoded by the tau cluster of the GST family may play a crucial role in maintaining the delicate balance between the production and scavenging of H_2O_2. This establishes a new equilibrium that enhances the functions of CAT and POD, thereby strengthening the plant's response to Vd [8]. This study identified nine enriched pathways in cotton response to VW through combined KEGG analysis of transcriptome and metabolome. The longitudinal and transverse pairwise analysis of DEGs in the transcriptome and the longitudinal comparison analysis of DAMs in the metabolome collectively enriched the glutathione metabolism pathway. The results of the nine-quadrant map and correlation clustering heatmap indicated that the metabolite spermine in the glutathione metabolism pathway was positively regulated by the annotated DEGs in this pathway. Furthermore, the construction of the glutathione metabolism pathway diagram revealed the involvement of spermine, spermidine, glutamic acid, L-cysteine, and 5-oxoproline in this pathway. Exogenous application of polyamines (spermine, spermidine, and spermine) significantly increased the activities of superoxide dismutase (SOD), catalase (CAT), and glutathione reductase (GR), maintaining their internal homeostasis of apricot and enhancing its resistance to black spot disease [63]. Glycine can increase soil absorption of heavy metals, alleviating plant toxicity caused by heavy metals [64]. Hydrogen sulfide and cysteine can alleviate cadmium-induced growth inhibition in Arabidopsis and enhance its tolerance to cadmium [65]. In the study of necrotic pathogens, it was found that the infected alfalfa significantly accumulated 5-oxoproline and malic acid, possibly enhancing its resistance to necrotic pathogens [66]. RNA-seq identified GST, DHAR, PRDX6, IDH1, and GGCT as participating in this pathway, with correlation analysis showing a significant association between putrescine and the gene-encoding GST. In this study, the DEG *GH_D11G2329* (*GSTF8*) was selected from the glutathione metabolism pathway. GSTs are a multifunctional enzyme encoded by a multigene superfamily that plays a detoxification role in various plants [67]. In the study of salt tolerance in the rice landrace HD961, a comprehensive analysis of the transcriptome and metabolome revealed that the glutathione metabolism pathway was significantly enriched, and a total of seven metabolites and 48 DEGs identified, most of which belong to the GST family. Among these, L-cysteine was the most active compound in response to salt stress. *GSTU2* showed a significant correlation with metabolites such as L-cysteine, spermidine, and L-ornithine. Additionally, there was a notable correlation between GSH and 16 genes. These findings suggest that these specific genes and glutathione metabolites play a crucial role in rice's response to salt stress [31]. Previous research has emphasized the vital importance of GSTs in combating peanut rust [23], cotton VW [8], and wheat *Powdery mildew* [68]. SA (salicylic acid) is one of the primary defense-related hormones involved in the response to various biotrophic and hemibiotrophic pathogens, and plays an auxiliary role against necrotrophic pathogens [69]. Vd initially behaves as a biotrophic pathogen during the early stages of infection but shifts to a necrotrophic lifestyle in the later stages. Therefore, the SA signaling pathway is required to confer resistance to Vd [13]. Additionally, the complex metabolic pathways of plant secondary metabolites often intersect with SA signaling. In the glutathione metabolism pathway, *GaGSTF9* may regulate reactive oxygen species content via catalyzing the reduction in tripeptide glutathione (GSH), thereby positively influencing cotton resistance to Vd and SA content [70]. Additionally, the constitutive overexpression of the gene encoding the key rate-limiting enzyme in spermine biosynthesis, *GhSAMDC* (S-adenosylmethionine decarboxylase), in *Arabidopsis thaliana* enhances resistance to Vd by activating SA signaling pathways [71]. Therefore, it is believed that the glutathione metabolism pathway plays an important role in early defense mechanisms of cotton against VW infection. In consideration of the relative disadvantages of traditional hybridization, self-cross, and backcross techniques adopted between TM-1 and Hai7124, such as long period, large population, and heavy workload, the candidate genes are even suggested to perform the genetic transformation for further cotton breeding. The more precise and efficient methods, including transgenic modification and gene editing by CRISPR-cas9 system, could break the reproductive isolation and trait linkage, finally achieving synchronous improvement in fiber yield, quality, and multiple resistances.

4. Materials and Methods

4.1. Cultivation of Materials and Preparation of Strains

The cotton varieties TM-1 (*G. hirsutum*) and Hai7124 (*G. barbadense*) were preserved in our laboratory (TM-1 and Hai7124 were provided by Dr. Fang Liu's research group at the Institute of Cotton Research). TM-1 and Hai7124 are two important cultivated cotton varieties. TM-1 exhibits stable agronomic traits and medium-level fiber quality, and it does not contain Bt, making it an excellent recipient for the introduction of exogenous genes. It is frequently used in constructing genetic linkage maps, serving as background material for developing substitution lines [72]. Hai7124 is characterized by its ultra-long or superfine quality fibers and exhibits strong resistance to VW [17]. The VW pathogen strain used was the highly pathogenic isolate V991 of Vd Kleb (V991 was provided by Professor Yingfan Cai from Henan University). The seeds of both cotton varieties were defuzzed using concentrated sulfuric acid, thoroughly washed with water, and air-dried. Uniform and fully developed seeds were selected and planted in a sterilized substrate composed of sand and vermiculite in a ratio of 3:2 in the greenhouse of Anyang Institute of Technology. The cultivation took place within a controlled greenhouse environment maintained at a temperature of 28 °C, adhering to a photoperiod of 16 h of light followed by 8 h of darkness, maintaining a relative humidity of 68%. Once the first true leaves of the cotton plants fully expanded, root tissues with consistent growth were collected as the wild-type (WT) control treatment. These roots were quickly transferred into collection tube that had been pre-cooled in liquid nitrogen to ensure the preservation of the samples. They were subsequently cryopreserved in liquid nitrogen and stored in an ultra-low temperature freezer set at −80 °C for future use. Additionally, we selected cotton plants exhibiting consistent growth for infection using the V991 strain (2 mL of V991 per plant), and root tissues were collected at days 1 and 2 post-infection. We rinsed the roots with tap water to remove soil, and then we collected the primary roots of the cotton plants below the hypocotyl. We quickly wrapped the root in aluminum foil and placed it into a pre-cooled collection tube with liquid nitrogen (0.5 g per replicate, consistent with the collection method for WT). One portion of the root tissues was allocated for transcriptome sequencing, while another portion was designated for metabolome sequencing.

The preserved pathogen strain was cultured on potato dextrose agar (PDA) plates. These plates were placed in an inverted position in a dark incubator with temperature controlled at 25 °C for 7 days. Subsequently, approximately 2 mm^2 fungal plaques were transferred to Czapek Dox liquid medium using sterilized toothpicks and placed at 25 °C in the dark incubator with a shaking speed of 150 r/min for 7 days. Prior to inoculation, the mycelium was filtered through four layers of gauze. The density of the resulting spore suspension was then measured using a hemocytometer. Then the suspension was diluted with ddH_2O to a concentration of 1×10^7 spores/mL, and the cotton roots were inoculated [73].

4.2. Extraction of RNA and Construction of Library

The FastPure Universal Plant Total RNA Isolation Kit (RC411) from Vazyme was used to isolate the total RNA. The RNA's concentration and quality were assessed using a NanoDrop 2000 spectrophotometer, and agarose gel electrophoresis was performed to check for RNA contamination and degradation. Samples that passed the quality standards were forwarded to Igenebook Biotechnology Co., Ltd., Wuhan, China for library construction and sequencing.

The library construction used the NEBNext® Ultra™ RNA Library Prep Kit (NEB, Ipswich, MA, USA), and PCR amplification was performed to obtain cDNA libraries. A Qubit 4.0 Fluorometer was employed to measure the concentration of the libraries, while the fragment sizes were assessed with a QSep400. The Qubit Fluorometer was again used to quantify the libraries. Once the libraries passed quality control, 18 cDNA libraries (from two cotton strains at 0, 1, and 2 days) were sequenced using the combinatorial probe-anchor synthesis (cPAS) technology.

4.3. Transcriptome Sequencing

The initial sequencing data were processed to filter out impurities to obtain clean reads. The reference genome of *Gossypium hirsutum* was downloaded, and alignment was performed using HISAT2 (https://www.cottongen.org/species/Gossypium_hirsutum/UTX-TM1_v2.1, accessed on 13 March 2024). The clean reads were assembled and expression levels were evaluated using StringTie version 2.0.4 software. RNA differential expression analysis between two different samples was performed by DESeq2 software (https://www.omicshare.com/tools/, accessed on 21 August 2024) (and by the edgeR package in R version 4.4.1 for comparisons between two samples). The genes/transcripts identified based on a false discovery rate (FDR) < 0.05 and absolute fold change ≥ 2 were considered differentially expressed genes/transcripts. OmicShare was employed to conduct functional enrichment analysis for KEGG pathways and GO categories (https://www.omicshare.com/tools/, accessed on 23 August 2024).

4.4. Metabolic Analysis (LC−MS/MS)

Root tissue samples were taken at 0, 1, and 2 DAI after being inoculated with V991 and subsequently sent to Igenebook Biotechnology Co., Ltd. for untargeted metabolite profiling. The extraction and analysis of the metabolites were performed through liquid chromatography/mass spectrometry (LC-MS) [74]. The raw data files were loaded into TraceFinder version 3.2.0 for analysis. Each metabolite was screened according to criteria including retention time and mass-to-charge ratio. To ensure accurate identification of metabolites, peak alignment was conducted on various samples. During the alignment process, we maintained a retention time deviation of 0.2 min and allowed for a mass deviation of 5 parts per million (ppm) to ensure accurate identification. Peak areas were quantified, and spectral matching was conducted against mzCloud alongside a locally established database (https://www.mzcloud.org/, accessed on 30 January 2024). In order to ensure accuracy in the data, the raw quantitative results were subjected to normalization processes. This step was crucial for facilitating the identification and relative quantification of metabolites present in the samples under investigation. The metabolites that were identified during this process were meticulously annotated with reference to several authoritative databases. Specifically, the KEGG database (https://www.genome.jp/kegg/pathway.html, accessed on 31 January 2024), the HMDB database (https://hmdb.ca/metabolites, accessed on 31 January 2024), and the LIPIDMaps database (http://www.lipidmaps.org/, accessed on 31 January 2024) were employed to ensure comprehensive and accurate annotations of the metabolites. DAMs were selected using a threshold of variable importance in projection (VIP) ≥ 1, a fold change of 2, and $p \leq 0.05$. The method chosen for calculating the p-value is ANOVA (https://cloud.metware.cn, accessed on 17 August 2024). Finally, pathway and other functional analyses were conducted to uncover the biological significance of the metabolites.

4.5. Gene Expression Analyses via qRT-PCR

In order to ensure the precision and dependability of the transcriptome data, a selection of 20 genes was randomly chosen for validation through real-time quantitative PCR (qRT-PCR). *GbUBQ7* was identified as the reference gene candidate based on the sequencing data from the transcriptome. The design of specific primers was accomplished using the Primer-BLAST tool available on the NCBI website, with primer sequences detailed in Table S4. The RNA samples used were the same as those returned from the transcriptome sequencing conducted by iGeneTech Co., Ltd., Seoul, Republic of Korea cDNA was synthesized from RNA through reverse transcription using the HiScript III RT SuperMix specifically designed for quantitative PCR, which includes a gDNA wiper to eliminate genomic DNA contamination. Subsequently, qRT-PCR assays were conducted with a ChamQ Universal SYBR qPCR Master Mix (Vazyme, Nanjing, China). The quantification of gene expression levels was carried out utilizing the $2^{-\Delta\Delta Ct}$ approach [75]. Three independent biological replicates along with three technical replicates were set up for both the control and experimental groups to ensure reproducibility and data reliability.

5. Conclusions

In this study, to identify key DEGs and DAMs related to resistance to VW in *G. hirsutum* and *G. barbadense*, and cultivate new disease-resistant varieties, RNA-seq and metabolomic analyses were conducted on root tissues of the resistant material Hai7124 and the susceptible material TM-1 at the early stages (0, 1, and 2 DAI) of V991 infection. The analysis of DEGs and DAMs using KEGG enrichment demonstrated that nine pathways were co-enriched, including glutathione metabolism, propionate metabolism, and pyrimidine metabolism. Through joint analysis, the glutathione metabolism pathway was identified as potentially significantly contributing to the early response of cotton to VW infection. A glutathione metabolism pathway map was constructed, revealing a significant positive correlation between the metabolite putrescine and DEGs within GSTs. Based on the comprehensive analysis, it is suggested that the DEG *GH_D11G2329* (*GSTF8*) could play a role in cotton response to VW infection. These results provided abundant information for screening the crucial VW-resistant genes and metabolites, thereby establishing a strong basis for further studying the molecular mechanism behind plant defense against VW infection.

Supplementary Materials: The supporting information can be downloaded at: https://www.mdpi.com/article/10.3390/ijms26010028/s1.

Author Contributions: B.X., Y.L., B.C., R.P., S.H. and P.L. designed the study; B.X., Y.L. and B.C. prepared the samples and generated the experiments; B.X. collected and organized the data and wrote the manuscript. Y.C., S.Z., Z.S., Q.L., X.D., X.L., L.H., B.W. and A.Z. provided suggestions and revised the paper. All authors have read and agreed to the published version of the manuscript.

Funding: This study was funded by the National Natural Science Foundation of China (32272179), Zhongyuan Scholars Workstation (224400510020), and Scientific and technological project in Henan Province (242102110262).

Institutional Review Board Statement: Not applicable.

Informed Consent Statement: Not applicable.

Data Availability Statement: The original contributions presented in the study are included in the article/Supplementary Materials, and further inquiries can be directed to the corresponding authors.

Conflicts of Interest: The authors declare no conflicts of interest.

References

1. Man, M.; Zhu, Y.; Liu, L.; Luo, L.; Han, X.; Qiu, L.; Li, F.; Ren, M.; Xing, Y. Defense Mechanisms of Cotton *Fusarium* and *Verticillium Wilt* and Comparison of Pathogenic Response in Cotton and Humans. *Int. J. Mol. Sci.* **2022**, *23*, 12217. [CrossRef] [PubMed]
2. Li, S.; Kong, L.; Xiao, X.; Li, P.; Liu, A.; Li, J.; Gong, J.; Gong, W.; Ge, Q.; Shang, H.; et al. Genome-wide artificial introgressions of *Gossypium barbadense* into *G. hirsutum* reveal superior loci for simultaneous improvement of cotton fiber quality and yield traits. *J. Adv. Res.* **2023**, *53*, 1–16. [CrossRef]
3. Egan, L.M.; Stiller, W.N. The Past, Present, and Future of Host Plant Resistance in Cotton: An Australian Perspective. *Front. Plant Sci.* **2022**, *13*, 895877. [CrossRef]
4. Palanga, K.K.; Liu, R.; Ge, Q.; Gong, J.; Li, J.; Lu, Q.; Li, P.; Yuan, Y.; Gong, W. Current advances in pathogen-plant interaction between *Verticillium dahliae* and cotton provide new insight in the disease management. *J. Cotton Res.* **2021**, *4*, 25. [CrossRef]
5. Xiong, X.-P.; Sun, S.-C.; Zhang, X.-Y.; Li, Y.-J.; Liu, F.; Zhu, Q.-H.; Xue, F.; Sun, J. GhWRKY70D13 Regulates Resistance to *Verticillium dahliae* in Cotton Through the Ethylene and Jasmonic Acid Signaling Pathways. *Front. Plant Sci.* **2020**, *11*, 69. [CrossRef]
6. Xiong, X.-P.; Sun, S.-C.; Zhu, Q.-H.; Zhang, X.-Y.; Liu, F.; Li, Y.-J.; Xue, F.; Sun, J. Transcriptome Analysis and RNA Interference Reveal GhGDH2 Regulating Cotton Resistance to Verticillium Wilt by JA and SA Signaling Pathways. *Front. Plant Sci.* **2021**, *12*, 654676. [CrossRef] [PubMed]
7. Fradin, E.F.; Thomma, B.P.H.J. Physiology and molecular aspects of *Verticillium wilt* diseases caused by *V. dahliae* and *V. alboatrum*. *Mol. Plant Pathol.* **2006**, *7*, 71–86. [CrossRef] [PubMed]
8. Li, Z.K.; Chen, B.; Li, X.X.; Wang, J.P.; Zhang, Y.; Wang, X.F.; Yan, Y.Y.; Ke, H.F.; Yang, J.; Wu, J.H.; et al. A newly identified cluster of glutathione S-transferase genes provides Verticillium wilt resistance in cotton. *Plant J.* **2019**, *98*, 213–227. [CrossRef] [PubMed]
9. Xu, L.; Zhu, L.; Tu, L.; Liu, L.; Yuan, D.; Jin, L.; Long, L.; Zhang, X. Lignin metabolism has a central role in the resistance of cotton to the *wilt fungus Verticillium dahliae* as revealed by RNA-Seq-dependent transcriptional analysis and histochemistry. *J. Exp. Bot.* **2011**, *62*, 5607–5621. [CrossRef]
10. Li, F.; Shen, H.; Wang, M.; Fan, K.; Bibi, N.; Ni, M.; Yuan, S.; Wang, X. A synthetic antimicrobial peptide BTD-S expressed in Arabidopsis thaliana confers enhanced resistance to *Verticillium dahliae*. *Mol. Genet. Genom.* **2016**, *291*, 1647–1661. [CrossRef] [PubMed]

11. Klosterman, S.J.; Atallah, Z.K.; Vallad, G.E.; Subbarao, K.V. Diversity, Pathogenicity, and Management of Verticillium Species. *Annu. Rev. Phytopathol.* **2009**, *47*, 39–62. [CrossRef]
12. Zhao, Y.-L.; Zhou, T.-T.; Guo, H.-S. Hyphopodium-Specific VdNoxB/VdPls1-Dependent ROS-Ca^{2+} Signaling Is Required for Plant Infection by *Verticillium dahliae*. *PLOS Pathog.* **2016**, *12*, e1005793. [CrossRef] [PubMed]
13. Zhu, Y.; Zhao, M.; Li, T.; Wang, L.; Liao, C.; Liu, D.; Zhang, H.; Zhao, Y.; Liu, L.; Ge, X.; et al. Interactions between *Verticillium dahliae* and cotton: Pathogenic mechanism and cotton resistance mechanism to Verticillium wilt. *Front. Plant Sci.* **2023**, *14*, 1174281. [CrossRef] [PubMed]
14. Ranga, A.D.; Kak, V.; Darvhankar, M.S. Genetic and Molecular Research of Resistance to Wilt in Cotton: A Concise Review. *Int. J. Curr. Microbiol. Appl. Sci.* **2020**, *9*, 2410–2422. [CrossRef]
15. He, L.; Han, Z.; Zang, Y.; Dai, F.; Chen, J.; Jin, S.; Huang, C.; Cheng, Y.; Zhang, J.; Xu, B.; et al. Advanced genes expression pattern greatly contributes to divergence in *Verticillium wilt* resistance between *Gossypium barbadense* and *Gossupium hirsutum*. *Front. Plant Sci.* **2022**, *13*, 979585. [CrossRef] [PubMed]
16. Yang, Z.; Gao, C.; Zhang, Y.; Yan, Q.; Hu, W.; Yang, L.; Wang, Z.; Li, F. Recent progression and future perspectives in cotton genomic breeding. *J. Integr. Plant Biol.* **2022**, *65*, 548–569. [CrossRef]
17. Chen, Z.J.; Scheffler, B.E.; Dennis, E.; Triplett, B.A.; Zhang, T.; Guo, W.; Chen, X.; Stelly, D.M.; Rabinowicz, P.D.; Town, C.D.; et al. Toward Sequencing Cotton (Gossypium) Genomes: Figure 1. *Plant Physiol.* **2007**, *145*, 1303–1310. [CrossRef]
18. Jin, Y.; Fan, L.; Zhang, Y.; Hu, W.; Han, X.; Yan, Q.; Yang, J.; Li, F.; Yang, Z. Functional divergence of GLP genes between *G. barbadense* and *G. hirsutum* in response to *Verticillium dahliae* infection. *Genomics* **2022**, *114*, 110470. [CrossRef]
19. Wang, Z.; Gerstein, M.; Snyder, M. RNA-Seq: A revolutionary tool for transcriptomics. *Nat. Rev. Genet.* **2009**, *10*, 57–63. [CrossRef]
20. Zhou, Z.; Gao, H.; Ming, J.; Ding, Z.; Lin, X.e.; Zhan, R. Combined Transcriptome and Metabolome analysis of Pitaya fruit unveiled the mechanisms underlying Peel and pulp color formation. *BMC Genom.* **2020**, *21*, 734. [CrossRef]
21. Fang, X.; Yan, P.; Guan, M.; Han, S.; Qiao, T.; Lin, T.; Zhu, T.; Li, S. Comparative Transcriptomics and Gene Knockout Reveal Virulence Factors of *Arthrinium phaeospermum* in *Bambusa pervariabilis* × *Dendrocalamopsis grandis*. *J. Fungi* **2021**, *7*, 1001. [CrossRef] [PubMed]
22. Bashyal, B.M.; Rawat, K.; Parmar, P.; Gupta, A.K.; Gupta, S.; Krishnan, S.G.; Choudhary, R.; Ercisli, S.; Kovacevic, A.; Aggarwal, R. Transcriptomic analysis of bakanae disease resistant and susceptible rice genotypes in response to infection by *Fusarium fujikuroi*. *Mol. Biol. Rep.* **2022**, *49*, 11959–11972. [CrossRef] [PubMed]
23. Rathod, V.; Hamid, R.; Tomar, R.S.; Patel, R.; Padhiyar, S.; Kheni, J.; Thirumalaisamy, P.P.; Munshi, N.S. Comparative RNA-Seq profiling of a resistant and susceptible peanut (*Arachis hypogaea*) genotypes in response to leaf rust infection caused by Puccinia arachidis. *3 Biotech* **2020**, *10*, 284. [CrossRef] [PubMed]
24. Hall, R.; Beale, M.; Fiehn, O.; Hardy, N.; Sumner, L.; Bino, R. Plant Metabolomics. *Plant Cell* **2002**, *14*, 1437–1440. [CrossRef] [PubMed]
25. Shen, S.; Zhan, C.; Yang, C.; Fernie, A.R.; Luo, J. Metabolomics-centered mining of plant metabolic diversity and function: Past decade and future perspectives. *Mol. Plant* **2023**, *16*, 43–63. [CrossRef] [PubMed]
26. Xiao, X.O.; Lin, W.; Feng, E.; Ou, X. Transcriptome and metabolome response of eggplant against Ralstonia solanacearum infection. *PeerJ* **2023**, *11*, e14658. [CrossRef] [PubMed]
27. Han, S.; Xu, X.; Yuan, H.; Li, S.; Lin, T.; Liu, Y.; Li, S.; Zhu, T. Integrated Transcriptome and Metabolome Analysis Reveals the Molecular Mechanism of Rust Resistance in Resistant (Youkang) and Susceptive (Tengjiao) Zanthoxylum armatum Cultivars. *Int. J. Mol. Sci.* **2023**, *24*, 14761. [CrossRef]
28. Kasote, D.M.; Jayaprakasha, G.K.; Ong, K.; Crosby, K.M.; Patil, B.S. Hormonal and metabolites responses in Fusarium wilt-susceptible and -resistant watermelon plants during plant-pathogen interactions. *BMC Plant Biol.* **2020**, *20*, 481. [CrossRef]
29. Yan, X.; Chen, S.; Pan, Z.; Zhao, W.; Rui, Y.; Zhao, L. AgNPs-Triggered Seed Metabolic and Transcriptional Reprogramming Enhanced Rice Salt Tolerance and Blast Resistance. *ACS Nano* **2022**, *17*, 492–504. [CrossRef] [PubMed]
30. Liu, H.; Wang, Y.; Chang, Q.; Li, Q.; Fang, J.; Cao, N.; Tong, X.; Jiang, X.; Yu, X.; Cheng, Y. Combined metabolome and transcriptome reveal HmF6′H1 regulating simple coumarin accumulation against powdery mildew infection in *Heracleum moellendorffii* Hance. *BMC Plant Biol.* **2024**, *24*, 507. [CrossRef]
31. Yang, S.; Liu, M.; Chu, N.; Chen, G.; Wang, P.; Mo, J.; Guo, H.; Xu, J.; Zhou, H. Combined transcriptome and metabolome reveal glutathione metabolism plays a critical role in resistance to salinity in rice landraces HD961. *Front. Plant Sci.* **2022**, *13*, 952595. [CrossRef]
32. Perez, L.M.; Mauleon, R.; Arick, M.A.; Magbanua, Z.V.; Peterson, D.G.; Dean, J.F.D.; Tseng, T.M. Transcriptome analysis of the 2,4-dichlorophenoxyacetic acid (2,4-D)-tolerant cotton chromosome substitution line CS-B15sh and its susceptible parental lines *G. hirsutum* L. cv. Texas Marker-1 and *G. barbadense* L. cv. Pima 379. *Front. Plant Sci.* **2022**, *13*, 910369. [CrossRef]
33. Li, T.; Wang, F.; Yasir, M.; Li, K.; Qin, Y.; Zheng, J.; Luo, K.; Zhu, S.; Zhang, H.; Jiang, Y.; et al. Expression Patterns Divergence of Reciprocal F1 Hybrids Between *Gossypium hirsutum* and *Gossypium barbadense* Reveals Overdominance Mediating Interspecific Biomass Heterosis. *Front. Plant Sci.* **2022**, *13*, 892805. [CrossRef]

34. Li, H.; Zhang, S.; Zhao, Y.; Zhao, X.; Xie, W.; Guo, Y.; Wang, Y.; Li, K.; Guo, J.; Zhu, Q.-H.; et al. Identification and Characterization of Cinnamyl Alcohol Dehydrogenase Encoding Genes Involved in Lignin Biosynthesis and Resistance to *Verticillium dahliae* in Upland Cotton (*Gossypium hirsutum* L.). *Front. Plant Sci.* **2022**, *13*, 840397. [CrossRef] [PubMed]
35. Zhang, T.; Zhao, Y.L.; Zhao, J.H.; Wang, S.; Jin, Y.; Chen, Z.Q.; Fang, Y.Y.; Hua, C.L.; Ding, S.W.; Guo, H.S. Cotton plants export microRNAs to inhibit virulence gene expression in a fungal pathogen. *Nat. Plants* **2016**, *2*, 16153. [CrossRef] [PubMed]
36. Zhang, J.; Ahmed, W.; Dai, Z.; Zhou, X.; He, Z.; Wei, L.; Ji, G. Microbial Consortia: An Engineering Tool to Suppress Clubroot of Chinese Cabbage by Changing the Rhizosphere Bacterial Community Composition. *Biology* **2022**, *11*, 918. [CrossRef]
37. Zhao, N.; Guo, A.; Wang, W.; Li, B.; Wang, M.; Zhou, Z.; Jiang, K.; Aierxi, A.; Wang, B.; Adjibolosoo, D.; et al. GbPP2C80 Interacts with GbWAKL14 to Negatively Co-Regulate Resistance to *Fusarium* and Verticillium wilt via MPK3 and ROS Signaling in Sea Island Cotton. *Adv. Sci.* **2024**, *11*, e2309785. [CrossRef] [PubMed]
38. Liu, S.; Zhang, X.; Xiao, S.; Ma, J.; Shi, W.; Qin, T.; Xi, H.; Nie, X.; You, C.; Xu, Z.; et al. A Single-Nucleotide Mutation in a GLUTAMATE RECEPTOR-LIKE Gene Confers Resistance to Fusarium wilt in *Gossypium hirsutum*. *Adv. Sci.* **2021**, *8*, 2002723. [CrossRef]
39. Sahoo, J.; Mahanty, B.; Mishra, R.; Joshi, R.K. Development of SNP markers linked to purple blotch resistance for marker-assisted selection in onion (*Allium cepa* L.) breeding. *3 Biotech* **2023**, *13*, 137. [CrossRef]
40. Kaur, B.; Bai, B.K.; Dhillon, G.S.; Kaur, J.; Sharma, A.; Srivastava, P.; Chhuneja, P.; Kaur, S. Mapping of *Aegilops speltoides* derived leaf rust and stripe rust resistance genes using 35K SNP array. *BMC Genom. Data* **2024**, *25*, 69. [CrossRef]
41. Maldonado dos Santos, J.V.; Ferreira, E.G.C.; Passianotto, A.L.d.L.; Brumer, B.B.; Santos, A.B.D.; Soares, R.M.; Torkamaneh, D.; Arias, C.A.A.; Belzile, F.; Abdelnoor, R.V.; et al. Association mapping of a locus that confers southern stem canker resistance in soybean and SNP marker development. *BMC Genom.* **2019**, *20*, 798. [CrossRef]
42. Kokane, S.B.; Bhose, S.; Kokane, A.; Gubyad, M.; Ghosh, D.K. Molecular detection, identification, and sequence analysis of 'Candidatus Liberibacter asiaticus' associated with Huanglongbing disease of citrus in North India. *3 Biotech* **2020**, *10*, 341. [CrossRef] [PubMed]
43. Cao, H.; Yang, Z.; Song, S.; Xue, M.; Liang, G.; Li, N. Transcriptome analysis reveals genes potentially related to maize resistance to *Rhizoctonia solani*. *Plant Physiol. Biochem.* **2022**, *193*, 78–89. [CrossRef] [PubMed]
44. Channappa, M.; Sharma, S.; Kulshreshtha, D.; Singh, K.; Bhardwaj, S.C.; Murugasamy, S.; Sindhu, A.; Vikas, V.K.; Aggarwal, R. Transcriptome profiling and differential gene expression analysis provides insights into Lr24-based resistance in wheat against *Puccinia triticina*. *3 Biotech* **2021**, *11*, 455. [CrossRef] [PubMed]
45. Samal, P.; Molla, K.A.; Bal, A.; Ray, S.; Swain, H.; Khandual, A.; Sahoo, P.; Behera, M.; Jaiswal, S.; Iquebal, A.; et al. Comparative transcriptome profiling reveals the basis of differential sheath blight disease response in tolerant and susceptible rice genotypes. *Protoplasma* **2021**, *259*, 61–73. [CrossRef]
46. Zeng, Q.; Man, X.; Dai, Y.; Liu, H. Pseudomonas spp. Enriched in Endophytic Community of Healthy Cotton Plants Inhibit Cotton Verticillium Wilt. *Front. Microbiol.* **2022**, *13*, 906732. [CrossRef] [PubMed]
47. Tan, B.A.; Daim, L.D.J.; Ithnin, N.; Ooi, T.E.K.; Md-Noh, N.; Mohamed, M.; Mohd-Yusof, H.; Appleton, D.R.; Kulaveerasingam, H. Expression of phenylpropanoid and flavonoid pathway genes in oil palm roots during infection by Ganoderma boninense. *Plant Gene* **2016**, *7*, 11–20. [CrossRef]
48. Li, C.; He, Q.; Zhang, F.; Yu, J.; Li, C.; Zhao, T.; Zhang, Y.; Xie, Q.; Su, B.; Mei, L.; et al. Melatonin enhances cotton immunity to Verticillium wilt via manipulating lignin and gossypol biosynthesis. *Plant J.* **2019**, *100*, 784–800. [CrossRef] [PubMed]
49. Hu, Q.; Min, L.; Yang, X.; Jin, S.; Zhang, L.; Li, Y.; Ma, Y.; Qi, X.; Li, D.; Liu, H.; et al. Laccase GhLac1 Modulates Broad-Spectrum Biotic Stress Tolerance via Manipulating Phenylpropanoid Pathway and Jasmonic Acid Synthesis. *Plant. Physiol.* **2018**, *176*, 1808–1823. [CrossRef]
50. Ling, Y.; Xiong, X.; Yang, W.; Liu, B.; Shen, Y.; Xu, L.; Lu, F.; Li, M.; Guo, Y.; Zhang, X. Comparative Analysis of Transcriptomics and Metabolomics Reveals Defense Mechanisms in Melon Cultivars against *Pseudoperonospora cubensis* Infection. *Int. J. Mol. Sci.* **2023**, *24*, 17552. [CrossRef] [PubMed]
51. Park, J.; Han, J.W.; Lee, N.; Kim, S.; Choi, S.; Lee, H.-H.; Kim, J.-E.; Seo, Y.-S.; Choi, G.J.; Lee, Y.-W.; et al. Sulfur metabolism-mediated fungal glutathione biosynthesis is essential for oxidative stress resistance and pathogenicity in the plant pathogenic fungus *Fusarium graminearum*. *mBio* **2024**, *15*, e0240123. [CrossRef]
52. Zhang, X.; Zhang, L.; Chen, L.; Lu, Y.; An, Y. Ectopic expression γ-glutamylcysteine synthetase of Vicia sativa increased cadmium tolerance in Arabidopsis. *Gene* **2022**, *823*, 146358. [CrossRef]
53. Zhang, H.; Ikram, M.; Li, R.; Xia, Y.; Zhao, W.; Yuan, Q.; Siddique, K.H.M.; Guo, P. Uncovering the transcriptional responses of tobacco (*Nicotiana tabacum* L.) roots to Ralstonia solanacearum infection: A comparative study of resistant and susceptible cultivars. *BMC Plant Biol.* **2023**, *23*, 620. [CrossRef] [PubMed]
54. Li, T.; Ren, G.; Jiang, D.; Liu, C. Dynamic Changes in Endophytic Microorganisms and Metabolites During Natural Drying of Licorice. *Front. Microbiol.* **2021**, *12*, 740721. [CrossRef] [PubMed]

55. Yu, D.; Boughton, B.A.; Hill, C.B.; Feussner, I.; Roessner, U.; Rupasinghe, T.W.T. Insights Into Oxidized Lipid Modification in Barley Roots as an Adaptation Mechanism to Salinity Stress. *Front. Plant Sci.* **2020**, *11*, 1. [CrossRef] [PubMed]
56. Yuan, H.; Zeng, X.; Shi, J.; Xu, Q.; Wang, Y.; Jabu, D.; Sang, Z.; Nyima, T. Time-Course Comparative Metabolite Profiling under Osmotic Stress in Tolerant and Sensitive Tibetan Hulless Barley. *BioMed Res. Int.* **2018**, *2018*, 9415409. [CrossRef]
57. Tamanna, N.; Mojumder, A.; Azim, T.; Iqbal, M.I.; Alam, M.N.U.; Rahman, A.; Seraj, Z.I. Comparative metabolite profiling of salt sensitive Oryza sativaand the halophytic wild rice Oryza coarctata under salt stress. *Plant-Environ. Interact.* **2024**, *5*, e10155. [CrossRef]
58. Zhu, J.; Tang, X.; Sun, Y.; Li, Y.; Wang, Y.; Jiang, Y.; Shao, H.; Yong, B.; Li, H.; Tao, X. Comparative Metabolomic Profiling of Compatible and Incompatible Interactions Between Potato and *Phytophthora infestans*. *Front. Microbiol.* **2022**, *13*, 857160. [CrossRef] [PubMed]
59. Zhao, T.; Tang, P.; Liu, C.; Zuo, R.; Su, S.; Zhong, Y.; Li, Y.; Yang, J. Multi-Omics Approach Reveals OsPIL1as a Regulator Promotes Rice Growth, Grain Development, and Blast Resistance. *J. Agric. Food Chem.* **2024**, *72*, 1822–1843. [CrossRef] [PubMed]
60. Zhao, P.; Gu, S.; Han, C.; Lu, Y.; Ma, C.; Tian, J.; Bi, J.; Deng, Z.; Wang, Q.; Xu, Q. Targeted and Untargeted Metabolomics Profiling of Wheat Reveals Amino Acids Increase Resistance to Fusarium Head Blight. *Front. Plant Sci.* **2021**, *12*, 762605. [CrossRef]
61. Sade, D.; Shriki, O.; Cuadros-Inostroza, A.; Tohge, T.; Semel, Y.; Haviv, Y.; Willmitzer, L.; Fernie, A.R.; Czosnek, H.; Brotman, Y. Comparative metabolomics and transcriptomics of plant response to Tomato *yellow leaf curl virus* infection in resistant and susceptible tomato cultivars. *Metabolomics* **2015**, *11*, 81–97. [CrossRef]
62. Mao, J.; Huang, L.; Chen, M.; Zeng, W.; Feng, Z.; Huang, S.; Liu, T. Integrated Analysis of the Transcriptome and Metabolome Reveals Genes Involved in Terpenoid and Flavonoid Biosynthesis in the Loblolly Pine (*Pinus taeda* L.). *Front. Plant Sci.* **2021**, *12*, 729161. [CrossRef] [PubMed]
63. Li, Y.; Ma, Y.; Zhang, T.; Bi, Y.; Wang, Y.; Prusky, D. Exogenous polyamines enhance resistance to *Alternaria alternata* by modulating redox homeostasis in apricot fruit. *Food Chem.* **2019**, *301*, 125303. [CrossRef]
64. Zhang, Y.; He, S.; Zhang, Z.; Xu, H.; Wang, J.; Chen, H.; Liu, Y.; Wang, X.; Li, Y. Glycine transformation induces repartition of cadmium and lead in soil constituents. *Environ. Pollut.* **2019**, *251*, 930–937. [CrossRef]
65. Fang, H.; Liu, Z.; Jin, Z.; Zhang, L.; Liu, D.; Pei, Y. An emphasis of hydrogen sulfide-cysteine cycle on enhancing the tolerance to chromium stress in Arabidopsis. *Environ. Pollut.* **2016**, *213*, 870–877. [CrossRef] [PubMed]
66. Fan, Q.; Creamer, R.; Li, Y. Time-course metabolic profiling in alfalfa leaves under Phoma medicaginis infection. *PLoS ONE* **2018**, *13*, e0206641. [CrossRef] [PubMed]
67. Jiang, S.; Chen, M.; He, N.; Chen, X.; Wang, N.; Sun, Q.; Zhang, T.; Xu, H.; Fang, H.; Wang, Y.; et al. MdGSTF6, activated by MdMYB1, plays an essential role in anthocyanin accumulation in apple. *Hortic. Res.* **2019**, *6*, 40. [CrossRef]
68. Wang, Q.; Guo, J.; Jin, P.; Guo, M.; Guo, J.; Cheng, P.; Li, Q.; Wang, B. GlutathioneS-transferase interactions enhance wheat resistance to powdery mildew but not wheat stripe rust. *Plant Physiol.* **2022**, *190*, 1418–1439. [CrossRef]
69. Dhar, N.; Chen, J.Y.; Subbarao, K.V.; Klosterman, S.J. Hormone Signaling and Its Interplay With Development and Defense Responses in Verticillium-Plant Interactions. *Front. Plant Sci.* **2020**, *11*, 584997. [CrossRef] [PubMed]
70. Gong, Q.; Yang, Z.; Chen, E.; Sun, G.; He, S.; Butt, H.I.; Zhang, C.; Zhang, X.; Yang, Z.; Du, X.; et al. A Phi-Class Glutathione S-Transferase Gene for Verticillium Wilt Resistance in Gossypium arboreum Identified in a Genome-Wide Association Study. *Plant. Cell Physiol.* **2018**, *59*, 275–289. [CrossRef] [PubMed]
71. Mo, H.J.; Sun, Y.X.; Zhu, X.L.; Wang, X.F.; Zhang, Y.; Yang, J.; Yan, G.J.; Ma, Z.Y. Cotton S-adenosylmethionine decarboxylase-mediated spermine biosynthesis is required for salicylic acid- and leucine-correlated signaling in the defense response to *Verticillium dahliae*. *Planta* **2016**, *243*, 1023–1039. [CrossRef]
72. Hu, Y.; Guo, W.Z.; Zhang, T.Z. Construction of a bacterial artificial chromosome library of TM-1, a standard line for genetics and genomics in Upland cotton. *J. Integr. Plant Biol.* **2009**, *51*, 107–112. [CrossRef] [PubMed]
73. Li, P.-t.; Rashid, M.H.o.; Chen, T.-t.; Lu, Q.-w.; Ge, Q.; Gong, W.-k.; Liu, A.-y.; Gong, J.-w.; Shang, H.-h.; Deng, X.-y.; et al. Transcriptomic and biochemical analysis of upland cotton (*Gossypium hirsutum*) and a chromosome segment substitution line from *G. hirsutum* × *G. barbadense* in response to *Verticillium dahliae* infection. *BMC Plant Biol.* **2019**, *19*, 19. [CrossRef] [PubMed]
74. Want, E.J.; Masson, P.; Michopoulos, F.; Wilson, I.D.; Theodoridis, G.; Plumb, R.S.; Shockcor, J.; Loftus, N.; Holmes, E.; Nicholson, J.K. Global metabolic profiling of animal and human tissues via UPLC-MS. *Nat. Protoc.* **2012**, *8*, 17–32. [CrossRef] [PubMed]
75. Xiang, B.; Hao, X.; Xie, Q.; Shen, G.; Liu, Y.; Zhu, X. Deletion of a Rare Fungal PKS CgPKS11 Promotes Chaetoglobosin A Biosynthesis, Yet Defers the Growth and Development of *Chaetomium globosum*. *J. Fungi* **2021**, *7*, 750. [CrossRef]

MDPI AG
Grosspeteranlage 5
4052 Basel
Switzerland
Tel.: +41 61 683 77 34

International Journal of Molecular Sciences Editorial Office
E-mail: ijms@mdpi.com
www.mdpi.com/journal/ijms

www.ingramcontent.com/pod-product-compliance
Lightning Source LLC
Chambersburg PA
CBHW040822250926
51850CB00007BA/1235
* 9 7 8 3 7 2 5 8 6 5 1 4 7 *